Heinrich von Weizsäcker
Gerhard Winkler

Stochastic Integrals

Advanced Lectures in Mathematics

Edited by Gerd Fischer and Manfred Knebusch

Jochen Werner
Optimization. Theory and Applications

Manfred Denker
Asymptotic Distribution Theory
in Nonparametric Statistics

Klaus Lamotke
Regular Solids and Isolated Singularities
(out of print)

Francesco Guaraldo, Patrizia Macrì,
Alessandro Tancredi
Topics on Real Analytic Spaces

Ernst Kunz
Kähler Differentials
(out of print)

Johann Baumeister
Stable Solution of Inverse Problems

Heinrich von Weizsäcker, Gerhard Winkler
Stochastic Integrals

Heinrich von Weizsäcker
Gerhard Winkler

Stochastic Integrals

An Introduction

Friedr. Vieweg & Sohn Braunschweig/Wiesbaden

AMS Subject Classification: 60 G xx, 60 G 44, 60 H xx, 60 H 05, 60 H 10

Vieweg is a subsidiary company of the Bertelsmann Publishing Group International

Produced by W. Langelüddecke, Braunschweig
Printed in the Federal Republic of Germany

ISSN 0932-7134

ISBN 3-528-06310-6

Preface

This text introduces at a moderate speed and in a thorough way the basic concepts of the theory of stochastic integrals and Itô calculus for semimartingales. There are many reasons to study this subject. We are fascinated by the contrast between general measure theoretic arguments and concrete probabilistic problems, and by the own flavour of a new differential calculus. For the beginner, a lot of work is necessary to go through this text in detail. As a reward it should enable her or him to study more advanced literature and to become at ease with a couple of seemingly frightening concepts. Already in this introduction, many enjoyable and useful facets of stochastic analysis show up.

We start out having a glance at several elementary predecessors of the stochastic integral and sketching some ideas behind the abstract theory of semimartingale integration. Having introduced martingales and local martingales in chapters 2 - 4, the stochastic integral is defined for locally uniform limits of elementary processes in chapter 5. This corresponds to the Riemann integral in one-dimensional analysis and it suffices for the study of Brownian motion and diffusion processes in the later chapters 9 and 12. Although only continuous semimartingales are needed as integrators for these applications, the integral and its calculus are developed also for non-continuous semimartingales since the noncontinuous theory is becoming more and more important. Predictability is introduced in chapter 6, mainly as a tool for the structure theory of semimartingales which culminates in the Dellacherie-Bichteler characterization theorem in chapter 10. In chapter 7, Itô's elegant algebraic calculus for differentials of semimartingales is developed. Chapter 8 contains the analytic part of this calculus the heart of which is Itô's formula. It extends the chain rule of classical analysis. This completes the main body of the technical tools.

In the second part of the book one sees this machinery at work. This starts with exponential semimartingales and a brief introduction to local time. Chapter 9 entirely is devoted to various paradigmatical aspects of Brownian motion. It is followed by Girsanov's change of measure technique in chapter 10. These two chapters open up prospects for stochastic differential equations and diffusions in the last two chapters.

In view of the almost infinite amount of material which we do not cover we regret most the limited number of examples. The reader may consult e.g. the monographs of Rogers-Williams (1987) and Karatzas-Shreve (1988) which are particularly rich in this respect. Also, our text definitely is not a special introduction to stochastic differential equations.

The measure theoretic part takes some space. We try to be pedantic in the details but we make sparing use of specific terminology. The host of new notions often obstructs the beginner's view of powerful results like Meyer's predictable desintegration theorem. On the whole, we follow the established lines. However, we hope that here and there also the expert will find a few new arguments. In particular, we do not assume the usual conditions on the underlying filtration (which even in the noncontinuous part requires a surprisingly small amount of extra work).

The reader we had in mind should have met Brownian motion. He should be able to handle conditional expectations and know the basic properties of $\boldsymbol{L}^1$- and $\boldsymbol{L}^2$-spaces. A first course on stochastic integration might be based on the techniques from chapters 2-4 and the integral from chapter 5. The applications to Brownian motion in chapter 9 - in the first place Lévy's characterization of Brownian motion - need only that part of the calculus in the chapters 7 and 8 which aim at continuous semimartingales and which refer to the integral from chapter 5. In chapter 7, therefore, we give separate proofs for this integral and for the extended integral from chapter 6. From chapter 8 only the sections 8.1 and 8.2 are needed. So one may select some of the items in chapter 9 or directly pass to the theory of diffusion processes in chapter 12. To this end it is not necessary to work through the proofs of chapter 11.

The work on this book was done by the first author at the universities of Kaiserslautern and (for three months) Berkeley and by the second author at the universities of Trier, Göttingen, Kaiserslautern, Erlangen and München.

We are grateful to a number of persons who contributed in various ways to this text. First of all, Eva Dengel typed an unknown number of versions of the manuscript. Manfred Krischke and Dieter Zimmermann helped us with the word processing. We are indebted to Michael Jerschow and Marc Yor for the first introduction to the subject and to Salah Mohammed and Michael Scheutzow for some useful hints. Finally, we thank our families for their patience and *gentle* mockery; the second author also wants to mention Mona Kolb and his (she-) cat Micki.

CONTENTS

CHAPTER 1

WARMING UP

This chapter first recalls some background information about the two basic processes Brownian motion and Poisson process. Then in section 1.2 a couple of preliminary versions of stochastic integration concepts are presented. They show that for many applications a straightforward approach is sufficient. In the third section we overview some of the main ideas of the later chapters in order to give the reader a flavour of the things to come.

In this chapter we freely use some notions from probability theory with which the reader perhaps is not perfectly familiar. One should not worry about this but try to grasp the ideas. All concepts will be thoroughly introduced in the later chapters where they are needed.

1.1 Brownian Motion and the Poisson Process

The main example in Itô's theory of stochastic integration - actually in the first versions of the theory the only example - is Brownian motion. This process has a long and fascinating history in biology, economics, physics and statistics which started with the observation of the wild fluctuations of small particles suspended in a fluid. In the year 1828 the botanist Robert Brown observed that pollen in water dissolves into a large number of small particles which perform highly irregular movements under the microscope. For 80 years several explanations of this phenomenon were proposed and discussed. The movement did not change if the particles were exposed to electric fields or light; it increased if the temperature increased or the inner friction of the surrounding medium decreased. The decisive break-through in the explanation came with a paper by Albert Einstein in 1905. However, already in 1900 the French mathematician Louis Bachelier in his thesis "Théorie de la spéculation" developed a number of aspects of "Brownian motion" independent of its natural science interpretation.

Einstein's motivation is remarkable. It reads in our translation:

»5. On the Motion of Particles Suspended in Stationary Fluids Postulated in the Molecularkinetic Theory of Heat;

by A. Einstein.

In this paper it will be shown that according to the molecular theory of heat, microscopically visible particles suspended in a fluid should be moved in such a fashion that these motions could be easily observed with a microscope. It is possible that this motion coincides with the so-called "Brownian molecular motion". The information about this available to me is too inprecise though, as that I might form an opinion.
If the motion considered here is genuinly observable, complete with the predicted regularity, then, the laws of classical thermodynamics must be considered inexact even for microscopically observable spaces. Further, it should then be possible to determine exactly the size of atoms. On the other hand, if the predictions prove to be incorrect, then this would provide a strong argument against the molecular theory of heat.«

Here the irregular movement of suspended particles serves as a crucial criterion for the acceptability of one theory of heat or the other. Einstein's idea is roughly that the particles must move because of the inequalities of the collisions with the molecules in the fluid - provided there are moving molecules. He made the idealizing assumption that there are so many elastic collisions, that in any two disjoint time intervals the changes of the positions are independent. Hence the directions of the movement in two small adjacent time intervals will be drastically different. This suggests that the particles cannot have velocities. Then he went on to show that the movements of the particles should be visible under a good microscope (although he did not want to commit himself to the prognosis that the motion he foresaw would be the same as the motion observed by Robert Brown). More precisely, he predicted the value of the diffusion constant D which is a measure for the intensity of the movement. This value was experimentally confirmed a few months later.

Going more deeply into the physics Einstein argued that under the idealizing independence assumption the probability of finding a particle in a certain domain provided it is released t time units before at a point x is given by a probability density $p(t,x,y)$ which fulfills the "forward" or "Fokker-Planck-equation" $\frac{\partial p}{\partial t} = \frac{1}{2}\cdot D\Delta_y p$ where $\Delta_y f(t,x,y) = \frac{\partial^2 f}{\partial y_1^2}(t,x,y) + \dots + \frac{\partial^2 f}{\partial y_3^2}(t,x,y)$ is the Laplacian in y.

The theory of partial differential equations tells us that for functions

$$p :]0,\infty[\times \mathbb{R}^d \times \mathbb{R}^d \longrightarrow]0,\infty[$$

the forward equation $\frac{\partial p}{\partial t} = \frac{1}{2}\cdot\Delta_y p$ has one and only one solution with the following properties:

(i) $\int f(y)\, p(t,x,y)\, dy \longrightarrow f(x)$ if $t \downarrow 0$ for continuous bounded f,

(ii) $p(t,x,\cdot)$ is a probability density,

(iii) p is infinitely often differentiable.

This solution is given by the **Brownian transition density function**

$$p(t,x,y) = (2\pi t)^{-\frac{1}{2}\cdot d} \exp\left(-\frac{|y-x|^2}{2t}\right)$$

where $|\cdot|$ denotes Euclidean norm. By symmetry, p also solves the backward equation $\frac{\partial p}{\partial t} = \frac{1}{2}\cdot\Delta_x p$.

There are several ways to study the time evolution of nondeterministic physical systems. Basically, there are analytical and probabilistic approaches. We are probabilists. For us the time evolution is given by a **stochastic process**, i.e. a probability space $(\Omega, \mathcal{F}, \mathbb{P})$ and a family $(X_t)_{t\geq 0}$ of random variables taking values in the real line or in an Euclidean space. Then $X_t(\omega)$ is the state at time t of the particle ω which is randomly choosen from Ω according to the probability law $\mathbb{P}$. The **path** $t \longmapsto X_t(\omega)$ describes the fate of the particle ω as time goes by. In this language the requirements on a stochastic process modelling the idealized Brownian motion can be comprized as follows:

Definition. A real-valued stochastic process $B = (B_t)_{t\geq 0}$ on a probability space $(\Omega, \mathcal{F}, \mathbb{P})$ is called a **Brownian motion** if the following holds:

(i) B has **independent increments**, i.e. for all times $0 \leq t_0 < \dots < t_n$ the random variables $B_{t_1} - B_{t_0}, \dots, B_{t_n} - B_{t_{n-1}}$ are independent.

(ii) $B_0 \equiv 0$ and for $t > s$ the increments $B_t - B_s$ have variance $t - s$.

(iii) The increments $B_t - B_s$ are normally distributed of mean zero.

(iv) $\mathbb{P}$-almost all paths $t \longmapsto B_t(\omega)$ are continuous.

For a model in higher dimensions one could take independent copies of one-dimensional Brownian motions.

We have indicated that the paths are not differentiable. It is by no means obvious that the conditions (i)-(iii) above are consistent with the requirement of *continuous* paths. N. Wiener's (1923) celebrated result answers this question positively. By $\mathscr{C}(\mathbb{R}_+)$ we denote the space of continuous real functions on $[0,\infty[$ and by $\mathscr{B}(\mathscr{C}(\mathbb{R}_+))$ the σ-field on this space generated by the projections $W_t : \mathscr{C}(\mathbb{R}_+) \longrightarrow \mathbb{R},\ \omega \longmapsto \omega(t)$.

Theorem (N. Wiener). There is one and only one probability measure $\mathbb{W}$ on the measurable space $(\mathscr{C}(\mathbb{R}_+), \mathscr{B}(\mathscr{C}(\mathbb{R}_+)))$ such that the process $W = (W_t)_{t\geq 0}$ is a Brownian motion.

The measure $\mathbb{W}$ on the space of continuous functions is called **Wiener measure** and the probability space $(\mathscr{C}(\mathbb{R}_+),\mathscr{B}(\mathscr{C}(\mathbb{R}_+)),\mathbb{W})$ is called **Wiener space.** The coordinate process W on Wiener space is called **standard Brownian motion.**

We stress that according to our definition not *all* paths of a Brownian motion are continous. But the proof of the theorem shows that under mild regularity assumptions (e.g. separability of the process, cf. section 2.3) the conditions (i)-(iii) imply the a.s. continuity (iv).

Some more remarks about Wiener space are in order here. The correspondence between $\mathbb{W}$ and the one-dimensional Brownian transition function is given by

$$\mathbb{W}\Big((W_{t_1}, \dots, W_{t_n}) \in B\Big)$$
$$= \int_B p(t_1,0,x_1)\cdot p(t_2 - t_1,x_1,x_2) \dots p(t_n - t_{n-1},x_{n-1},x_n)\ dx_1\, dx_2 \dots dx_n.$$

This implies uniqueness since $\mathbb{W}$ is completely determined by its finite dimensional marginals (cf. the remarks below). One of the many possible ways to carry out the construction of $\mathbb{W}$ can be found in Billingsley (1979), section 37.

Usually the symbol $\mathscr{B}$ indicates that we are dealing with the **Borel-σ-field** on some metric space, i.e. the smallest σ-field which contains the open sets. The correct topology on $\mathscr{C}(\mathbb{R}_+)$ is that of uniform convergence on compact intervals. Plainly, this topology is induced by the metric

$$d(\omega_1,\omega_2) = \sum_n \frac{d_n(\omega_1,\omega_2)}{2^n(1 + d_n(\omega_1,\omega_2))}, \quad d_n(\omega_1,\omega_2) = \sup_{t\in[0,n]} |\omega_1(t) - \omega_2(t)|.$$

We shall not deal with the metric in the sequel, but let us shortly comment on it. The metric is easily seen to be complete and separable. Hence $\mathscr{C}(\mathbb{R}_+)$ is a Polish space. Let us for the present denote the σ-field generated by the projections W_t by σ and the Borel-σ-field by $\mathscr{B}$. Then σ is contained in $\mathscr{B}$ since the projections are continuous. For the converse note that for each $\omega^* \in \Omega$ the map

$$\omega \longmapsto d_n(\omega,\omega^*) = \sup_{q\in[0,n]\cap\mathbb{Q}} |W_q(\omega) - W_q(\omega^*)|$$

is measurable w.r.t. σ and hence the map $\omega \longmapsto d(\omega,\omega^*)$ too. A closed set F in $\mathscr{C}(\mathbb{R}_+)$ contains a dense sequence (ω_k) and thus

$$\omega \longmapsto d(\omega,F) = \inf_k d(\omega,\omega_k)$$

is measurable w.r.t. σ and hence also the zero-set F of this map. In summary, the Borel-σ-field for the metric d is generated by the projections W_t.

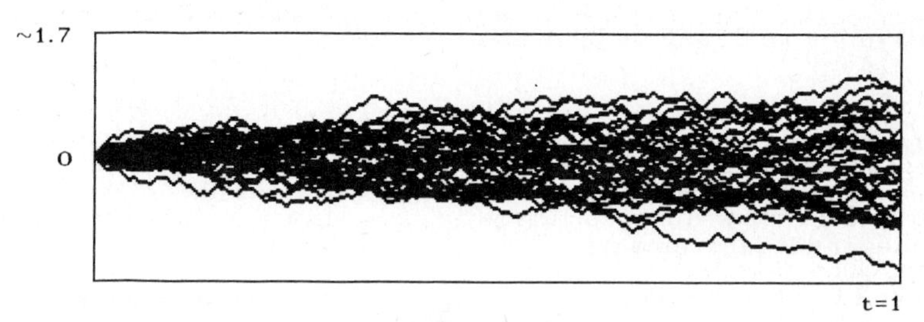

Fig. 1.1.1

Figure 1.1.1 shows a sample of 35 (approximately) Brownian paths on the unit interval. The unit on the y-axis is one tenth. The next picture shows the trace of a single Brownian path in two dimensions before it leaves for the first time a given rectangle. (To be honest we selected this from a large collection of less beautiful pictures.) Note that a portion of the *whole* path would produce a totally black rectangle since a typical Brownian path in two dimensions visits each small disk infinitely often.

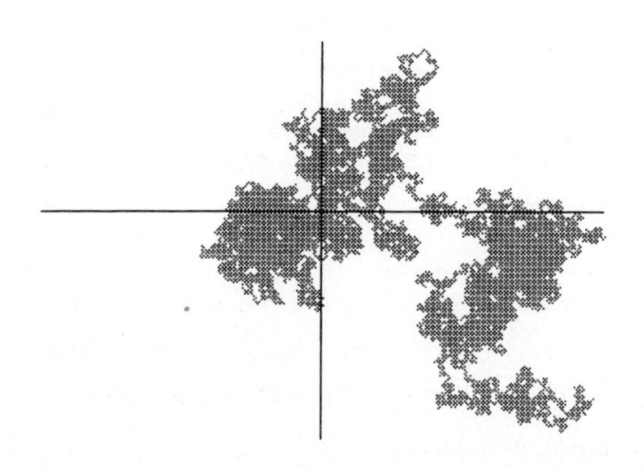

Fig. 1.1.2

Let us stress again that the role of Brownian motion is not restricted to the physical context sketched above. It shows up nearly everywhere where highly irregular behaviour has to be modelled. We already mentioned Louis Bachelier who in his "Théorie de la spéculation" found important properties of Brownian motion like the diffusion equation and the distribution of the maximum of the process up to time t before Einstein and Wiener (B. Mandelbrot (1983), chapter XII reviews the history of the reception of Bachelier's ideas). The universality of Brownian motion has several mathematical reasons. Two of them are closely related to stochastic integration and will be explained in chapter 9: Firstly, each continuous transition function which is homogeneous in space and time can be written in the form $p(\sigma^2 t, x, y - \mu t)$; secondly, each continuous martingale - the most important class of processes we will work with - can be derived from a Brownian motion by transforming the time axis for each ω separately in a suitable way.

Since we will deal also with non-continuous processes, we will comment on a non-continuous analogue of Brownian motion.

Definition. A real-valued stochastic process $N = (N_t)_{t\geq 0}$ on a probability space $(\Omega, \boldsymbol{F}, \mathbb{P})$ is called a **Poisson process** if the following holds:
(i) N has independent increments.
(ii) $N_0 \equiv 0$ and for $t > s$ the increments $N_t - N_s$ have variance $t - s$.
(iii') The increments $N_t - N_s$ are Poisson distributed of mean $t - s$.

In contrast to Brownian motion such a process can be constructed by elementary means (cf. Billingsley(1979)). Let $T_1, T_2, \ldots$ be independent and identically distributed waiting times with an exponential law of mean 1. Let $S_0 \equiv 0$ and $S_n = T_1 + \ldots + T_n$, $n \geq 1$, be the times when the n-th event occurs. Then the event

$$\Omega^* = \left\{\omega : S_0(\omega) < S_1(\omega) < S_2(\omega) < \ldots, \sup_n S_n(\omega) = \infty\right\}$$

has probability one and we may define on this set

$$N_t(\omega) = \max\left\{n \geq 0 : S_n(\omega) \leq t\right\}.$$

Outside Ω^* we set $N_t(\omega) = 0$. Since $\{N_t \geq n\} \cap \Omega^* = \{S_n \leq t\} \cap \Omega^*$ each N_t is a random variable and thus we have defined a stochastic process. The properties above can be verified by elementary calculations. Note that for all $\omega \in \Omega^*$ the path starts at zero, has jumps of size 1 only, is right-continuous and constant on $\{S_n \le t < S_{n+1}\}$. We write $\boldsymbol{D}(\mathbb{R}_+)$ for the space of those functions on $\mathbb{R}_+$ which are right-continuous and have left-hand limits. $\mathcal{B}(\boldsymbol{D}(\mathbb{R}_+))$ denotes the σ-field on this space generated by the projections N_t. In this notation the analogue of Wiener's theorem reads

Theorem. There is one and only one probability measure on the measurable space $(\boldsymbol{D}(\mathbb{R}_+), \mathcal{B}(\boldsymbol{D}(\mathbb{R}_+)))$ such that the process $N = (N_t)_{t \ge 0}$ is Poisson process.

Again $\mathcal{B}(\boldsymbol{D}(\mathbb{R}_+))$ is the Borel-σ-field for a suitable metric . Usually one takes the Skorohod metric (cf. Billingsley (1968)).The measure on $\mathcal{B}(\boldsymbol{D}(\mathbb{R}_+))$ has on cylinder sets a form like that noted above for $\mathbb{W}$ but with the **Poisson transition function**

$$q(t,m,n) = e^{-t} \cdot \frac{t^{n-m}}{(n-m)!}, \quad n \ge m, \; m, n \in \mathbb{N}_0,$$

replacing the Brownian transition function. Fig. 1.1.3 shows a typical path of the Poisson process. The Poisson process is the starting point for many models in insurance, queueing theory and science. If we insist that the variables should be centered we set $M_t = N_t - t$ and get thus the **Poisson martingale** M. If the mean in (iii') is required to be $\alpha(t-s)$ with $\alpha > 0$ instead of $t-s$ then we arive at the Poisson process or martingale with parameter α.

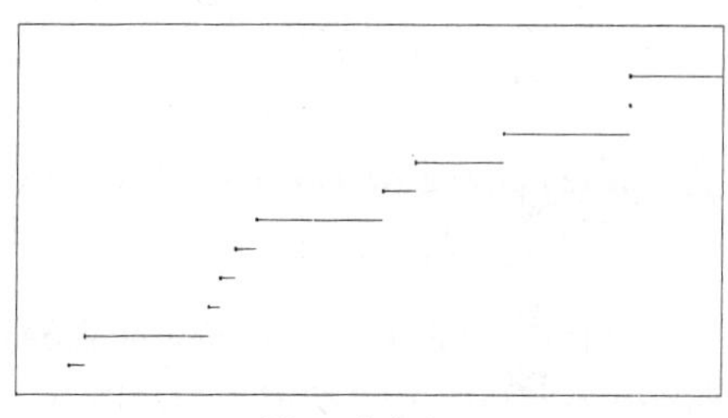

Fig. 1.1.3

The Poisson martingale fits better into the theory of stochastic integration. Fig. 1.1.4 shows a typical path (for $\alpha \sim ½$).

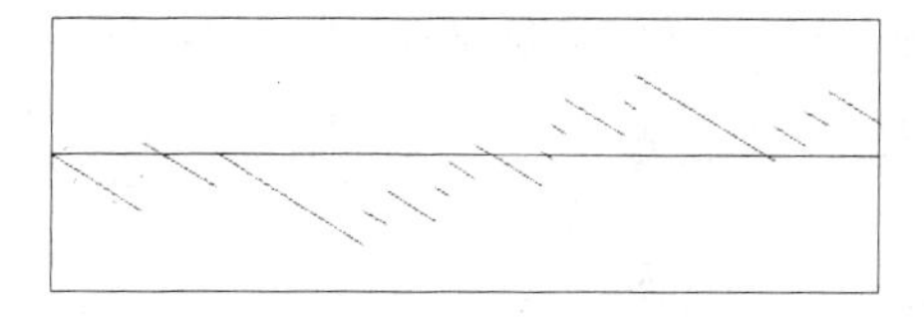

Fig.1.1.4

1.2 Simple Versions of the Stochastic Integral

Stochastic integrals $\int_0^t H\,dM$ were introduced by K. Itô in the 40-s in order to arrive at a probabilistically more intuitive understanding of diffusion processes. Up to then they where mathematically only tractable by Kolmogorov's partial-differential-equation description of the transition functions. These integrals have random functions as integrators and possibly also as integrands.

Random integrals though were used long before K. Itô introduced what we today call a stochastic integral. Many problems - arising for instance in physics or engineering - can be treated with special stochastic integrals which may be introduced with a minimum of theory. Let us consider three predecessors of the Itô integral which are of great practical and historical interest. The first one is a straightforward randomization of the usual Stieltjes integral, the second one extends this integral to more general integrating processes by formal integration by parts and the third integral only admits deterministic integrands but fairly general integrating processes. Although simple to construct - it already exhibits some of the features of the Itô integral.

Let us start from the very beginning. Students are used to integrals of the form $\int h(t)\,d\mu(t)$ with a measure μ - say on the Borel sets of the real line. In our context it is convenient to write them as Stieltjes integrals $\int h(t)\,dm(t)$ with the measure defining function $m(t) = \mu(]-\infty,t])$. (If μ is a probability measure then m is the cumulative distribution function). The domain of standard integration theory in $\mathbb{R}_+$ is indicated by the following theorem which is basically due to F. Riesz (1914). Two notions should be introduced in advance: A function $m : [0,t] \longrightarrow \mathbb{R}$ has **finite variation** if

$$\int_0^t |dm| = \sup\Big\{ \sum_{i=0}^{n} |m(t_{i+1}) - m(t_i)| \;:\; 0 \le t_0 < \ldots < t_{n+1} \le t,\ n \in \mathbb{N}\Big\} < \infty$$

and a **finite signed measure** μ on the Borel sets of $[0,t]$ is just the difference $\mu = \mu_1 - \mu_2$ of two finite nonnegative measures μ_1 and μ_2. The theorem is stated only for measures which do not charge the origin.

Theorem. There is a 1-1-1 correspondence between

(i) right-continuous functions $m : [0,t] \longrightarrow \mathbb{R}$ satisfying $m(0) = 0$ and which are of finite variation,

(ii) finite signed measures μ on the Borel sets of $]0,t]$,

(iii) continuous linear functionals I on the space $\{h \in \mathscr{C}([0,t]) : h(0) = 0\}$ endowed with the supremum norm.

The correspondence is given by the relations

$$\int_0^t h(s)\, dm(s) = \int_0^t h\, d\mu = I(h),$$

$$m(b) - m(a) = \mu(]a,b]) .$$

If part (a) is used as a definition then the integral is called a Lebesgue-Stieltjes integral. It will be dicussed in more detail in section 5.1. Here it serves as the starting point for the first simple version of a stochastic integral.

(A) Random Parameters. If the integrand or the integrating function or both depend on an additional parameter ω and if

$$\Big(\int_0^t h(s,\cdot)\, dm(s,\cdot)\Big)(\omega) = \int_0^t h(s,\omega)\, dm(s,\omega)$$

exists for each ω then we get a family of Stieltjes integrals. If the parameter is random then the integral is random too.

Let us consider a very simple example of a random integrator. Let $(\Omega, \boldsymbol{F}, \mathbb{P})$ be a probability space and let $N : \mathbb{R}_+ \times \Omega \longrightarrow \{0,1,\ldots\}$ be a Poisson process. Then the paths of N, i.e. the functions $t \longmapsto N(t,\omega)$ are monotone and right-continuous and hence one can integrate with respect to them. Similarly, the process M defined by $M(t,\omega) = N(t,\omega) - t$, the Poisson martingale, has paths which are of finite variation on each bounded interval. Thus the stochastic integral with respect to M makes sense either. For example, if $m(t,\omega) = N_t(\omega)$ with the Poisson process N then

$$\Big(\int_0^t h\, dN\Big)(\omega) = \sum_{i=1}^{N_t(\omega)} h(T_i(\omega),\omega),$$

where $T_i(\omega)$ is the time where the path ω jumps to i.

Despite the simplicity of its construction, this random integral is in some sense more than just a collection of usual integrals. For instance, one could ask if the integral process has a version with nice measurability properties. Such questions will be tackled in section 5.3.

(B) Stochastic Integration Via Integration by Parts and the Ornstein-Uhlenbeck Process. Suppose for simplicity that the deterministic functions h(t) and m(t) are continuous. If m does not have finite variation then the Stieltjes integral $\int_0^t h\,dm$ is not defined. On the other hand, if h has finite variation and we write down the formula for integration by parts, namely

$$\int_0^t h\,dm = h(t)\,m(t) - h(0)\,m(0) - \int_0^t m\,dh$$

then the right-hand side perfectly makes sense. Similarly, let X and Y be stochastic processes with continuous paths and assume that every path of X has finite variation (but the paths of Y possibly may not). Then we can define a stochastic integral path by path setting:

$$\int_0^t X(s,\omega)\,Y(ds,\omega) = X(t,\omega)\,Y(t,\omega) - X(0,\omega)\,Y(0,\omega) - \int_0^t Y(s,\omega)\,X(ds,\omega).$$

Again this integral - often called stochastic Stieltjes integral - is itself a stochastic process. As we will learn in section 5.3 the most interesting continuous stochastic processes - among them Brownian motion B - do not have paths of finite variation. Hence integrals like $\int_0^t B(s)\,dB(s)$ cannot be defined as stochastic Stieltjes integrals.

An important example - the Ornstein-Uhlenbeck process - *can* be treated with this integral (at least in its original form). Although in agreement with experiment, Brownian motion is a highly idealized model for real phenomena. In particular, it is far from Newtonian mechanics since the particles don't have velocities. Every model in classical mechanics, however, should rest upon Newton's principles. One first step in reducing physical Brownian motion to the Hamilton-Jacobi formalism was done by L.S. Ornstein and G.E. Uhlenbeck (1930) in their paper "On the theory of Brownian motion". Let us have a look at their "Brownian motion".

Consider a particle of mass m moving in three-dimensional Euclidean space and assume that no external field of force is present. We study the projections of the particle's position onto the coordinate axes separately. Let $x(t)$ be one of these components. In accordance with physical reality, we assume that the velocity $v(t) = \frac{dx}{dt}(t)$ exists and is differentiable. Assume further that a frictional force $-m\beta v$ acts on the particle; here β is a constant with the dimension of inverse time (frequency). Then Newton's laws would give the differential equations

$$\begin{aligned} \frac{dx}{dt} &= v, \\ m\frac{dv}{dt} &= -m\beta\, v, \qquad (1.2.1) \\ x(0) = x_o,\ v(0) &= v_o\,. \end{aligned}$$

Suppose now that the particle is exposed to the bombardment of much smaller particles. Formal application of Newton's principles and a suitable choice of parameters in Brownian motion then gives the **Langevin equation**

$$m\frac{dv}{dt} = -m\beta\, v + m\,\frac{\text{"}dB\text{"}}{dt}\,. \qquad (1.2.2)$$

The additional term on the right-hand side reflects the irregular movement of the small particles. We write quotation marks since Brownian paths are nowhere differentiable. This "stochastic differential equation" contains besides time t the random parameter ω. Therefore the initial conditions are now random variables $x(0,\cdot)$ and $v(0,\cdot)$.

Let us formally apply variation of constants and set:

$$v(t,\cdot) = c(t,\cdot)\, e^{-\beta t}$$

where $c(0,\cdot) = v(0,\cdot)$ and

$$\frac{dc}{dt} = \frac{\text{"}dB\text{"}}{dt}\, e^{\beta t}\,.$$

By formal integration we get

$$c(t,\cdot) - c(0,\cdot) = \int_o^t e^{\beta s}\, dB(s,\cdot)$$

and hence the solution should have the form

$$v(t,\cdot) = v(0,\cdot)\, e^{-\beta t} + e^{-\beta t}\int_o^t e^{\beta s}\, dB(s,\cdot).$$

Now formal integration by parts yields

$$v(t,\cdot) = v(0,\cdot)\, e^{-\beta t} + e^{-\beta t}\Big(B(t,\cdot)e^{\beta t} - \int_o^t B(s,\cdot)\beta e^{\beta s}\, ds\Big)$$

and the integral in the brackets is an ordinary Riemann-integral. We may also give a precise meaning to the stochastic differential equation

$$dv = -\beta v + dB$$

as an integral equation, namely

$$v(t,\cdot) = v(0,\cdot) - \int_0^t \beta\, v(s,\cdot)\, ds + \int_0^t dB(s,\cdot)\,.$$

The last integral of course is $B(t,\cdot) - B(0,\cdot)$. The process v is called the **Ornstein-Uhlenbeck velocity process.** In the figures on the left we compare the direction field of the Langevin equation (1.2.1) and a discrete approximation of (1.2.2). In each figure one smooth solution of (1.2.1) is drawn in.

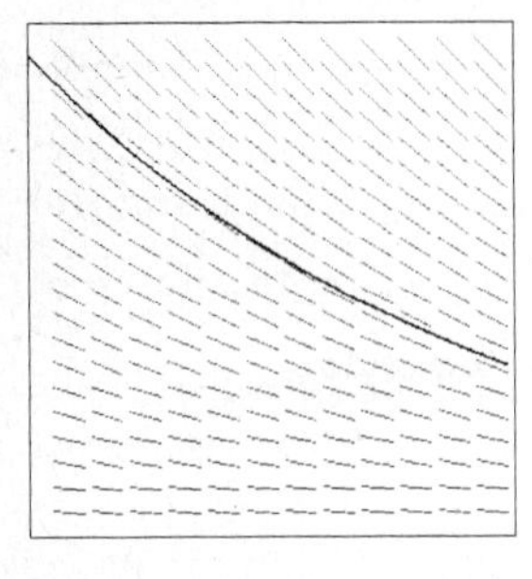

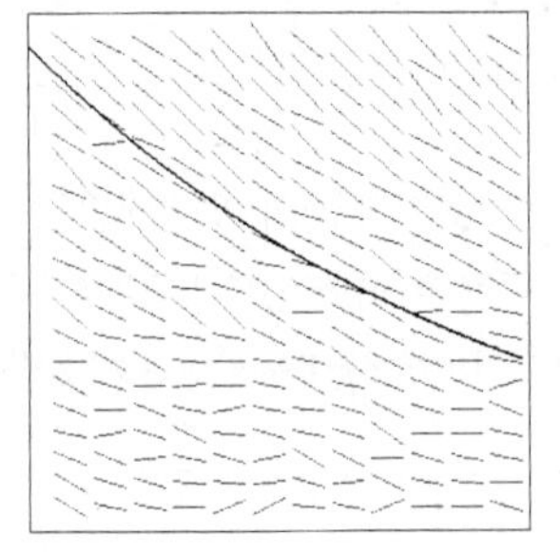

Fig. 1.2.1

Note that the mean or expectation $\mathbb{E}(v(t,\cdot))$ can be computed from the above formulas as

$$\mathbb{E}\big(v(t,\cdot)\big) = e^{-\beta t}\, \mathbb{E}\big(v(0,\cdot)\big)$$

i.e. the initial mean velocity is damped exponentially. Note further that the **Ornstein-Uhlenbeck position process**

$$x(t,\cdot) = x(0,\cdot) + \int_0^t v(s,\cdot)\, ds$$

is no Markov process in contrast to the velocity process. In fact, its transition behaviour depends on the past: If x is known on the time interval $]t - \delta, t[$ then $v(t,\cdot)$ is known and hence the direction into which x is going to move next. However, this direction cannot be identified from $x(t,\cdot)$ alone.

A detailed discussion of the Ornstein-Uhlenbeck process in the light of physical intuition is part of Nelson (1967). In recent years the (infinite dimensional) Ornstein-Uhlenbeck process was the object of research because it arises in a natural way if one studies smoothness of solutions of differential equations in dependence of smoothness of the coefficients. The keyword is "Malliavin calculus"; but this topic is beyond the scope of the present text.

[illegible] g the stochastic Stieltjes integral we refer to Letta [illegible] Dellacherie-Meyer (1980), chap. VIII.1. In section 7.3 [illegible] the general stochastic integral will become clear.

(C) A Stoch[illegible]egral for Deterministic Integrands. The integral touched upon in the preceding paragraphs is a random variable - hence "stochastic". On the other hand, the construction was carried out path by path in a most classical way. Let us now have a look at another type of integral. Only deterministic functions will be integrated (but they need not have finite variation). This restriction to nonrandom integrands allows a fairly simple construction. The integrating processes, however, may be of rather general type. It serves as a powerful tool in the study of stationary processes and time series. It is worth noting, that here the integral is not only used as a tool to solve stochastic differential equations but shows up in its own right.

Only $\boldsymbol{L}^2$-techniques will be used in this section. If $(\Omega, \boldsymbol{F}, \mu)$ is a measure space then one has the Hilbert space $\boldsymbol{L}^2(\Omega, \boldsymbol{F}, \mu)$ of μ-equivalence classes of real- or complex-valued square-integrable functions. The scalar product and the norm are given by

$$\langle f,g\rangle = \int f\cdot\bar{g}\, d\mu \text{ and } \|f\|_2^2 = \int_\Omega |f|^2\, d\mu$$

(where $\bar{z}$ denotes the complex conjugate of a complex number z).

Construction of the integral. Let I denote any interval - bounded or not - on the real line. We consider **second-order processes** $X = (X_t)_{t\in I}$, i.e. each X_t is a complex-valued and square-integrable random variable. (It does not cause any complications and will turn out convenient for applications to consider complex-valued processes.) The process X has **orthogonal increments** if

$$\mathbb{E}\Big((X_{t_2} - X_{t_1})(\bar{X}_{t_4} - \bar{X}_{t_3})\Big) = 0 \text{ whenever } t_1 < t_2 < t_3 < t_4.$$

Many important processes like Brownian motion and the Poisson martingale have orthogonal increments. To every such process X a real-valued increasing function m on I may be associated such that

$$m(t) - m(s) = \mathbb{E}\Big(|X_t - X_s|^2\Big) \quad \text{whenever } s < t. \tag{1.2.3}$$

We may simply fix some $t_0 \in I$ and set

$$m(t) = \mathbb{E}\Big(|X_t - X_{t_0}|^2\Big) \quad \text{for} \quad t \geq t_0$$

and

$$m(t) = -\mathbb{E}\left(|X_{t_0} - X_t|^2\right) \quad \text{for} \quad t \le t_0.$$

If for instance X is standard Brownian motion and $t_0 = 0$ then $m(t) = t$; similarly, for a Poisson martingale with intensity α we get $m(t) = \alpha t$. Since the function m increases it defines by the Riesz representation theorem a non-negative measure μ on the Borel sets $\mathscr{B}(I)$ of I by

$$\mu\big(]s,t]\big) = m(t+) - m(s+).$$

Observe that each function m satisfying (1.2.3) defines the same measure μ. For standard Brownian motion and the Poisson martingale with parameter one μ is Lebesgue measure λ on I.

We are going to define the integral $\int_I f\, dX$ as a random variable on $(\Omega, \boldsymbol{F}, \mathbb{P})$ - and in fact as an element of $\boldsymbol{L}^2(\Omega, \boldsymbol{F}, \mathbb{P})$ - for all functions f in $\boldsymbol{L}^2(I, \mathscr{B}(I), \mu)$. The integral is first defined for step functions

$$f(t) = \sum_{i=0}^{n-1} f_i \, \mathbf{1}_{]t_i, t_{i+1}]} \quad \text{where} \quad t_0 < t_1 < \ldots < t_n$$

and each t_i is a point of continuity for m (and $\mathbf{1}_A$ is the characteristic function of the set A). It is given pathwise as

$$\int_I f(s)\, dX_s(\omega) = \sum_{i=0}^{n-1} f_i\big(X_{t_{i+1}}(\omega) - X_{t_i}(\omega)\big). \tag{1.2.4}$$

It is well-defined and linear on the linear space $\boldsymbol{E}$ of these step functions. Now the essential point is that the map

$$f \longmapsto \int_I f\, dX$$

is an isometry between $\boldsymbol{E}$ considered as a subspace of $\boldsymbol{L}^2(I, \mathscr{B}(I), \mu)$ and its image in $\boldsymbol{L}^2(\Omega, \boldsymbol{F}, \mathbb{P})$. In fact, representing step functions f and g by means of common partitions $t_0, \ldots, t_n$ and using orthogonality of the increments we get

$$\int_\Omega \Big(\int_I f\, dX \cdot \overline{\int_I g\, dX}\Big)\, d\mathbb{P} = \mathbb{E}\Big(\sum_{i,j=0}^{n-1} f_i\, \bar{g}_j \big(X_{t_{i+1}} - X_{t_i}\big)\big(\bar{X}_{t_{j+1}} - \bar{X}_{t_j}\big)\Big)$$

$$= \Big(\sum_{i=0}^{n-1} f_i\, \bar{g}_i\, \mathbb{E}\big(\big|X_{t_{i+1}} - X_{t_i}\big|^2\big) = \sum_{i=0}^{n-1} f_i\, \bar{g}_i \big(m_{t_{i+1}} - m_{t_i}\big) = \int_I f\, \bar{g}\, d\mu .$$

For the last equality we used that m is continuous at all points t_i .

We may now extend the integral to the closure of $\boldsymbol{E}$ in $\boldsymbol{L}^2(\mu)$, i.e. to all of $\boldsymbol{L}^2(\mu)$. (The assumption that μ does not charge the jumping points of the functions in $\boldsymbol{E}$ does not invalidate this density statement.)

In summary, we have defined the integral $\int f\,dX$ in $\boldsymbol{L}^2(\mathbb{P})$ for all functions in $\boldsymbol{L}^2(\mu)$, the map $f \longmapsto \int f\,dX$ is a linear isomorphism and for step functions it has the form given in (1.2.4).

If $\mathbb{P}$ is Wiener measure $\mathbb{W}$ we have established a linear isomorphism between $\boldsymbol{L}^2(\mathbb{R}_+, \mathcal{B}(\mathbb{R}_+), \lambda)$ and $\boldsymbol{L}^2(\mathcal{C}(\mathbb{R}_+), \mathcal{B}(\mathcal{C}(\mathbb{R}_+)), \mathbb{W})$ which is called the **Wiener integral**. The integral w.r.t to the Poisson martingale is also included. It is not yet clear that the present approach results in the same integral as the pathwise construction in paragraph (A). This will become clear in chapter 7 (cf. corollary 7.1.9).

In the cases $I = [0,\infty[$ or $I = \mathbb{R}$ one sets

$$Y_t = \int_0^t f\,dX = \int \mathbf{1}_{[0,t]}\, f\,dX$$

thus defining a stochastic process Y which is easily seen to have again orthogonal increments.

One of the far-reaching applications is

The spectral representation of weakly stationary processes. These are complex second-order processes $Y = (Y_t)_{t\in\mathbb{R}}$ such that the covariance of any two variables Y_s and Y_t only depends on $t - s$. Let us assume that all Y_t have expectation zero. Then we may define the covariance function K by

$$K(h) = \langle Y_{t+h}, Y_t \rangle = \int Y_{t+h}\,\bar{Y}_t\,d\mathbb{P}.$$

The celebrated Bochner-Khintchine theorem tells us that

$$K(h) = \int_{-\infty}^{\infty} e^{ih\lambda}\,dm(\lambda)$$

for an increasing and *bounded* function m on $\mathbb{R}$. The nonnegative and finite measure μ induced by m is called the **spectral measure** of the process Y.

Example. Let us have a look at the "derivative $\frac{dB}{dt}$" of Brownian motion - which caused the trouble with the Langevin equation in (B) - from this point of view. This will also give an explanation of the name **white noise** for $\frac{dB}{dt}$. We fix any positve δ and consider the difference quotients

$$D^{(\delta)}(t) = \frac{1}{\delta}(B(t+\delta) - B(t)).$$

Since $\mathbb{E}((B(t) - B(s))^2) = t - s$ the covariance function of $D^{(\delta)}$ is

$$K(h) = \langle D^{(\delta)}(t+h), D^{(\delta)}(t)\rangle = \begin{cases} 0 & \text{if } |h| \geq \delta \\ \delta^{-2}(\delta - |h|) & \text{if } |h| < \delta \end{cases}.$$

We readily find the spectral measure setting $d\mu(\lambda) = dm(\lambda) = m'(\lambda)d\lambda$ with the **spectral density**

$$m'^{(\delta)}(\lambda) = \frac{1}{2\pi}\int_{-\delta}^{\delta} e^{-i\lambda h}\,\delta^{-2}(\delta - |h|)\,dh = \frac{1-\cos(\lambda\delta)}{\lambda^2\,\delta^2}.$$

If δ is small then $m'^{(\delta)}$ is nearly the constant 1. If therefore the limit of $D^{(\delta)}(t)$ for $\delta \to 0$ would exist in $\boldsymbol{L}^2$ then the limiting process should behave like a stationary process with constant spectral density (which cannot exist since spectral measures are finite). The fact that then all frequencies λ would appear with the same weight is the main reason for the name white noise. Note, however, that these considerations also apply to the Poisson martingale.

Let us continue with the discussion of the spectral representation. In the $\boldsymbol{L}^2$-theory of stationary processes one studies exclusively those features of stochastic processes which are inherent in the covariance function or - in Hilbert space terminology - in the scalar product of variables Y_t and Y_{t+h}. Therefore one tries to replace Y by an especially handy process with the same scalar products. To start with, one associates to the process Y with random variables Y_t in $\boldsymbol{L}^2(\Omega, \boldsymbol{F}, \mathbb{P})$ a family $(\varphi(Y_t))_{t\geq 0}$ of functions in $\boldsymbol{L}^2(I, \mathscr{B}(I), \mu)$ setting

$$\varphi(Y_t) = e^{it\cdot}$$

(note that μ is not necessarily normalized and hence $\varphi(Y)$ is not really a stochastic process; but all stochastic concepts which are of interest in the $\boldsymbol{L}^2$-theory - like the covariance function - have a Hilbert space analogue - like the scalar product). The map φ extends canonically to the linear span of the functions Y_t. We want to extend it further to the closure of this span in $\boldsymbol{L}^2(\mathbb{P})$ as an isometry into (and in fact onto) $\boldsymbol{L}^2(\mu)$. It is sufficient to observe that

φ preserves scalar products:

$$\langle\varphi(Y_{t+h}), \varphi(Y_t)\rangle = \int_{\infty}^{-\infty} e^{i(t+h)\lambda}\, \overline{e^{it\lambda}}\, dm(\lambda)$$

$$= \int_{-\infty}^{\infty} e^{ih\lambda}\, dm(\lambda) = K(h) = \langle Y_{t+h}, Y_t\rangle.$$

Because the functions $e^{it\cdot}$ are total in $\boldsymbol{L}^2(\mu)$ the map φ is even onto and thus an isometric isomorphism.

Let us have a closer look at the inverse mapping. Define a process Z by

$$Z_a = \varphi^{-1}\left(\mathbf{1}_{]-\infty,a]}\right).$$

By definition, the process Z is of second order. Moreover, it has orthogonal increments. In fact, for $a < b < c < d$ we compute

$$\mathbb{E}\left((Z_b - Z_a)\overline{(Z_d - Z_c)}\right) = \langle Z_b - Z_a\,,\, Z_d - Z_c\rangle$$

$$= \langle\varphi(Z_b) - \varphi(Z_a),\ \varphi(Z_d) - \varphi(Z_c)\rangle$$

$$= \int_{-\infty}^{\infty} \mathbf{1}_{]a,b]}(\lambda)\ \mathbf{1}_{]c,d]}(\lambda)\ d\mu(\lambda) = 0.$$

The increasing function of Z is at all points of continuity equal to m:

$$\mathbb{E}\left(|Z_b - Z_a|^2\right) = \int_{-\infty}^{\infty} \mathbf{1}_{]a,b]}(\lambda)\ d\mu = m(b) - m(a)\ .$$

It induces the same spectral measure μ as m. Therefore the integral $\int_{-\infty}^{\infty} f\, dZ$ is defined exactly on $\boldsymbol{L}^2(\mu)$ according to the discussion following (1.2.4). Moreover, this integral is nothing but the inverse of the operator φ:

$$\varphi^{-1}(f) = \int_{-\infty}^{\infty} f\, dZ\ .$$

In fact, this identity is easily checked for characteristic functions of intervals

$$\varphi\left(\int_{-\infty}^{\infty} \mathbf{1}_{]a,b]}\, dZ\right) = \varphi\left(Z_b - Z_a\right) = \mathbf{1}_{]-\infty,b]} - \mathbf{1}_{]-\infty,a]} = \mathbf{1}_{]a,b]}$$

and since φ^{-1} is linear and continuous the identity holds for general f.

In particular, for $f = e^{it\cdot}$ we get

$$Y_t = \varphi^{-1}\left(e^{it\cdot}\right) = \int_{-\infty}^{\infty} e^{ita}\, dZ_a.$$

This equality is called the **spectral representation** of the stationary process Y.

Once a stationary process is given in its spectral representation, it is easy to execute linear transformations of the process. They are frequently called **linear filters** and are of great importance in technical applications.

Linear Filters. Suppose we want to differentiate the process Y, i.e. we want to find the derivative

$$Y_t' = \lim_{h \to 0} \frac{1}{h}\left(Y_{t+h} - Y_t\right)$$

where the limit is taken in the $\mathcal{L}^2$-sense. Using the spectral representation we get for the difference quotient

$$\frac{1}{h}\left(Y_{t+h} - Y_t\right) = \int_{-\infty}^{\infty} e^{ita} \frac{1}{h}\left(e^{iha} - 1\right) dZ_a$$

and read off that Y_t' exists if and only if

$$\frac{1}{h}\left(e^{iha} - 1\right) \text{ converges in } \mathcal{L}^2(\mu) \text{ as } h \longrightarrow 0$$

which in turn is equivalent to $\int_{-\infty}^{\infty} a^2 \, d\mu(a) < \infty$. Then we get that

$$Y_t' = \int_{-\infty}^{\infty} e^{ita} \, ia \, dZ_a \, .$$

If we want to replace the process Y' by the process $(e^{it\cdot})_t$ on a suitable measure space we must compute the covariance function of Y':

$$K^{(')}(h) = \langle Y_{t+h}', \overline{Y_t'} \rangle$$

$$= \mathbb{E}\left(\int_{-\infty}^{\infty} e^{i(t+h)a} ia \, dZ_a, \overline{\int_{-\infty}^{\infty} e^{ita} ia \, dZ_a} \right)$$

$$= \int_{-\infty}^{\infty} e^{i(t+h)a} ia \cdot e^{-ita} \overline{ia} \, d\mu(a) = \int_{-\infty}^{\infty} e^{iha} a^2 \, d\mu(a)$$

and thus the spectral measure of Y' is given by $a^2 d\mu(a)$. Induction shows that the n-th derivative of Y has spectral measure $a^{2n} d\mu(a)$ if a^{2n} is μ-integrable.

This may be rephrased as follows: if the "input" Y is fed into the **linear filter "differentiation"** then the "output" X is of the form

$$X_t = \int_{-\infty}^{\infty} e^{ita} H(a) \, dZ_a \, , \int_{-\infty}^{\infty} |H(a)|^2 \, d\mu < \infty;$$

where $H(a) = ia$. Replacing in the above computation of the covariance ia by a general H(a) we see that the covariance of X is $\int_{-\infty}^{\infty} e^{ita} |H(a)|^2 \, d\mu(a)$ and hence the spectral measure of X is given by $|H(a)|^2 d\mu(a)$. This can be used to design linear filters just by a suitable choice of the **system function** H. If for instance H has support in a small interval around the origin - i.e. "the associated filter lets only small frequencies pass" - then the filter is called a **low pass filter**. Similarly, if H lives on two small intervals symmetrically positioned on the left and right of the origin, it produces a **band pass filter**.

Of particular interest is the case where the system function has the form

$$H(a) = \int_{-\infty}^{\infty} e^{iat} h(t)\, dt, \; h \in L^2(\lambda).$$

Since H is bounded and continuous the linear filter is always defined. The output is

$$X_t = \int_{-\infty}^{\infty} e^{ita} H(a)\, dZ_a = \int_{-\infty}^{\infty} \int_{-\infty}^{\infty} e^{ia(t+s)} h(s)\, ds\, dZ_a .$$

One can show that the order of integration may be interchanged. Hence

$$X_t = \int_{-\infty}^{\infty} \int_{-\infty}^{\infty} e^{ia(t+s)} h(s)\, dZ_a\, ds = \int_{-\infty}^{\infty} h(s)\, Y_{s+t}\, ds.$$

This is the **linear filter of integration** corresponding to the **weight function** h. Plainly, one could start the procedure with the weight function and compute the system function afterwards.

The word linear filters is used also for the corresponding electronical units.

A good reference for more detailed information still is J.L. Doob's monograph "Stochastic Processes".

1.3 A Guided Tour through Semimartingale Integration

The theory of Itô integration for semimartingales is ill-famed for requiring a lot of technical details. Our chapters 4 to 8 abound in them. This section overviews the most important landmarks.

(A) Continuous Time Martingales . Mathematically speaking the two main features of the Brownian motion process are

(i) the independent increments,

i.e. the probability distribution of $M(t,\cdot) - M(s,\cdot)$ is not affected by any information about what the process does up to time s. This property is shared by the Poisson process and the Poisson martingale.

(ii) the continuity of the paths.

We consider now stochastic processes on a probability space $(\Omega, \boldsymbol{F}, \mathbb{P})$. Note the following two facts:

Lemma 1. If M is a process with independent increments and zero expectation then M is a **martingale**.

Lemma 2. If M is a martingale whose paths are a.s. continuous and of finite variation on bounded intervals then M is a.s. constant in time.

These results show - whatever the meaning of the word martingale may be - that the paths of Brownian motion are not a.s. of locally finite variation (i.e. of finite variation on bounded intervals). Therefore the stochastic integral with respect to Brownian motion cannot be introduced as a Lebesgue-Stieltjes integral, i.e. stochastic integration is outside the range of Riesz' theorem. For deterministic integrands the construction from paragraph (C) in section 1.2 might apply but now we consider random integrands. So the plan for this section is to give some indication what one actually does in order to define these integrals and moreover to show that there still is an analogue of Riesz' theorem for stochastic integrals where now the functions of locally finite variation are replaced by semimartingales.

Definition. A **filtration** is a family $(\boldsymbol{F}_t)_{t \geq 0}$ of sub-σ-fields of the σ-field $\boldsymbol{F}$ such that $\boldsymbol{F}_s$ is contained in $\boldsymbol{F}_t$ for $s < t$.

The σ-field $\boldsymbol{F}_t$ contains all those 'events' E - i.e. measurable subsets of Ω - for which the question whether or not E occurs (i.e. whether or not the point ω which represents the actual course of the fate of the process lies in E) can be answered before time t.

Any filtration defines an increasing family $(L^2(\Omega, F_t, \mathbb{P}))_{t\geq 0}$ of subspaces of the Hilbert space $L^2(\Omega, F, \mathbb{P})$.

Definition. A process $M : \mathbb{R}_+ \times \Omega \longrightarrow \mathbb{R}$ is a (L^2-)**martingale** if and only if for $s < t$ the random variable $M(s,\cdot)$ is the orthogonal projection of $M(t,\cdot)$ to the space $L^2(\Omega, F_s, \mathbb{P})$. This projection operator is the conditional expectation and is denoted by $\mathbb{E}(\cdot \mid F_s)$.

Proofs of lemma 1 and lemma 2. For this introduction we assume that all variables $M_s = M(s,\cdot)$ are square integrable. As for lemma 1, the assumption may be rephrased saying that the increment $M_t - M_s$ is independent of the σ-field F_s generated by $\{M_{s'} : s' \leq s\}$ and therefore for every F_s-measurable L^2-random variable Z one has

$$\mathbb{E}(Z(M_t - M_s)) = \mathbb{E}(Z)\mathbb{E}(M_t - M_s) = \mathbb{E}(Z)(0 - 0) = 0$$

which is the orthogonality of $M_t - M_s$ to the space $L^2(\Omega, F_s, \mathbb{P})$. Since moreover $M_s \in L^2(\Omega, F_s, \mathbb{P})$ this shows that M_s is the orthogonal projection of M_t to this space. Thus M is a martingale.
Now assume for the proof of lemma 2 that M has continuous paths of locally finite variation. Then we could use Lebesgue-Stieltjes integration for these paths. In particular, we could use the following change of variable rule.

Lemma 3. Let m be continuous and of locally finite variation. Let φ be a $\mathscr{C}^1$-function on $\mathbb{R}$. Then

$$\varphi(m(t)) - \varphi(m(0)) = \int_0^t \varphi'(m(s))\, dm(s).$$

Plugging in $\varphi(y) = y^2$ and assuming for simplicity $M_0 = 0$ we would get for our martingale by Riemann approximation

$$\mathbb{E}(M_t^2) = \mathbb{E}\left(2\int_0^t M_s\, dM_s\right) = \lim \mathbb{E}\left(2 \sum M_{t_i} \left(M_{t_{i+1}} - M_{t_i}\right)\right) = 0 .$$

Here we have interchanged the order of taking the limit and of taking expectation. This is easily justified by standard probabilistic arguments like stopping before the values become too large (this concept will be explained in chapter 4). Moreover, we have used that M_{t_i} is in the space $L^2(\Omega, F_{t_i}, \mathbb{P})$. It is orthogonal to $M_{t_{i+1}} - M_{t_i}$ since M is a martingale. But $\mathbb{E}(M_t^2) = 0$ implies that $M_t^2 = 0$ almost surely, i.e. M is the zero process. This proves lemma 2.

(B) The Construction of the Martingale Integral. So let us give an outline of how one defines stochastic integrals for martingale integrators. One starts with a filtration $(\boldsymbol{F}_t)$ and a martingale M with respect to this filtration.

Technical note. The available texts on this subject all assume that the filtration satisfies the 'usual conditions' of P.A. Meyer but we shall verify that nothing of the following really needs these conditions. So we do not formulate them here.

One proceeds in three steps.

Step 1 : The elementary integral. A bounded process H is called elementary if all its paths are left-continuous step functions with the number of steps bounded as a function of ω and if H is **adapted** to the filtration, i.e. each random variable H_t is $\boldsymbol{F}_t$-measurable. An elementary process H can be written in the form

$$H = \sum_{i=1}^{n} H_i \, 1_{]T_i, T_{i+1}]}$$

where the (random) jumping times T_i are **stopping times**, i.e. $\{T_i < t\} \in \boldsymbol{F}_t$ for all t. For every elementary process H and every process M one can define the integral $\int_0^\infty H \, dM$ by pathwise integration:

$$\int_0^\infty \sum_{i=1}^{n} H_i \, 1_{]T_i, T_{i+1}]} \, dM = \sum_{i=1}^{n} H_i \left(M_{T_{i+1}} - M_{T_i} \right) .$$

Here everything (except n) depends on ω. The 'stopping theorem for martingales' implies that on the right all terms are orthogonal in $\boldsymbol{L}^2$ if M is a $\boldsymbol{L}^2$-bounded martingale with right-continuous paths. Hence we get the $\boldsymbol{L}^2$-estimate

$$\left\| \int_0^\infty H \, dM \right\|_2^2 = \sum \left\| H_i \left(M_{t_{i+1}} - M_{t_i} \right) \right\|_2^2$$

$$\leq \|H\|_\infty^2 \sum \left\| M_{t_{i+1}} - M_{t_i} \right\|_2^2 \leq \|H\|_\infty^2 \left\| M_{T_{n+1}} \right\|_2^2 .$$

Thus the elementary integral defines a linear operator $H \longmapsto \int_0^\infty H dM$ which is continuous w.r.t. the norms $\|\cdot\|_\infty$ and $\|\cdot\|_2$. Therefore we can pass to the

Step 2: Extension of the integral to $\boldsymbol{L}^2$-bounded martingales M and to uniform limits of elementary processes H.

Step 3: Delocalization. If there is a sequence (T_n) of stopping times converging a.s. to infinity and if for each n the process $(M_{t\wedge T_n})_{t\geq 0}$ is a $\mathcal{L}^2$-bounded martingale then M is called a **local $\mathcal{L}^2$-martingale**. If in addition each process $(H_{t\wedge T_n})_{t\geq 0}$ is a uniform limit of elementary processes then the definition of the integrals $\int_0^t H\, dM$ can be given for all t which are dominated by T_n for some n - i.e. for all t. The main new feature after this delocalization is the fact that statements about convergence which under the assumption of step 2 hold in $\mathcal{L}^2$ now hold only as convergence in probability. A crucial fact is

Theorem 1. The integral process $(\int_0^t H\, dM)_{t\geq 0}$ is again a local $\mathcal{L}^2$-martingale.

The integral developed so far is sufficient for many applications. It corresponds to the Riemann integral for regular functions in standard one-dimensional integration. Let us combine it with the pathwise Stieltjes integrals into one single concept.

Definition. A process X is called a **semimartingale** with respect to the given filtration if it can be written in the form X = M + A where M is a right-continuous local $\mathcal{L}^2$-martingale and A is adapted right-continuous and a.s. of locally finite variation. Given such a decomposition we write

$$\int_0^t H\, dX = \int_0^t H\, dM + \int_0^t H\, dA.$$

(C) An Application of the Poisson Integral. As we saw in the last section the integral w.r.t. the Poisson process or martingale can easily be introduced. Here is an application of theorem 1 and the concept of martingales to the Poisson process. It shows that the theory of martingale integrals is useful even if there is no difficulty in the definition.

Theorem (PASTA: Poisson Arrivals See Time Averages). Let $T_1, T_2, \ldots$ be the jumping times of a Poisson process. Let H be a bounded process adapted to the filtration with left-continuous paths. Consider

$$\lim_n \frac{1}{n} \sum_{i=1}^{n} H_{T_i} \quad \text{and} \quad \lim_t \frac{1}{t} \int_0^t H_s\, ds\,.$$

Then if either of the two limits exists a.s. then the other does and they are a.s. equal.

The result applies in particular to any deterministic continuous function H on $\mathbb{R}_+$ and it is really surprising that the Poisson points are a.s. well enough distributed so that one can read off the integral mean from this countable collection of points. The result simplifies some tricky questions in queuing theory. The proof consists basically in observing that a difference of the form $\sum_{i \le n} G_{T_i} - \int_0^{T_n} G_s ds$ is an integral with respect to the Poisson martingale and hence a martingale by theorem 1. Thus martingale convergence theory can be applied.

(D) The Quadratic Variation of a Semimartingale and Itô's Formula. If M is a local $\boldsymbol{L}^2$-martingale i.e. if it satisfies the assumption in step 3 then martingale theory shows that almost every path of M has left limits, i.e. it is a regular function. Similarly, functions of finite variation are regular. Since regular functions on compact intervals are uniform limits of step functions it is not surprising that the left-continuous version (X_t^-) of a semimartingale X can be integrated, i.e.

$$\int_0^t X_s^- \, dX_s = \lim \sum X_{t_i}\left(X_{t_{i+1}} - X_{t_i}\right) \text{ in probability}$$

makes sense. (We have removed the "$-$ " from the factor $X_{t_i}^-$ on the right-hand side; this is easily justified.) The limit is taken along a sequence of partitions of the interval [0,t] whose length converges to 0. Because of the algebraic identity

$$X_t^2 - X_0^2 = \sum X_{t_{i+1}}^2 - X_{t_i}^2$$
$$= \sum \left(X_{t_{i+1}} - X_{t_i}\right)^2 + 2X_{t_i}\left(X_{t_{i+1}} - X_{t_i}\right)$$

we see that the sum of squares of the increments converge to

$$X_t^2 - X_0^2 - 2\int_0^t X_s^- \, dX_s \,.$$

This process is called the quadratic variation process of X and is denoted by $[X,X]_t$ or for continuous X simply by $\langle X \rangle_t$. For Brownian motion B one has $\langle B \rangle_t = t$ or $\int_0^t B \, dB = \frac{1}{2}(B_t^2 - t)$. Note that for continuous X the process $\langle X \rangle$ would vanish if the transformation formula of lemma 3 would apply. More generally one has instead of lemma 3

Theorem 2. (Itô-formula). Let φ be a $\mathscr{C}^2$-function on $\mathbb{R}$ and let X be a semimartingale with a.s. continuous paths. Then

$$\varphi(X_t) - \varphi(X_0) = \int_0^t \varphi'(X_s)\, dX_s + \tfrac{1}{2} \int_0^t \varphi''(X_s)\, d\langle X\rangle_s .$$

The idea of the proof is of course to sum up the second order Taylor approximation

$$\varphi(X_s) - \varphi(X_r) \approx \varphi'(X_r)(X_s - X_r) + \tfrac{1}{2}\varphi''(X_r)(X_s - X_r)^2 .$$

There is also a version of Itô's formula for noncontinuous X. But it looks more complicated. As an illustration of Itô's formula let us check in a complicated way that an affine function of an a.s. continuous local martingale X is again an a.s. continuous local martingale: If φ is affine then $\varphi'' = 0$ and hence the second integral vanishes and the first integral defines a local martingale according to theorem 1. In exactly the same way one verifies in higher dimensions that a harmonic function of a vector valued Brownian motion is again a local martingale. This indicates that there is a close connection between these stochastic integrals and elliptic equations.

Another application is the following

Corollary. A $\mathscr{C}^2$-function of a continuous semimartingale is a semimartingale.

(E) Predictability and Meyer's Desintegration Theorem. Now let us return to the abstract integration theory. Here it is convenient to change the way to think of processes. We now really consider the processes as functions on the product space $\mathbb{R}_+ \times \Omega$.

Definition. A process is called **predictable** if it is measurable with respect to the predictable σ-field on $]0,\infty[\times \Omega$ which is generated by all continuous adapted processes or equivalently by the stochastic intervals

$$]0,T] = \{(t,\omega) : 0 < t \le T(\omega)\}$$

where T is a stopping time.

The Poisson process is not predictable ! There is a gap between the σ-field generated by the right-continuous adapted processes and the left-continuous adapted processes, respectively.

The following result is at the heart of the more advanced theory of stochastic integration. Its difficulty lies exactly in the fact that the right-open stochastic intervals generate a larger σ-field than the predictable one.

Theorem 3 (P.A. Meyer (1963)**).** Let μ be a finite nonnegative measure on the predictable σ-field. If there is a right-continuous a.s. increasing process A such that $A_0 = 0$ and $\mu([0,T]) = \mathbb{E}(A_T)$ for every stopping time T then there is even a *predictable* process with the same properties.

Another illustration of the role of predictability is given by the following result.

Proposition. Every right-continuous predictable martingale is a.s. continuous.

This explains for example that the Poisson martingale cannot be predictable. An application of this is the

Corollary. The process A in Meyer's theorem is unique a.s..

Indeed if A and A' are as in the theorem then the difference is right-continuous, of locally finite variation and it satisfies $\mathbb{E}(A_T - A'_T) = 0$ for every stopping time. This implies that it is a (local) martingale and up to localization lemma 2 shows that it vanishes a.s..

(F) A Stochastic Analogue of Riesz' Theorem. We close the circle of the ideas in this section by explaining some of the ideas behind the following result of Dellacherie-Bichteler-Mokobodzki. It shows that semimartingales are for stochastic integration what functions of finite variation are for classical integration.

Theorem 4. Let X be a right-continuous adapted process such that the elementary integral given by step 1 is continuous in the sense that

$$\|H^n\|_\infty \longrightarrow 0 \text{ implies } \int_0^t H^n \, dX \longrightarrow 0 \text{ in probability for every } t.$$

Then X is a semimartingale.

The first step of the proof is to note that two equivalent probability measures on the underlying space have the same semimartingales. This is a beautiful result related to what is called the Cameron-Martin-Maruyama-Girsanov formula. Therefore one may choose the measure $\mathbb{P}$ freely in its equivalence class. A functional analytic argument then shows that $\mathbb{P}$ can be chosen such that sup $|X_t|$ is $\mathbb{P}$-integrable. Assuming this integrability we define the set function μ on the predictable sets by

$$\mu([0,T]) = \mathbb{E}(X_T) = \mathbb{E}\Big(\int_0^\infty \mathbf{1}_{[0,T]} \, dX\Big).$$

Very much like one sees that on the real line every right-continuous function of finite variation defines a σ-additive set function, one shows that μ is a signed measure and hence the theorem of Meyer gives a process of finite variation A such that

$$\mathbb{E}(A_T) = \mu([0,T]) = \mathbb{E}(X_T)$$

for all stopping times T. Therefore the difference process M = X - A satisfies $\mathbb{E}(M_T) = 0$ for all T. This implies that it is a martingale and X = M + A is the desired representation of X as a semimartingale.

CHAPTER 2

PROCESSES AND FILTRATIONS

A **stochastic process** on a probability space $(\Omega,\boldsymbol{F},\mathbb{P})$ is a family $(X_t)_{t\in I}$ of random variables on Ω. In this book, the index set I is always a subset of $[0,\infty]$. The process X may also be viewed as a single function of two variables t and ω. Stopping times and filtrations are introduced in order to deal with measurability questions arising from the interplay between t an ω.

Let us introduce some conventions. Random variables are real-valued or extended real-valued; nevertheless, we will also consider random variables taking values in Euclidean n-space. Most frequently, the index set I of a process will be either a finite set $\{0,\dots,n\}$ or $\{0,\dots\}$ or $\mathbb{R}_+$. We will write $X(t,\cdot)$ instead of $X_t(\cdot)$ if convenient. The functions $t \longmapsto X(t,\omega)$ where $\omega \in \Omega$ are called the **paths** of X. We adopt the following way of speaking: If a certain property of functions on I is shared by all (resp. $\mathbb{P}$-almost all) paths of X we ascribe this property to X. Thus we call a process **left-continuous** (resp. **a.s. left-continuous**) if all (resp. almost all) of its paths are left-continuous functions on I, etc. . If X and Y are two processes we say that X **equals** Y **a.s.** if $\mathbb{P}$-almost all paths of the two processes coincide on all of I. Some authors use the somewhat heavy term indistinguishability. Others call X and Y **versions** of each other. If only $\mathbb{P}(X_t = Y_t) = 1$ for each t then X and Y are **modifications** of each other.

2.1 Filtrations

Let us introduce the first basic concept.

2.1.1 Definition. Suppose that $(\Omega,\boldsymbol{F})$ is a measurable space and $I \subset [0,\infty]$ is an index set. A **filtration** is a family $(\boldsymbol{F}_t)_{t\in I}$ of sub-σ-fields of $\boldsymbol{F}$ such that $\boldsymbol{F}_s \subset \boldsymbol{F}_t$ whenever $s \le t$.

Most of the time the index set is either $\mathbb{R}_+$ or $\mathbb{N}$; in these cases we simply write $t \geq 0$ and $n \geq 1$ instead of $t \in \mathbb{R}_+$ and $n \in \mathbb{N}$.

The σ-field $\boldsymbol{F}_t$ may be viewed as representing the knowledge of a hypothetical person at time t - i.e. the totality of all events for which at time t this person can decide whether they occur or not. So, a filtration mirrors the increase of information as time increases.

The most important filtrations are those generated by stochastic processes. The σ-field generated by a family $\boldsymbol{A}$ of sets or functions is denoted by $\sigma(\boldsymbol{A})$.

2.1.2 Definition. Let $X = (X_t)_{t \geq 0}$ be a stochastic process.

(a) Call

$$\boldsymbol{F}_t = \sigma\left(X_s : s \leq t\right)$$

the **σ-field of the past of X up to time t**. The family $(\boldsymbol{F}_t)_{t \geq 0}$ is called the **own** or **natural filtration** of X.

(b) The process X is **adapted** to a filtration $(\boldsymbol{F}_t)_{t \geq 0}$ or **$(\boldsymbol{F}_t)$-adapted** if for every $t \geq 0$ the random variable X_t is $\boldsymbol{F}_t$-measurable.

Sometimes in the literature an adapted process is called "nonanticipating" in accordance with the interpretation that all the information which can be obtained by observing X up to time t is contained in $\boldsymbol{F}_t$. Clearly, the natural filtration of a process X is the smallest filtration to which X is adapted. Often, it is convenient or even necessary to enlarge the natural filtration. Consider for instance two processes X and Y simultaneously. Then

$$\boldsymbol{F}_t = \sigma(X_s, Y_s : s \leq t)$$

gives the appropriate filtration which in general will be larger than the natural filtrations both of X and Y.

In the case of continuous time, frequently one adds to $\boldsymbol{F}_t$ the information of an infinitesimal glance into the future.

2.1.3 Definition. Given a filtration $(F_t)_{t \ge 0}$ let

$$F_t^+ = \bigcap_{\varepsilon > 0} F_{t+\varepsilon}$$

for every $t \ge 0$. A filtration is called **right-continuous** if $F_t = F_t^+$ for all t.

Plainly, the family $(F_t^+)_{t \ge 0}$ is itself a right-continuous filtration and $F_t \subset F_t^+$. The following example illustrates the meaning of the "+".

2.1.4 Example. (a) Let Ω be the space $\mathscr{C}(\mathbb{R}_+)$ of continuous functions on $\mathbb{R}_+$, denote by Π the coordinate process $\Pi_t(\omega) = \omega(t)$ and by $(F_t)_{t \ge 0}$ its natural filtration. Then for every F_t-measurable function f and $\omega \in \Omega$ the value $f(\omega)$ depends only on the restricted path $\omega|[0,t]$. If f is F_t^+-measurable then $f(\omega)$ is determined by $\omega|[0,t+\varepsilon]$ for each $\varepsilon > 0$ but not necessarily by $\omega|[0,t]$.
(b) For the natural filtration (F_t) of a Markov process usually the two filtrations (F_t) and (F_t^+) are "almost identical" (Blumenthal's 0-1 law cf. corollary 9.7.2). If $\mathbb{P}$ describes e.g. standard Brownian motion and (F_t) is its natural filtration then for every F_t^+-measurable f there is a F_t-measurable $\tilde{f}$ with $\mathbb{P}(f = \tilde{f}) = 1$.
(c) On the other hand, suppose that $\mathbb{P}$ is the law of the (non-Markovian) process $X_t = C \cdot B_t$ where B is a standard Brownian motion starting at 0 and the nonnegative random "scale" variable C is independent of B. Then using quadratic variation (see example 5.6.6) the value of C can be computed from the restriction of almost every path of X to $[0,\varepsilon]$ for every $\varepsilon > 0$. This implies that C is $\mathbb{P}$-a.s. equal to a (F_0^+) measurable random variable. But $\mathbb{P}(X_0 = 0) = 1$ and hence C is not $\mathbb{P}$-a.s. equal to a F_0-measurable random variable if it is not $\mathbb{P}$-a.s. constant.

Finally, we mention the "usual conditions" on a filtration. As the name indicates they are a standard assumption in many texts on our subject.

2.1.5 Definition. Let $(\Omega, F, \mathbb{P})$ be a probability space.
(a) For every σ-field $G \subset F$ let ${}^oG$ be the nullset augmentation of G, i.e. the σ-field generated by G and the system of all F-measurable $\mathbb{P}$-nullsets.
(b) A filtration $(F_t)_{t \ge 0}$ is said to satisfy the **usual conditions** if

(i) $\boldsymbol{F}_t = \boldsymbol{F}_t^+$: the filtration is right-continuous;

(ii) $\boldsymbol{F}_t = {}^{\circ}\boldsymbol{F}_t$: every $\mathbb{P}$-nullset $N \in \boldsymbol{F}$ belongs to all σ-fields $\boldsymbol{F}_t$, $t \geq 0$;

(iii) every subset of a $\mathbb{P}$-nullset is $\boldsymbol{F}_t$-measurable.

Every filtration can be augmented such as to satisfy any combination of the three properties (i), (ii), (iii). However, this augmentation in general destroys some other properties of the filtration. We do not assume the usual conditions. In the case of (i) and (ii) this implies a surprisingly small amount of extra work and we do not cover material in which (iii) is essential.

The construction of an adapted process gives a slightly stronger result if one chooses $(\boldsymbol{F}_t)_{t\geq 0}$ instead of $(\boldsymbol{F}_t^+)_{t\geq 0}$. But during the work, the larger filtration $(\boldsymbol{F}_t^+)_{t\geq 0}$ is a very useful tool. This is true in particular in connection with stopping times to which we turn in the next section.

That we do not want to assume (ii) has the following reason (cf. D. Stroock and S. Varadhan (1979)) : In some situations one would like to use the same filtration with two nonequivalent probability measures at the same time. Suppose e.g. that $\Omega = \mathscr{C}(\mathbb{R}_+)$ and the filtration $(\boldsymbol{F}_t)$ are the same as in Wiener space and $\mathbb{W}^x$ is the distribution of a Brownian motion which starts a.s. at some point $x \neq 0 \in \mathbb{R}$. Then the system of all $\mathbb{W}^x$-nullsets induces the full Borel-σ-field on the subset

$$\Omega_0 = \{\omega \in \Omega : \omega(0) = 0\}.$$

Adding the $\mathbb{W}^x$-nullsets to $\boldsymbol{F}_0$ destroys the filtration on Ω_0 which is the space on which ordinary Brownian motion lives. So, if one studies nonequivalent probability distributions this nullset enlargement of the filtration may not be feasible.

As a consequence we have to distinguish e.g. between continuous processes and a.s. continuous processes. But the reader should keep in mind that there is always the obvious "nullset elimination argument": If a probabilistic statement holds, say, for every continuous process with some additional properties then one can expect it to carry over to the almost surely continuous case. One just substitutes the original space Ω by the set $\Omega\backslash N$ with the induced filtration and measure (where N is the exceptional set). On $\Omega\backslash N$ the continuous version of the result in question can be applied and usually the conclusion stays valid if one shifts back to Ω.

2.2 Stopping Times

Suppose one wants to stop a stochastic process at the instant when a given event occurs - say when it crosses a certain level for the third time. Then the decision to stop has to be taken for each path individually and it can only be based on the behaviour of the path in the past and may not depend on any information about the future. This is made precise by the following

2.2.1 Definition. Suppose that $(\Omega, \boldsymbol{F})$ is a measurable space with a filtration $(\boldsymbol{F}_t)_{t\geq 0}$. We call a map $T : \Omega \longrightarrow [0,\infty]$

(a) a **strict $(\boldsymbol{F}_t)$-stopping time** if $\{T \leq t\} \in \boldsymbol{F}_t$ for all $t \geq 0$,

(b) a **$(\boldsymbol{F}_t)$-stopping time** if $\{T \leq t\} \in \boldsymbol{F}_t^+$ for all $t \geq 0$.

The specification "$(\boldsymbol{F}_t)$-" is omitted unless the choice of the filtration is not clear from the context.

Note. Frequently, in the literature the term "stopping time" is used for our strict stopping times and stopping times in the sense of 2.2.1(b) are called "wide sense stopping times". But in stochastic integration the latter are much more important and therefore we let them have the simpler name. Plainly, this distinction is superfluous if $(\boldsymbol{F}_t)_{t\geq 0}$ is right-continuous.

If the filtration $(\boldsymbol{F}_t)_{t\in I}$ has a discrete index set I then a map $T : \Omega \longrightarrow I \cup \{\infty\}$ is a stopping time if it satisfies either of the conditions "$\{T \leq t\} \in \boldsymbol{F}_t$ for all $t \in I$" or "$\{T = t\} \in \boldsymbol{F}_t$ for all $t \in I$" which are easily seen to be equivalent for discrete index sets (cf. also the comment following lemma 2.2.2).

Every strict stopping time is a stopping time. Obviously, a constant random variable $T \equiv t$ is a strict stopping time. Stopping times are denoted by capital letters in contrast to real numbers which we denote by small letters. Thus a real number will not be confused with a constant stopping time.

Stopping times can be defined in terms of the original filtration $(\boldsymbol{F}_t)$ instead of $(\boldsymbol{F}_t^+)$ as the next lemma shows.

2.2.2 Lemma. Let $(\boldsymbol{F}_t)_{t\geq 0}$ be a filtration. A map $T : \Omega \longrightarrow [0,\infty]$ is a stopping time if and only if $\{T < t\} \in \boldsymbol{F}_t$ for all $t \geq 0$.

Proof. Asume that $\{T < t\} \in \boldsymbol{F}_t$ for all $t \geq 0$. Then

$$\{T \leq t\} = \bigcap_{n \in \mathbb{N}} \{T < t + \tfrac{1}{n}\} \in \bigcap_{n \in \mathbb{N}} \boldsymbol{F}_{t+1/n} = \boldsymbol{F}_t^+ \quad \text{for all } t \geq 0 .$$

Conversely, assume $\{T \leq t\} \in \boldsymbol{F}_t^+$ for all $t \geq 0$. Then

$$\{T < 0\} = \varnothing \in \boldsymbol{F}_0$$

and for $t > 0$ one has

$$\{T < t\} = \bigcup_{n \in \mathbb{N}} \{T \leq t - \tfrac{1}{n}\} \in \boldsymbol{F}_t$$

since $\boldsymbol{F}^+_{t-\frac{1}{n}} \subset \boldsymbol{F}_t$ for all $n > 0$. ■

Comment. (a) Since $\boldsymbol{F}_t^{++} = \boldsymbol{F}_t^+$ the filtrations $(\boldsymbol{F}_t)$ and $(\boldsymbol{F}_t^+)$ have the same stopping times (but not the same strict stopping times). In particular, T is a stopping time if $\{T < t\} \in \boldsymbol{F}_t^+$ for all t.

(b) Another condition characterizing stopping times is

$$\{T < t\} \in \sigma\Big(\bigcup_{s<t} \boldsymbol{F}_s\Big) \text{ for all } t > 0.$$

This condition actually is equivalent to "$\{T \leq t\} \in \boldsymbol{F}_t$ for all t" in the discrete case. We did not take it as a definition only because the σ-field $\sigma\big(\bigcup_{s<t} \boldsymbol{F}_s\big)$ is not needed in the remaining text except in the optional chapter 6.

Given a stopping time T one introduces a mathematical equivalent of "the information gathered up to time T":

2.2.3 Definition. To every random time $T : \Omega \to [0,\infty]$ we associate the **collections** $\boldsymbol{F}_T$ and $\boldsymbol{F}_T^+$ of **events known up to time T** (the "T-past")

$$\boldsymbol{F}_T = \{A \in \boldsymbol{F} : A \cap \{T \leq t\} \in \boldsymbol{F}_t \text{ for all } t \geq 0\}$$

$$\boldsymbol{F}_T^+ = \{A \in \boldsymbol{F} : A \cap \{T \leq t\} \in \boldsymbol{F}_t^+ \text{ for all } t \geq 0\}.$$

Note that $\boldsymbol{F}_T$ is defined also in the discrete case. Plainly, $\boldsymbol{F}_T = \boldsymbol{F}_t$ and $\boldsymbol{F}_T^+ = \boldsymbol{F}_t^+$ for a constant stopping time $T \equiv t$.

2.2.4 Lemma. (a) Let T be a strict stopping time. Then $\boldsymbol{F}_T$ is a σ-field and a random variable f is $\boldsymbol{F}_T$-measurable if and only if $f \cdot \mathbf{1}_{\{T \leq t\}}$ is $\boldsymbol{F}_t$-measurable for all $t \geq 0$. In particular, T itself is $\boldsymbol{F}_T$-measurable.

(b) If T is a stopping time then $\mathcal{F}_T^+$ is a σ-field and a random variable f is $\mathcal{F}_T^+$-measurable if and only if $f \cdot \mathbf{1}_{\{T<t\}}$ is $\mathcal{F}_t$-measurable for all t.

Proof. Part (a) is a straightforward application of the definitions. The proof of (b) just repeats the arguments in the proof of lemma 2.2.2. ∎

The next lemma compares the σ-fields associated with several stopping times.

2.2.5 Lemma. (a) If $S \le T$ are stopping times then $\mathcal{F}_S^+ \subset \mathcal{F}_T^+$.

(b) If $(S_n)_{n\ge1}$ is a sequence of stopping times then $S = \inf_{n\ge1} S_n$ is a stopping time and one has

$$\mathcal{F}_S^+ = \bigcap_{n=1}^{\infty} \mathcal{F}_{S_n}^+.$$

(c) If $(S_n)_{n\ge1}$ is a sequence of stopping times then $S = \sup_{n\ge1} S_n$ is a stopping time.

Proof. (a) Choose $A \in \mathcal{F}_S^+$ and $t \ge 0$. Then

$$A \cap \{T \le t\} = \big(A \cap \{S \le t\}\big) \cap \{T \le t\} \in \mathcal{F}_t^+.$$

Hence $A \in \mathcal{F}_T^+$.

(b) The random variable $S = \inf S_n$ is a stopping time since

$$\{S < t\} = \bigcup_{n=1}^{\infty} \{S_n < t\} \in \mathcal{F}_t.$$

From (a) follows $\mathcal{F}_S^+ \subset \bigcap_{n=1}^{\infty} \mathcal{F}_{S_n}^+$. For the converse, choose $A \in \bigcap_{n=1}^{\infty} \mathcal{F}_{S_n}^+$. Then by lemma 2.2.4(b),

$$A \cap \{S < t\} = A \cap \bigcup_{n=1}^{\infty} \{S_n < t\} = \bigcup_{n=1}^{\infty} A \cap \{S_n < t\} \in \mathcal{F}_t$$

and hence $A \in \mathcal{F}_S^+$. This completes the proof of (b).

(c) For $S = \sup S_n$ and $t \ge 0$ the identity

$$\{S \le t\} = \bigcap_{n=1}^{\infty} \{S_n \le t\} \in \mathcal{F}_t^+.$$

and thus (c) holds. ∎

The definition of $\mathcal{F}_T$ is somewhat formal. Therefore we give a more explicit description of $\mathcal{F}_T$ in an important special case. On most canonical spaces of paths the σ-fields $\mathcal{F}_T$ are generated by the paths up to time T. To be more specific we take $\Omega = \mathscr{C}(\mathbb{R}_+)$. We shall not use example 2.2.6 in the sequel.

2.2.6 Example. Suppose that $\Omega = \mathscr{C}(\mathbb{R}_+)$, Π_t is the evaluation map $\omega \longmapsto \omega(t)$, $\boldsymbol{F} = \sigma(\Pi_t : t \geq 0)$ and $(\boldsymbol{F}_t)_{t\geq 0}$ is the natural filtration of $\Pi = (\Pi_t)_{t\geq 0}$. Then for each strict stopping time T one has

$$\boldsymbol{F}_T = \sigma(\Pi_{t\wedge T} : t \geq 0).$$

Proof. We show here just the inclusion "$\subset$". For the reverse inclusion some technical details must be clarified and so the proof is deferred to the end of the next section.

Define $a_T : \mathscr{C}(\mathbb{R}_+) \longrightarrow \mathscr{C}(\mathbb{R}_+)$ by $a_T(\omega)_t = \omega_{t\wedge T(\omega)}$ or $\Pi_t \circ a_T = \Pi_{t\wedge T}$. Let $f : \mathscr{C}(\mathbb{R}_+) \longrightarrow \mathbb{R}$ be a $\boldsymbol{F}_T$-measurable function. We show that $f = f \circ a_T$. Since f is $\sigma(\Pi_t : t \geq 0)$-measurable it then follows that it is measurable w.r.t. $\sigma(\Pi_t \circ a_T : t \geq 0) = \sigma(\Pi_{t\wedge T} : t \geq 0)$ which proves the inclusion.

Fix a point $\omega_0 \in \Omega$. Let $t_0 = T(\omega_0)$ and $c = f(\omega_0)$. The variable $f\cdot\mathbf{1}_{\{T\leq t_0\}}$ is $\boldsymbol{F}_{t_0}$-measurable since f is $\boldsymbol{F}_T$-measurable. On the other hand, $\boldsymbol{F}_{t_0}$ is generated by the variables Π_s, $s \leq t_0$, and for each $s \leq t_0$,

$$\Pi_s(\omega_0) = \Pi_{s\wedge T(\omega_0)}(\omega_0) = \Pi_s(a_T(\omega_0)) .$$

Therefore, all $\boldsymbol{F}_{t_0}$-measurable functions coincide at the two points ω_0 and $a_T(\omega_0)$. In particular, $a_T(\omega_0)$ is also in the $\boldsymbol{F}_{t_0}$-measurable set $\{T \leq t_0\}$. Hence

$$f(a_T(\omega_0)) = f\cdot\mathbf{1}_{\{T\leq t_0\}}(a_T(\omega_0)) = f\cdot\mathbf{1}_{\{T\leq t_0\}}(\omega_0) = f(\omega_0)$$

which completes the proof. ■

2.3 Stopping a Process

Our first aim is to discuss an important class of stopping times: the random times when a stochastic process enters a given set. Clearly, the fact that a process $(X_t)_{t\geq 0}$ has visited a set G before time t depends only on the history of the process up to time t and thus the first entrance time of X into G should be a stopping time. However, there remains a measurability problem which is trivial only under additional assumptions on X and G. The concept of a separable process is useful in this context.

2.3.1 Definition. An $\mathbb{R}^d$-valued stochastic process $(X_t)_{t\geq 0}$ is called **separable** if there is a countable set U of indices such that for each $\omega \in \Omega$ the set $\{(u,X_u(\omega)) : u \in U\}$ is dense in the set $\{(t,X_t(\omega)) : t \geq 0\}$. Any such set U is a **separability set** for X.

In other words, the value of the path at time t can be approximated by values of the path in $U \cap]t-\varepsilon,t+\varepsilon[$ for each $\varepsilon > 0$ and $t \geq 0$.

We could equally well work with the smaller class of processes which have a fixed countable dense subset of $\mathbb{R}_+$ - say $\mathbb{Q}_+$ - as a separability set. The more general concept of definition 2.3.1 is better known because J.L. Doob proved that every process has a separable modification. We do not make use of Doob's theorem. For us, the important fact is that the first time when a separable adapted process enters an open domain is a stopping time (see proposition 2.3.4 below) and that separability is easy to verify (cf. proposition 2.3.2). Also it is convenient that for separable processes $(X_t)_{t\geq 0}$ suprema and infima can be taken over whole t-intervals without destroying the measurability.

2.3.2 Proposition. If at every point t each path of a process is either right-continuous or left-continuous then the process is separable.

Proof. $\mathbb{Q}_+$ is a separability set. ■

The entrance times mentioned above are defined as follows.

2.3.3 Definition. Let $X = (X_t)_{t \geq 0}$ be a $\mathbb{R}^d$-valued process and let G be a subset of $\mathbb{R}^d$. Then **the entrance time** $T^G : \Omega \longrightarrow [0,\infty]$ of X into G is defined by

$$T^G(\omega) = \inf\left\{t \geq 0 : X_t(\omega) \in G\right\}.$$

It may happen that a path never hits the set in question. Then the entrance time is infinite following the convention $\inf \emptyset = +\infty$.

2.3.4 Proposition. The entrance time of a d-dimensional separable and $(\mathcal{F}_t)$-adapted process into an open set is a stopping time.

Proof. Let X denote the process in question, let G be an open set in $\mathbb{R}^d$ and let U be a (countable) set of separability for X. For $t \geq 0$ one has

$$\left\{T^G < t\right\} = \bigcup_{s<t} \left\{X_s \in G\right\} = \bigcup_{s \in U \cap [0,t[} \left\{X_s \in G\right\} \in \mathcal{F}_t.$$

The restriction to the countably many indices in U is possible since G is open. Hence T^G is a stopping time by lemma 2.2.2. ■

Using nontrivial results of abstract measure theory (e.g. G. Choquet's capacitability theorem) one can show that the entrance time into *any* Borel set is a stopping time if the filtration satisfies the usual conditions 2.1.5 and if the process X is progressively measurable (cf. definition 2.3.8 below). This is where the completeness condition 2.1.5 (iii) would be essential. For us proposition 2.3.4 is sufficient and it would be inadequate to invoke that general theory here.

It is at least plausible that if a process X enters G immediately after time t with a jump then this cannot be predicted from the behaviour of X up to time t. For reasons like this in general one cannot expect the entrance times to be *strict* stopping times. Just for the sake of illustration we show that the entrance times of *continuous* processes into *closed* sets *are* strict stopping times. In this case they are even " predictable" in a precise sense. Later on, this concept of predictable stopping times will appear mainly in the optional chapter 6. The reader may skip the proof of proposition 2.3.5 in a first round.

2.3.5 Proposition. Suppose that the d-dimensional process X is $(\boldsymbol{F}_t)$-adapted and has continuous paths. Then the entrance time T^C into a closed set C is a strict stopping time.
It is predictable in the following sense: There is an "announcing" sequence $(T_n)_{n\geq 1}$ of stopping times such that $T_n < T^C$ on $\{T^C > 0\}$ and $T_n \uparrow T^C$.

Proof. Let G_n be the open set of all points in $\mathbb{R}^d$ whose distance from C is strictly smaller than $\frac{1}{n}$. We show

$$T^{G_n} < T^C \text{ on } \{0 < T^C < \infty\} \text{ and } \sup_{n\geq 1} T^{G_n} = T^C. \tag{2.3.1}$$

The first statement is obvious. For the second one, fix $\omega \in \Omega$ and assume that

$$t^* = \sup_{n\geq 1} T^{G_n}(\omega) < \infty$$

(if $t^* = \infty$ nothing has to be proved). Choose a sequence (t_n) converging to t^* such that $X_{t_n}(\omega) \in G_n$ for all n. Since the sets G_n decrease even $X_{t_r}(\omega) \in G_n$ for each $r \geq n$ and consequently (the bar indicates closure)

$$X_{t^*}(\omega) \in \bigcap_{n\geq 1} \overline{G_n} = C .$$

Hence $T^C \leq t^*$ which proves (2.3.1). By proposition 2.3.4

$$\{T^C \leq t\} = \bigcap_{n\geq 1} \{T^{G_n} < t\} \in \boldsymbol{F}_t \text{ for all } t > 0$$

and

$$\{T^C = 0\} = \{X_0 \in C\} \in \boldsymbol{F}_0.$$

Hence T^C is a strict stopping time. Moreover, $T_n = n \wedge T^{G_n}$ defines a sequence with the desired properties. ■

Our second goal is to show that the evaluation of an adapted process X at a stopping time is consistent with the measurable structure induced by the stopping time. The state of the process at the moment of stopping is described by the random variable X_T.

2.3.6 Definition. Let $(X_t)_{t\geq 0}$ be a stochastic process and $T : \Omega \longrightarrow [0,\infty]$ be a stopping time. We set $X_T(\omega) = X_{T(\omega)}(\omega)$ if $T(\omega) < \infty$. If both $T(\omega) = \infty$ and $\lim_{t\to\infty} X_t(\omega)$ exists let $X_T(\omega)$ be this limit. Otherwise X_T is not defined. The symbol $X_T \mathbf{1}_{\{T<\infty\}}$ is used for the function which agrees with X_T on $\{T < \infty\}$ and vanishes on $\{T = \infty\}$.

The property we are looking for can be expressed in two equivalent ways.

2.3.7 Proposition. For a $(\boldsymbol{F}_t)$-adapted process X the following are equivalent:
(a) X_T is $\boldsymbol{F}_T^+$-measurable for every finite stopping time T,
(b) the process $(X_{t\wedge T})_{t\geq 0}$ is $(\boldsymbol{F}_t)$-adapted for every stopping time T.

Proof. Suppose that (a) holds. If T is any stopping time and $t \geq 0$ then

$$X_{t\wedge T} = X_t \cdot \mathbf{1}_{\{T\geq t\}} + X_{t\wedge T}\, \mathbf{1}_{\{t\wedge T<t\}}.$$

The first term is $\boldsymbol{F}_t$-measurable since $\{T \geq t\} \in \boldsymbol{F}_t$ by lemma 2.2.2. The random time $t \wedge T$ is a finite stopping time by lemma 2.2.5(b). Thus $X_{t\wedge T}$ is $\boldsymbol{F}_{t\wedge T}^+$-measurable by our assumption and hence the second term is $\boldsymbol{F}_t$-measurable by 2.2.4(b). So $(X_{t\wedge T})_{t\geq 0}$ is $(\boldsymbol{F}_t)$-adapted.
Conversely, let T be a finite stopping time and assume that $(X_{t\wedge T})_{t\geq 0}$ is $(\boldsymbol{F}_t)$-adapted. Then $X_T \cdot \mathbf{1}_{\{T<t\}} = X_{t\wedge T} \cdot \mathbf{1}_{\{T<t\}}$ is $\boldsymbol{F}_t$-measurable for every t and hence X_T is $\boldsymbol{F}_T^+$-measurable. ■

In order to verify these conditions progressive measurability is useful.

2.3.8 Definition. A process X is said to be **progressively measurable** (with respect to the filtration $(\boldsymbol{F}_t)$) if for every $t \geq 0$ the restriction $X|[0,t] \times \Omega$ is measurable with respect to the product-σ-field $\mathscr{B}([0,t]) \otimes \boldsymbol{F}_t$.

Simple sufficient criteria for progressive measurability can be formulated in terms of the right-hand and the left-hand limits of processes. The following definition takes into account that these limits may not exist everywhere.

Let $(X_t)_{t\geq 0}$ be a (realvalued) process. We define two processes X^+ and X^- by

$$X_t^+ = \begin{cases} \limsup\limits_{s\to t, s>t, s\in\mathbb{Q}} X_s & \text{if this is finite} \\ 0 & \text{otherwise} \end{cases}, \tag{2.3.2}$$

$$X_t^- = \begin{cases} \limsup\limits_{s\to t, s<t, s\in\mathbb{Q}_+} X_s & \text{if } \sup\limits_{s<t, s\in\mathbb{Q}_+} |X_s| < \infty \\ 0 & \text{otherwise} \end{cases}. \tag{2.3.3}$$

Remark. The process X^- is separable (with separability set $\mathbb{Q}_+$). This will be convenient and thus we take into account the slight asymmetry in the definition of X^+ and X^-, respectively.

The following gives simple sufficient criteria for this property.

2.3.9 Proposition. Let $X = (X_t)_{t\geq 0}$ be an $(\boldsymbol{F}_t)$-adapted process. Then:
(a) The process X^+ is progressively measurable for the filtration $(\boldsymbol{F}_t^+)_{t\geq 0}$.
(b) If X is either right-continuous or left-continuous then it is progressively measurable w.r.t. $(\boldsymbol{F}_t)_{t\geq 0}$.

Proof. (a) Let $((t_n^k)_{1\leq k\leq k_n})_{n\geq 1}$ be an enumeration of all finite increasing subfamilies of $\mathbb{Q}_+$. For each n define the process X^n by

$$X^n(t,\omega) = \sum_{k=1}^{\infty} X(t_n^{k+1},\omega)\, \mathbf{1}_{]t_n^k,t_n^{k+1}]}(t). \tag{2.3.4}$$

Fix $t \geq 0$. For every n let r_n be the largest index such that $t_n^{r_n} < t$. The restriction of X^n to $[0,t_n^{r_n}] \times \Omega$ is $\mathscr{B}([0,t_n^{r_n}])\otimes \boldsymbol{F}_t$-measurable. Moreover, $X^+ = \limsup\limits_{n\to\infty} X^n$ where this limsup is finite and $X^+ = 0$ otherwise. Since $t_n^{r_n} \longrightarrow t$ we conclude that $X^+|_{[0,t[\times\Omega}$ is $\mathscr{B}([0,t[)\otimes \boldsymbol{F}_t$ -measurable. On the other hand the random variable X_t^+ is $\boldsymbol{F}_t^+$-measurable. Thus X^+ is progressively measurable for $(\boldsymbol{F}_t^+)_{t\geq 0}$. It is progressively measurable even for $(\boldsymbol{F}_t)_{t\geq 0}$ if X_t^+ is $\boldsymbol{F}_t$-measurable for every $t \geq 0$. This is true in particular if X is right-continuous because then $X^+ = X$.
The argument for left-continuous X is similar but simpler, replacing $X(t_n^{k+1},\omega)$ in (2.3.4) by $X(t_n^k,\omega)$. This completes the proof. ■

In lemma 6.1.3 we shall see that X^- even satisfies a stronger condition than progressive measurability. Progressive measurability implies that X_T is measurable for strict stopping times T.

2.3.10 Proposition. Suppose that the process X is progressively measurable and that T is a strict stopping time. Then $\mathbf{1}_{\{T<\infty\}} X_T$ is measurable w.r.t. the σ-field $\boldsymbol{F}_T$.

Proof. We may assume $T < \infty$ everywhere. The map

$$\varphi : \Omega \longrightarrow [0,t] \times \Omega,\ \omega \longmapsto (T(\omega)\wedge t,\omega)$$

is measurable w.r.t. $\boldsymbol{F}_t$ and $\mathscr{B}([0,t])\otimes \boldsymbol{F}_t$. Choose $B \in \mathscr{B}(\mathbb{R})$ and $t \geq 0$. Then

$$\{X_T \in B\} \cap \{T \leq t\} = \{X \in B\} \cap \{T \leq t\} = \{X \circ \varphi \in B\} \cap \{T \leq t\} \in \boldsymbol{F}_t$$

which completes the proof. ■

This result is useful also for general stopping times.

2.3.11 Proposition. Let the process $(X_t)_{t\geq 0}$ be $(\boldsymbol{F}_t)$-adapted.
(a) The variable $X_T^+ \cdot \mathbf{1}_{\{T<\infty\}}$ is $\boldsymbol{F}_T^+$ -measurable for every stopping time T.
(b) Let X be right-continuous. Then $(X_{t\wedge T})_{t\geq 0}$ is $(\boldsymbol{F}_t)_{t\geq 0}$-adapted for every stopping time T.

Proof. (a) Let T be a stopping time. Then it is a strict stopping time for the filtration $(\boldsymbol{F}_t^+)_{t\geq 0}$. Since X^+ is progressively measurable for this filtration the random variable $X_T^+ \cdot \mathbf{1}_{\{T<\infty\}}$ is $\boldsymbol{F}_T^+$-measurable by proposition 2.3.9(a).
(b) If X is right-continuous then $X^+ = X$ and hence X_T is $\boldsymbol{F}_T^+$-measurable by (a) for every finite stopping time. Since X is $(\boldsymbol{F}_t)$-adapted this implies the assertion by proposition 2.3.7. ■

Finally, we complete the argument in example 2.2.6:
Let T be a strict stopping time. Then $t \wedge T$ is a finite strict stopping time and therefore by the propositions 2.3.9(b) and 2.3.10 the following inclusion holds if Π is (left-)continuous:

$$\sigma(\Pi_{t\wedge T} : t \geq 0) \subset \sigma\Big(\bigcup_{t\geq 0} \boldsymbol{F}_{t\wedge T}\Big) \subset \boldsymbol{F}_T.$$

CHAPTER 3

MARTINGALES

Let B be a Brownian motion. We shall study integrals with integrator B - say $X_t = \int_0^t B_s dB_s$. This integral is a new stochastic process. Actually, Itô calculus implies $X_t = \frac{1}{2}(B_t^2 - t)$. This process in turn might appear as an integrator - say in $Y_t = \int_0^t B_s dX_s$ - and so on. Thus it is desirable to develop stochastic integration for a class of integrators rich enough to contain processes like X or Y. The appropriate class to start with is the class of (local) martingales.

Therefore, some basic concepts from martingale theory are reviewed in this chapter. Basically, it is sufficient to know three central results: the martingale convergence theorem, the stopping theorem and an inequality which we refer to as "Doob's $\boldsymbol{L}^2$-inequality". There are many alternative approaches to these results. We believe that the classical line of Doob still is the clearest one.

Throughout this chapter, a probability space $(\Omega, \boldsymbol{F}, \mathbb{P})$ with a filtration $(\boldsymbol{F}_t)_{t \in I}$ is given where $I \subset \mathbb{R}$. If f is an integrable random variable then $\mathbb{E}(f)$ denotes the expectation and $\mathbb{E}(f | \boldsymbol{G})$ the conditional expectation w.r.t. a sub-σ-field $\boldsymbol{G}$ of $\boldsymbol{F}$.

3.0.1 Definition. A **martingale** w.r.t. $(\boldsymbol{F}_t)_{t \in I}$ or a **$(\boldsymbol{F}_t)$-martingale** is a $(\boldsymbol{F}_t)$-adapted stochastic process $M = (M_t)_{t \in I}$ where each random variable M_t is integrable and $\mathbb{E}(M_t | \boldsymbol{F}_s) = M_s$ whenever $s \le t$.
If "$\ge$" holds then M is a **submartingale**,
if "$\le$" holds then M is a **supermartingale**.

The qualification "w.r.t. $(\boldsymbol{F}_t)$" is omitted wherever the choice of the filtration is unambiguous.

The fundamental significance of martingales was underlined by a number of remarkable results of J.L. Doob. Concerning the name we quote from Elliott (1982): »The term martingale has an interesting history. A martingale is, in fact, part of a horses's harness which prevents the horse from raising its head too high. Through horse racing the word became a gambling term, and the mathematical definition ... can be thought of as representing a fair game of chance, where the conditional expectation of the reward at a later time equals one's present reward.« Let us add that a martingale is also a device in a sailing boat keeping the sail tight. The probabilistic concept of a martingale seems to have been first studied by L. Bachelier.

3.0.2 Examples. (a) Let $(\boldsymbol{F}_t)$ be a filtration and let M be an integrable $(\boldsymbol{F}_t)$-adapted process. Suppose that the increment $M_t - M_s$ is independent of $\boldsymbol{F}_s$ for $s < t$ and has mean 0. Then M is a $(\boldsymbol{F}_t)$-martingale. In fact

$$\mathbb{E}(M_t \mid \boldsymbol{F}_s) = \mathbb{E}(M_s \mid \boldsymbol{F}_s) + \mathbb{E}(M_t - M_s \mid \boldsymbol{F}_s) = M_s \text{ if } s \leq t .$$

In particular, every Brownian motion is a martingale with respect to its own filtration. More generally, a $(\boldsymbol{F}_t)$-Brownian motion is a $(\boldsymbol{F}_t)$-martingale: Given a filtration $(\boldsymbol{F}_t)_{t \geq 0}$, a Brownian motion B is called a $(\boldsymbol{F}_t)$-**Brownian motion** if the increment $B_t - B_s$ is independent of $\boldsymbol{F}_s$ for $s < t$ and - for technical reasons - all paths are right-continuous. (We will see in lemma 4.3.5 that this right-continuity is not an essential restriction.) If e.g. $B^1, \ldots, B^d$ are independent (strictly continuous) Brownian motions and

$$\boldsymbol{F}_t = \sigma\{ B_s^i : 0 \leq s \leq t, 1 \leq i \leq d \}$$

then each B^i is a $(\boldsymbol{F}_t)$-Brownian motion.

(b) Let N be a Poisson process and $(\boldsymbol{F}_t)_{t \geq 0}$ be its natural filtration. Then N is $(\boldsymbol{F}_t)$-adapted and the random variable $N_t - N_s$ has a Poisson distribution with expectation $t - s$ and is independent of $\boldsymbol{F}_s$ for $t > s$. If N has these properties w.r.t. a general filtration $(\boldsymbol{F}_t)$ then N is called a $(\boldsymbol{F}_t)$-**Poisson process**. The process $(N_t - t)_{t \geq 0}$ is a $(\boldsymbol{F}_t)$-martingale and hence called Poisson martingale.

(c) Suppose that $\mathbb{P}_0$ and $\mathbb{P}_1$ are two probability measures on $\boldsymbol{F}$ such that M is a $(\boldsymbol{F}_t)$-martingale w.r.t. both of them. Then M is also a martingale under any convex combination

$$\mathbb{P}_\alpha = \alpha\mathbb{P}_1 + (1-\alpha)\mathbb{P}_0,\ 0 \le \alpha \le 1.$$

In fact, for $F \in \boldsymbol{F}_s$ and $t > s$ one has

$$\begin{aligned}\int_F M_t \, d\mathbb{P}_\alpha &= \alpha\int_F M_t \, d\mathbb{P}_1 + (1-\alpha)\int_F M_t \, d\mathbb{P}_0 \\ &= \alpha\int_F M_s \, d\mathbb{P}_1 + (1-\alpha)\int_F M_s \, d\mathbb{P}_0 \\ &= \int_F M_s \, d\mathbb{P}_\alpha .\end{aligned}$$

Suppose e.g. $\Omega = \mathscr{C}(\mathbb{R}_+)$. Let $\mathbb{P}_1$ be the law of a Brownian motion $B = (B_t)_{t\ge0}$ (i.e. $\mathbb{P}_1$ is Wiener measure) and let $\mathbb{P}_0$ be the law of $(aB_t)_{t\ge0}$ i.e. Brownian motion on a scale enlarged by the factor a^{-1}. Then the canonical projection process Π is a martingale both under $\mathbb{P}_0$ and $\mathbb{P}_1$. The measure $\mathbb{P}_{1/2}$ describes Brownian motion on a random scale which is determined by a tossed coin. This process is a martingale with *dependent* increments.
Here is another description of $\mathbb{P}_{1/2}$ which relates $\mathbb{P}_{1/2}$ more explicitely to example 2.1.4 . If B lives on some general probability space $(\Omega', \boldsymbol{F}', Q')$ and C is a random variable on Ω' independent of B such that

$$Q'(C = 1) = \tfrac{1}{2} = Q'(C = a)$$

then B and aB have the law $\mathbb{P}_1$ and $\mathbb{P}_0$, respectively, and $C\cdot B$ has the law $\mathbb{P}_{1/2}$.

Submartingales arise naturally as convex functions of martingales.

3.0.3 Lemma. Suppose that M is a martingale, that $\varphi : \mathbb{R} \longrightarrow \mathbb{R}$ is convex and each random variable $\varphi(M_t)$ is integrable. Then the process $(\varphi(M_t))_{t\ge0}$ is a submartingale.

Proof. Jensen's inequality for conditional expectations yields for $s \le t$ that

$$\mathbb{E}(\varphi(M_t) \mid \boldsymbol{F}_s) \ge \varphi\big(\mathbb{E}(M_t \mid \boldsymbol{F}_s)\big) = \varphi(M_s)$$

which proves the lemma. ■

3.1 Martingales with a Finite Index Set: Doob's Inequalities

We recall some basic facts about discrete-time martingales and submartingales which are at the root of our later work. We give the proofs since they are short and illuminating.

The following simple "Pythagoras type" identities of $\boldsymbol{L}^2$-martingales explain the importance of quadratic variation in stochastic integration.

3.1.1 Lemma. Let $(\mathcal{G}_i)_{0\le i\le n}$ denote a filtration and let $(M_i)_{0\le i\le n}$ be a $(\mathcal{G}_i)_{0\le i\le n}$-martingale in $\boldsymbol{L}^2$. Then

(a) $$\mathbb{E}((M_j - M_i)^2 \mid \mathcal{G}_i) = \mathbb{E}(M_j^2 - M_i^2 \mid \mathcal{G}_i) \text{ for } i \le j ,$$

(b) $$\mathbb{E}(M_n^2) = \mathbb{E}(M_0^2) + \mathbb{E}\left(\sum_{i=0}^{n-1} (M_{i+1} - M_i)^2\right).$$

Proof. (a) The identity is verified by the computation

$$\begin{aligned}\mathbb{E}(M_j - M_i)^2 \mid \mathcal{G}_i) &= \mathbb{E}(M_j^2 \mid \mathcal{G}_i) - 2M_i \cdot \mathbb{E}(M_j \mid \mathcal{G}_i) + M_i^2 \\ &= \mathbb{E}(M_j^2 \mid \mathcal{G}_i) - M_i^2 = \mathbb{E}(M_j^2 - M_i^2 \mid \mathcal{G}_i).\end{aligned}$$

(b) By (a), we get the identity

$$\sum_{i=0}^{n-1} \mathbb{E}((M_{i+1} - M_i)^2) = \sum_{i=0}^{n-1} \mathbb{E}(M_{i+1}^2 - M_i^2) = \mathbb{E}(M_n^2) - \mathbb{E}(M_0^2)$$

which implies (b). ∎

The (super-)martingale property extends from constant times to stopping times.

3.1.2 Proposition. (Stopping theorem, discrete time) Let $(M_i)_{0\le i\le n}$ be a $(\mathcal{G}_i)_{0\le i\le n}$-supermartingale and let S and T be stopping times (with values in $\{0,\dots,n\}$) such that $S \le T$. Then

$$\mathbb{E}(M_T \mid \mathcal{G}_S) \le M_S$$

with equality if M is a martingale.

Proof. It is easy to check that in this discrete case a random variable X is $\mathcal{G}_S$-measurable if and only if $X \cdot 1_{\{S=k\}}$ is $\mathcal{G}_k$-measurable for every $k \in \{0,\ldots,n\}$. Therefore, the variable M_S is $\mathcal{G}_S$-measurable and we have to prove that for every $A \in \mathcal{G}_S$

$$\mathbb{E}\left(M_T 1_A 1_{\{S=k\}}\right) \le \mathbb{E}\left(M_k 1_A 1_{\{S=k\}}\right) \text{ for every } k \in \{0, \ldots ,n\}.$$

We show this by induction on the largest value m of T. If $m = 0$ then $S = T = 0$ and nothing has to be proved. For the induction step from $m - 1$ to m we may assume $k < m$ since $S \le T \le m$ implies

$$M_T 1_A 1_{\{S=m\}} = M_m 1_A 1_{\{S=m\}} .$$

If $k < m$ then

$$A \cap \{S = k, T = m\} = \left(A \cap \{S = k\}\right) \setminus \{T \le m-1\} \in \mathcal{G}_{m-1}$$

and therefore the supermartingale property implies

$$\begin{aligned}
\mathbb{E}\left(M_T 1_A 1_{\{S=k\}}\right) &= \mathbb{E}\left(M_T 1_A 1_{\{S=k,T\le m-1\}}\right) + \mathbb{E}\left(M_m 1_A 1_{\{S=k,T=m\}}\right) \\
&\le \mathbb{E}\left(M_T 1_A 1_{\{S=k,T\le m-1\}}\right) + \mathbb{E}\left(M_{m-1} 1_A 1_{\{S=k,T=m\}}\right) \\
&= \mathbb{E}\left(M_{T\wedge(m-1)} 1_A 1_{\{S=k\}}\right) \\
&\le \mathbb{E}\left(M_k 1_A 1_{\{S=k\}}\right)
\end{aligned}$$

where we have used the induction hypothesis for the last inequality. ■

The following inequalities control the fluctuation of the process - thus justifying the name martingale in its original meaning from a mathematical point of view.

3.1.3 Lemma. (**Doob's maximal inequalities**) Let $(\mathcal{G}_i)_{0\le i\le n}$ be a filtration.

(a) If $(M_i)_{0\le i\le n}$ is a $(\mathcal{G}_i)$-submartingale then

$$c\, \mathbb{P}(\{\max_{0\le i\le n} M_i \ge c\}) \le \int_{\{\max_{0\le i\le n} M_i \ge c\}} M_n \, d\mathbb{P} \text{ for every } c > 0.$$

(b) If $(M_i)_{0\le i\le n}$ is a nonnegative $(\mathcal{G}_i)$-supermartingale then

$$c\, \mathbb{P}(\{\max_{0\le i\le n} M_i \ge c\}) \le \mathbb{E}(M_0) \qquad \text{for every } c > 0 .$$

Proof. Set

$$S = \begin{cases} \min\{i : M_i \geq c\} & \text{if } \max\limits_{0\leq i\leq n} M_i \geq c \\ n & \text{otherwise} \end{cases}.$$

Then $\{S \leq k\} \in \sigma(M_0, \dots, M_k)$, i.e. S is a stopping time such that $0 \leq S \leq n$ and $\{\max M_i \geq c\} = \{M_S \geq c\} \in \boldsymbol{F}_S$. Moreover

$$c\,\mathbb{P}\{\max M_i \geq c\} \leq \int_{\{\max M_i \geq c\}} M_S \, d\mathbb{P}.$$

By the stopping theorem the right hand side can be estimated in case (a) by

$$\int_{\{\max M_i \geq c\}} M_n \, d\mathbb{P}$$

and in case (b) by

$$\int_\Omega M_S \, d\mathbb{P} \leq \mathbb{E}(M_0).$$

This completes the proof. ■

3.1.4 Theorem. (**Doob's L^2-inequality**) Let $(M_i)_{0\leq i\leq n}$ be a $(\boldsymbol{G}_i)_{0\leq i\leq n}$-martingale or a nonnegative submartingale in $\boldsymbol{L}^2$. Then

$$\mathbb{E}\Big(\max_{0\leq i\leq n} M_i^2\Big) \leq 4\mathbb{E}(M_n^2).$$

We shall see in example 3.2.11 that this estimate is sharp.

Proof. In both cases $(|M_i|)_{0\leq i\leq n}$ is a submartingale. In the martingale case this follows from the general inequality $|\mathbb{E}(X \mid \boldsymbol{G})| \leq \mathbb{E}(|X| \mid \boldsymbol{G})$, in the submartingale case from $M_i = |M_i|$. Set

$$V = \max_{0\leq i\leq n} |M_i| \quad \text{and} \quad U = |M_n|.$$

In the following computation, use Fubini's theorem for the second and fourth and Doob's maximal inequality 3.1.3(a) for the third step:

$$\begin{aligned} \mathbb{E}(V^2/2) &= \int_\Omega \Big(\int_{\mathbb{R}_+} y\, \mathbf{1}_{\{y\leq V(\omega)\}}(y)\, dy\Big)\, d\mathbb{P}(\omega) \\ &= \int_{\mathbb{R}_+} y\, \mathbb{P}(\{y \leq V\})\, dy \\ &\leq \int_{\mathbb{R}_+} \int_\Omega \mathbf{1}_{\{y\leq V(\omega)\}}(y)\, U(\omega)\, d\mathbb{P}(\omega)\, dy \\ &= \mathbb{E}\Big(U \cdot \int_0^{V(\cdot)} dy\Big) \leq \|U\|_2\, \|V\|_2. \end{aligned}$$

Here $\|V\|_2$ denotes the L^2-norm $(\mathbb{E}(V^2))^{1/2}$. If $\|V\|_2 \neq 0$ then

$$\|V\|_2 = \mathbb{E}(V^2)/\|V\|_2 \leq 2\|U\|_2.$$

If $V = 0$ a.s. then the assertion holds also and the proof is complete. ■

In proving convergence theorems it is of interest how often a path crosses a fixed interval. Consider a process $M = (M_i)_{i\in F}$ where F is some finite subset of the real line. A path $(M_i(\omega))_{i\in F}$ crosses the interval $]a,b[$ say from the left to the right if $M_i(\omega) \leq a < b \leq M_j(\omega)$ for some indices $i < j$. Such a run is called an **upcrossing**. The gross number of upcrossings is therefore the largest integer $N_F(a,b) = m$ such that we can find indexes $i_1 < \dots < i_{2m}$ in F with

$$M_{i_{2k-1}} \leq a < b \leq M_{i_{2k}} \quad \text{for } k = 1, \dots, m.$$

3.1.5 Proposition. (**Doob's upcrossing inequality**) Let F be a finite subset of $\mathbb{R}_+$ and let $(M_i)_{i\in F}$ be a $(\boldsymbol{G}_i)_{i\in F}$-supermartingale. The gross number $N_F(a,b)$ of upcrossings for an interval $]a,b[$, $a < b$, is then a random variable and

$$\mathbb{E}\left(N_F(a,b)\right) \leq \frac{1}{b-a} \max_{i\in F} \mathbb{E}\left(0 \vee -(M_i - a)\right).$$

Proof. We may assume $F = \{0,\dots,n\}$. Set $p = [\frac{n}{2}] + 1$. Define $T_0 \equiv 0$ and recursively for $k \in \{1, \dots, p\}$

$$T_{2k-1} = \begin{cases} \min\{j \geq T_{2k-2} : M_j \leq a\} & \text{if there is such a } j \\ n & \text{otherwise} \end{cases}$$

$$T_{2k} = \begin{cases} \min\{j \geq T_{2k-1} : M_j \geq b\} & \text{if there is such a } j \\ n & \text{otherwise} \end{cases}$$

It is easily verified that $T_0 \leq T_1 \leq T_2 \leq \dots \leq T_{2p}$ are stopping times. Consider the difference $M_{T_{2k}} - M_{T_{2k-1}}$. It corresponds to the k-th upcrossing of $(M_i)_{i\in F}$ over $]a,b[$ if $T_{2k} < n$. For at most one index k one has $T_{2k-1} < T_{2k} = n$ and in this case

$$M_{T_{2k}} - M_{T_{2k-1}} \geq M_n - a.$$

For the remaining indices k one gets $T_{2k-1} = T_{2k} = n$ and

$$M_{T_{2k}} - M_{T_{2k-1}} = 0.$$

In summary,

$$\sum_{k=1}^{p} M_{T_{2k}} - M_{T_{2k-1}} \geq (b - a)\, N_F(a,b) + \min(0, M_n - a).$$

The left-hand side has nonpositive expectation by the stopping theorem. Taking expectations we thus get

$$\mathbb{E}\big((b - a)\, N_F(a,b)\big) \leq \mathbb{E}\big(\max(0, -(M_n - a))\big)$$

which implies the assertion. ■

3.2 Convergence Theorems and Continuous Time Martingales

In this section we collect a few properties of martingales with a continuous index set. We also treat the slightly more general case of submartingales which however will be used only in section 6.6. But except for theorem 3.2.4(a) the proofs for martingales and for submartingales are essentially the same. The key to the pathwise behaviour of a submartingale is the following preparatory lemma which is based on Doob's upcrossing inequality for supermartingales.

3.2.1 Lemma. Let $(M_t)_{t \in U}$ by a submartingale with a countable index set $U \subset \mathbb{R}_+$. Assume $\inf_U \mathbb{E}(M_t) > -\infty$. Then there is a set $\Omega_0 \in \sigma\{M_t : t \in U\}$ such that $\mathbb{P}(\Omega_0) = 1$ and for each $\omega \in \Omega_0$ the following holds: For every monotone (increasing or decreasing) sequence $(t_n)_{n \geq 1}$ in U the sequence $(M_{t_n}(\omega))_{n \geq 1}$ converges in $[-\infty, \infty]$.

Proof. A sequence $(c_n)_{n \in \mathbb{N}}$ of real numbers does not converge in $[-\infty, \infty]$ if and only if it has more than one limit point or equivalently, if there is a pair of rationals a, b and a subsequence (c_{n_k}) such that $c_{n_{2k}} < a < b < c_{n_{2k+1}}$ for all k. Let Ω_0 be the set of those points for which each of the countably many random variables

$$N(a,b) = \sup_{\substack{F \subset U \\ F \text{ finite}}} N_F(a,b), \quad a, b \in \mathbb{Q},$$

is finite at ω. (The number $N_F(a,b)$ is defined before proposition 3.1.5.) In order to apply the upcrossing inequality we replace M by the supermartingale

$-M$ which satisfies $\sup_U \mathbb{E}(-M_t) < \infty$. Since U is countable the expectations $\mathbb{E}(N(a,b))$ are finite by proposition 3.1.5 and monotone convergence. Hence $\mathbb{P}(\Omega_0) = 1$. Let $\omega \in \Omega_0$ and let $(t_n)_{n\geq 1}$ be a monotone sequence in U and suppose that the sequence $(M_{t_n}(\omega))_{n\geq 1}$ does not converge. Then the introductory remark implies that $N(a,b)(\omega)$ is infinite for some $a,b \in \mathbb{Q}$ in contradiction to the definition of Ω_0. This completes the proof. ■

For convergence in the mean the following observation is important. Recall that a family $(X_i)_{i\in J}$ of random variables on a probability space $(\Omega, \boldsymbol{F}, \mathbb{P})$ is **uniformly integrable** if

$$\lim_{a\to\infty} \sup_{i\in J} \int_{\{|X_i|>a\}} |X_i| \, d\mathbb{P} = 0 .$$

The useful fact is that for a uniformly integrable sequence a.s. convergence implies $\boldsymbol{L}^1$-convergence (Billingsley (1979), theorem 16.13).

3.2.2 Lemma. For each integrable random variable Y the set

$$\left\{\mathbb{E}(Y|\boldsymbol{G}) \;\middle|\; \boldsymbol{G} \text{ is a sub-}\sigma\text{-field of } \boldsymbol{F}\right\}$$

is uniformly integrable.

Proof. For $\varepsilon > 0$ choose $\delta > 0$ such that

$$\int_A |Y| \, d\mathbb{P} < \varepsilon \text{ if } \mathbb{P}(A) < \delta ,$$

and let $a > \frac{1}{\delta}\mathbb{E}(|Y|)$. For every σ-field $\boldsymbol{G} \subset \boldsymbol{F}$ we have, writing X instead of $\mathbb{E}(Y \mid \boldsymbol{G})$, that

$$|X| \leq \mathbb{E}(|Y| \;|\; \boldsymbol{G}) .$$

This implies

$$\mathbb{P}(|X| > a) \leq \frac{1}{a}\mathbb{E}(|X|) \leq \frac{1}{a}\mathbb{E}(|Y|) < \delta$$

and hence

$$\int_{\{|X|>a\}} |X| \, d\mathbb{P} \leq \int_{\{|X|>a\}} |Y| \, d\mathbb{P} < \varepsilon .$$

This estimate does not depend on $\boldsymbol{G}$ and hence the uniform integrability is proved. ■

Frequently, martingales are constructed by conditioning.

3.2.3 Definition. An $(\boldsymbol{F}_t)$-martingale $(M_t)_{t\in I}$ is **closed by the random variable** Y if

$$M_t = \mathbb{E}(Y|\boldsymbol{F}_t) \text{ for all } t \geq 0.$$

Similarly, a submartingale is closed by Y if "≤" holds.

According to lemma 3.2.2 every closed martingale is uniformly integrable. The converse is part of the following theorem.

3.2.4 Theorem. Let $M = (M_t)_{t\in U}$ be a submartingale with a index set $U \subset \mathbb{R}$. Assume either that U is countable or that the process M is separable. Then the following holds:

(a) (**Decreasing submartingale convergence theorem**) Suppose that

$$\inf_{U} \mathbb{E}(M_t) > -\infty.$$

Let

$$\boldsymbol{F}_{-\infty} = \bigcap_{t\in U} \boldsymbol{F}_t.$$

As t *decreases* through U the family $(M_t)_{t\in U}$ converges a.s. and in $\boldsymbol{L}^1$ to a $\boldsymbol{F}_{-\infty}$-measurable random variable $M_{-\infty}$ such that $(M_t)_{t\in\{-\infty\}\cup U}$ is a submartingale.

(b) (**Increasing submartingale convergence theorem**) There is a random variable M_∞ such that $(M_t)_{t\in U}$ converges a.s. to M_∞ as t *increases* through U. If M is uniformly integrable then the convergence is also in $\boldsymbol{L}^1(\mathbb{P})$ and M_∞ closes the submartingale M.

(c) (**$\boldsymbol{L}^2$-martingale convergence theorem**) If M is a $\boldsymbol{L}^2$-bounded martingale then in (a) and (b) the convergence is also in $\boldsymbol{L}^2$.

Proof. (a) A family in a metric space with index set U converges as t decreases through U if and only if a limit exists along every decreasing cofinal sequence (t_n) in U. (Two such sequences cannot have different limits since one can merge the two sequences into a single one).

For the a.s. convergence assume first that U is countable. Lemma 3.2.1 shows that

$$\lim_{t\downarrow,\, t\in U} M_t(\omega)$$

exists a.s. in $[-\infty, +\infty]$. The limit is given by a $\boldsymbol{F}_{-\infty}$-measurable random variable $M_{-\infty}$. The $\boldsymbol{L}^1$-convergence proved below will show that $M_{-\infty}$ is a.s.

finite. If M is separable and U_0 is a countable set of separability then this a.s. convergence stays valid since limits along U_0 are also limits along the full index set.

Now let us turn to $\mathcal{L}^1$-convergence. By the introductory remark we need to consider only sequences. If M is a martingale then uniform integrability (lemma 3.2.2) implies the assertion.

In the *submartingale case* we prove that M has the Cauchy property in $\mathcal{L}^1$. Let $\varepsilon > 0$ and choose t_1 such that

$$\mathbb{E}(M_{t_1}) \leq \inf_{U} \mathbb{E}(M_t) + \varepsilon .$$

Denote by (N_t) the martingale $(\mathbb{E}(M_{t_1} \mid \mathcal{F}_t))$. We already know that a decreasing martingale converges in $\mathcal{L}^1$, i.e. there is some index t_2 such that for all $t \leq t_2$

$$\|N_t - N_{t_2}\|_1 < \varepsilon .$$

Moreover $M_t \leq \mathbb{E}(M_{t_1} \mid \mathcal{F}_t) = N_t$ and therefore

$$\|M_t - N_t\|_1 = \mathbb{E}(N_t - M_t) = \mathbb{E}(M_{t_1}) - \mathbb{E}(M_t) < \varepsilon .$$

These two estimates give

$$\|M_t - N_{t_2}\|_1 < 2\varepsilon$$

for all $t \leq t_2$, and hence $\|M_t - M_{t'}\|_1 < 4\varepsilon$ for all $t, t' \leq t_2$. This proves the Cauchy property of (M_t), i.e. (M_t) converges in $\mathcal{L}^1$.

In particular, the limit variable $M_{-\infty}$ is integrable and, as t decreases along U,

$$\int_A M_{-\infty}\, d\mathbb{P} = \lim \int_A M_t\, d\mathbb{P} \leq \int_A M_{t_0}\, d\mathbb{P}$$

for every $A \in \bigcap_{t \in U} \mathcal{F}_t$ and every $t_0 \in U$. Thus

$$M_{-\infty} \leq \mathbb{E}(M_{t_0} \mid \bigcap_{t \in U} \mathcal{F}_t)$$

and $M_{-\infty}$ extends M to a submartingale.

(b) In the increasing case we may assume that U has a smallest element t_0. Then by the submartingale property

$$\inf_{t \in U} \mathbb{E}(M_t) \geq \mathbb{E}(M_{t_0}) > -\infty .$$

Thus the a.s. convergence follows from lemma 3.2.1 as in (a). Uniform integrability then implies $\mathcal{L}^1$-convergence and M_∞ is seen to close M as in (a).

(c) If M is a martingale bounded in $\mathcal{L}^2$ then it follows from lemma 3.1.1 (a) that

(3.2.1) $$0 \leq \|M_t - M_s\|_2^2 = \mathbb{E}(M_t^2) - \mathbb{E}(M_s^2) \text{ for } s \leq t.$$

Since $(\mathbb{E}(M_t^2))_{t \in U}$ is bounded and increasing it is a Cauchy net in $\mathbb{R}$ and this identity shows that $(M_t)_{t \in U}$ is a Cauchy net in $\boldsymbol{L}^2(\mathbb{P})$, both in the decreasing and in the increasing case. Together with almost sure convergence this implies that the $\boldsymbol{L}^2$-limits are $M_{-\infty}$ and M_∞ respectively. ■

Most parts of the preceding result could have been derived directly from the maximal inequality 3.1.3 without referring to the upcrossing inequality (via lemma 3.2.1). This is no longer true for the next theorem concerning the regularity of the paths of a (sub-)martingale. Note that right- and left-continuity of the paths, respectively, involves uncountably many limit statements. First we recall the concept of regular functions.

3.2.5 Definition. A function $f : \mathbb{R}_+ \longrightarrow \mathbb{R}$ is **regular** if for each $t > 0$ the limits

$$f(t+) = \lim_{s \to t, s > t} f(s) \quad \text{and} \quad f(t-) = \lim_{s \to t, s < t} f(s)$$

exist, and if $f(0+)$ exists.

Even if a process is separable its paths are not regular in general. However for martingales this implication holds a.s..

3.2.6 Theorem. Let $M = (M_t)_{t \geq 0}$ be a submartingale.

(a) If M is separable then it is a.s. regular.

(b) Suppose that the filtration $(\boldsymbol{F}_t)_{t \geq 0}$ is right-continuous. If the map $t \longmapsto M_t \in \boldsymbol{L}^1$ is right-continuous then M has a right-continuous modification. For a martingale and a right-continuous filtration the map $t \longmapsto M_t \in \boldsymbol{L}^1$ automatically is right-continuous.

(c) If the map $t \longmapsto M_t \in \boldsymbol{L}^1(\mathbb{P})$ is left-continuous then M^- is a separable a.s. left-continuous modification of M.

Proof. (a) In order to prove a.s. regularity we need to consider only a bounded interval $[0,t_1]$. On $[0,t_1]$ the function $t \longmapsto \mathbb{E}(M_t)$ is bounded. Therefore a.s. all one-sided limits along a separability set U exist in $[-\infty,\infty]$ according to lemma 3.2.1. We show that a.s. all these limits are finite by proving that almost every path is bounded on $[0,t_1]$.

Write $N_t = \mathbb{E}(M_{t_1} \mid \boldsymbol{F}_t)$ and let F be a finite subset of $[0,t_1]$. Then lemma 3.1.3(a) applied to the submartingale $|N|$ yields for $c > 0$

$$\mathbb{P}\Big(\sup_{t \in F} |N_t| \geq c \Big) \leq \frac{1}{c} \mathbb{E}\big(|M_{t_1}|\big)$$

and 3.1.3(b) applied to the nonnegative supermartingale $N_t - M_t$ (recall $M_t \leq \mathbb{E}(M_{t_1} \mid \boldsymbol{F}_t) = N_t$) gives

$$\mathbb{P}\Big(\sup_{t \in F} |N_t - M_t| \geq c \Big) \leq \frac{1}{c} \mathbb{E}\big(N_0 - M_0\big) .$$

The right hand sides do not depend on F and hence the sup can be taken over $U \cap [0,t_1]$ and then over $[0,t_1]$. As $c \longrightarrow \infty$ this shows the boundedness of the paths on $[0,t_1]$.

(b) We want to take

$$(3.2.2) \qquad \lim_{t' \downarrow t, t' \in \mathbb{Q}} M_{t'}(\omega)$$

as a modification. In order to deal in a measurable way with those cases in which this limit does not exist we recall from lemma 3.2.1 that for each $s \in \mathbb{Q}_+$ there is a set $\Omega_s \in \boldsymbol{F}_s$ such that $\mathbb{P}(\Omega_s) = 1$ and all one-sided limits of the restricted process $(M_{s'})_{s'<s, s' \in \mathbb{Q}}$ exist on Ω_s. According to the argument in (a) we may also assume that these limits are finite on Ω_s. Consider the random time

$$T(\omega) = \sup\big\{s \in \mathbb{Q} : \omega \in \Omega_s\big\} .$$

Then $\{T < t\} = \bigcup_{s \in \mathbb{Q}, s<t} \Omega \backslash \Omega_s \in \boldsymbol{F}_t$ for every $t \geq 0$ i.e. T is a stopping time. Moreover $T = \infty$ a.s. and for $t < T(\omega)$ the limit (3.2.2) exists. Therefore

$$\tilde{M}_t(\omega) = \begin{cases} \lim_{t' \downarrow t, t' \in \mathbb{Q}} M_{t'}(\omega) & \text{for} \quad t < T(\omega) \\ 0 & \text{for} \quad T(\omega) \leq t \end{cases}$$

defines a right-continuous process $\tilde{M}$ which is adapted since $(\boldsymbol{F}_t)$ is right-continuous. Because of the right-continuity in the $\boldsymbol{L}^1$-sense, $\tilde{M}_t = M_t$ a.s. and hence $\tilde{M}$ is a modification of M.

If M is a martingale then the $\boldsymbol{L}^1$ -right-continuity follows from the decreasing martingale convergence theorem 3.2.4(a).

(c) The process M^- from (2.3.3) is separable with separability set $\mathbb{Q}_+$. The increasing martingale convergence theorem 3.2.4(b) and the $\boldsymbol{L}^1$-left-continuity imply that

$$M_t = \lim_{s \uparrow t, s \in \mathbb{Q}_+} M_s \text{ a.s..}$$

Moreover

$$\sup_{s<t, s\in\mathbb{Q}_+} |M_s| < \infty \text{ a.s.}$$

by part (a). Hence $M_t^- = M_t$ a.s. i.e. M^- is a modification. Every regular path of M^- is left-continuous since in the definition of M^- the left limit is taken whenever possible. ■

Remark. If $(\mathcal{F}_t)_{t\geq 0}$ is right-continuous and $t \longmapsto M_t$ is $\mathcal{L}^1$-continuous then both (b) and (c) can be applied. The example $M_t = N_t - t$ where N_t is a Poisson process shows that in general the left-continuous and the right-continuous modifications cannot be forced to coincide since there is no a.s. continuous modification.

The next theorem is fundamental for stochastic integration. Its proof makes strong use of right-continuity. Together with part (b) of the preceding theorem this is one of the reasons why many authors assume the filtration $(\mathcal{F}_t)_{t\geq 0}$ to be right-continuous. However, there are also other ways to arrive at right-continuous martingales. For instance, we will see in chapter 10 that for the natural filtration of Brownian motion every martingale has a right-continuous and even a.s. continuous modification. An analogue of theorem 3.2.7 below for left-continuous martingales is valid only for "predictable" stopping times (cf. theorem 6.4.10).

3.2.7 Theorem. (Stopping theorem) Suppose that M is a right-continuous $(\mathcal{F}_t)$-submartingale. Then it is also a $(\mathcal{F}_t^+)$-submartingale. Let S and T be stopping times such that $S \leq T$. Suppose either that T is bounded or that M is closed by the random limit variable M_∞. Then

$$M_S \leq \mathbb{E}(M_T \mid \mathcal{F}_S^+) . \tag{3.2.4}$$

In particular $M_S \leq \mathbb{E}(M_\infty \mid \mathcal{F}_S^+)$ and

$$\mathbb{E}(M_0) \leq \mathbb{E}(M_T) . \tag{3.2.5}$$

Proof. First note that by right-continuity and theorem 3.2.4(a) M is also a submartingale with respect to the right-continuous filtration $(\mathcal{F}_t^+)$. For the proof of (3.2.4) let us first assume that S and T are stopping times taking their values in a finite set $F = \{t_0, \ldots, t_m\}$ (where possibly $t_m = +\infty$). In this case the result follows from the corresponding result 3.1.2 for finite parameter martingales applied to the $(\mathcal{F}^+_{t_i})_{0\le i\le m}$-submartingale $(M_{t_i})_{0\le i\le m}$. Now suppose that S and T are general stopping times such that $S \le T$ and that M_∞ closes M. Let (F_n) be an increasing sequence of finite subsets of $[0,\infty]$ such that each F_n contains ∞ and the union of the F_n is dense in $[0,\infty]$. Let

$$S^n(\omega) = \min\{t \in F_n : S(\omega) < t\}$$

and

$$T^n(\omega) = \min\{t \in F_n : T(\omega) < t\}.$$

Then S^n and T^n are (even strict) stopping times taking values in the finite set F_n since e.g.

$$\{S^n \le t\} = \bigcup_{s\le t, s\in F_n} \{S < s\} \in \mathcal{F}_t.$$

Also $S^n \le S^m \le T^m$ for $n \ge m$ and hence by the discrete time stopping theorem

$$M_{S^n} \le \mathbb{E}(M_{T^m} \mid \mathcal{F}^+_{S^n}) \text{ for } n \ge m.$$

Keep now m fixed. Use $S^n \downarrow S$, right-continuity, the decreasing martingale convergence theorem 3.2.4(a) and lemma 2.2.5(b) to get

$$\begin{aligned} M_S &= \lim_{n\to\infty} M_{S^n} \le \lim_{n\to\infty} \mathbb{E}\left(M_{T^m} \mid \mathcal{F}^+_{S^n}\right) \\ &= \mathbb{E}\left(M_{T^m} \mid \bigcap_{n=1}^{\infty} \mathcal{F}^+_{S^n}\right) = \mathbb{E}\left(M_{T^m} \mid \mathcal{F}^+_S\right). \end{aligned}$$

According to the discrete case the sequence $(M_{T^m})_{m\in\mathbb{N}}$ is a submartingale such that $\mathbb{E}(M_0) \le \inf_m \mathbb{E}(M_{T^m})$. Therefore M_{T^m} converges to M_T not only pointwise but also in $\mathcal{L}^1(\mathbb{P})$. Thus we get the desired inequality (3.2.4):

$$M_S = \lim_{m\to\infty} \mathbb{E}\left(M_{T^m} \mid \mathcal{F}^+_S\right) = \mathbb{E}\left(M_T \mid \mathcal{F}^+_S\right) \text{ a.s.}.$$

The second inequality (3.2.5) is obtained by choosing $S \equiv 0$ and taking ordinary expectations.

If $T \le t_0 < \infty$ then consider the closed submartingale $(M_t)_{t\le t_0}$. ■

Consider a $(\mathcal{F}_t)$-Brownian motion B starting in 0 and let T be the first time at which the Brownian path hits -1. Then $B_T = -1$ a.s. and thus $\mathbb{E}(B_T) < 0 = \mathbb{E}(B_0)$. This shows that Brownian motion is not closed and that theorem 3.2.7 in general is not valid for nonclosed martingales.

The theorem implies that stopping is consistent with the (sub-)martingale property. This is made precise in different ways in the following corollary.

3.2.8 Corollary. (**Optional sampling**) Let M be a right-continuous submartingale.

(a) Suppose that M is closed and let T_n, $n \geq 1$, be pointwise increasing stopping times. Then the process $(M_{T_n})_{n\geq 1}$ is a submartingale w.r.t. the filtration $(\boldsymbol{F}^+_{T_n})_{n\geq 1}$.

(b) For each stopping time T the process $(M_{t\wedge T})_{t\geq 0}$ is a $(\boldsymbol{F}^+_t)$-submartingale and a $(\boldsymbol{F}_t)$-submartingale.

(c) If M is closed and T is a stopping time then $(M_{t\wedge T})_{t\geq 0}$ is closed by M_T and for every other stopping time S one has

$$M_{S\wedge T} \leq \mathbb{E}(M_T \mid \boldsymbol{F}^+_S) . \tag{3.2.6}$$

If M is a uniformly integrable martingale then so is $M_{\cdot\wedge T}$.

Proof. (a) follows immediately from theorem 3.2.7.

(b) By proposition 2.3.11(b) the process $(M_{t\wedge T})_{t\geq 0}$ is $(\boldsymbol{F}_t)$-adapted. Consider $A \in \boldsymbol{F}^+_s$. Then for every $t \geq 0$

$$A \cap \{T \geq s\} \cap \{T < t\} = \begin{cases} \emptyset & \text{if } t \leq s \\ A \cap \{s \leq T < t\} \in \boldsymbol{F}_t & \text{if } s < t \end{cases}$$

which implies $A \cap \{T \geq s\} \in \boldsymbol{F}^+_T$. Lemma 2.2.5 yields

$$A \cap \{T \geq s\} \in \boldsymbol{F}^+_{s\wedge T} .$$

For $s < t$ the stopping times $s \wedge T$ and $t \wedge T$ are bounded and hence the stopping theorem applies and yields

$$\begin{aligned} \int_A M_{s\wedge T}\, d\mathbb{P} &= \int_{A\cap\{T\geq s\}} M_{s\wedge T}\, d\mathbb{P} + \int_{A\cap\{T<s\}} M_T\, d\mathbb{P} \\ &\leq \int_{A\cap\{T\geq s\}} M_{t\wedge T}\, d\mathbb{P} + \int_{A\cap\{T<s\}} M_T\, d\mathbb{P} \qquad (3.2.7) \\ &= \int_A M_{t\wedge T}\, d\mathbb{P} . \end{aligned}$$

Thus

$$M_{s\wedge T} \leq \mathbb{E}\left(M_{t\wedge T} \mid \boldsymbol{F}^+_s\right),$$

i.e. $M_{\cdot\wedge T}$ is a $(\boldsymbol{F}^+_t)$-submartingale. By proposition 2.3.7 it is $(\boldsymbol{F}_t)$-adapted and hence also a $(\boldsymbol{F}_t)$-submartingale.

(c) If M is closed then one can rewrite (3.2.7) with T in place of $t \wedge T$ to get

$$\int_A M_{s \wedge T} \, d\mathbb{P} \le \int_A M_T \, d\mathbb{P}$$

for every $A \in \mathcal{F}_s^+$, i.e. M_T closes the submartingale $M_{\cdot \wedge T}$. If one applies the stopping theorem to $(M_{t \wedge T})_{t \ge 0}$ and the pair S, ∞ of stopping times one gets $\mathbb{E}(M_T \mid \mathcal{F}_S^+) = M_{S \wedge T}$. ■

For martingales (3.2.5) implies $\mathbb{E}(M_T) = \mathbb{E}(M_0)$ for every bounded stopping time. It is useful to observe that this even characterizes martingales.

3.2.9 Proposition. (**Characterization of martingales**) Suppose that M is an integrable right-continuous process adapted to the filtration $(\mathcal{F}_t)_{t \ge 0}$. Then M is an $(\mathcal{F}_t)$-martingale if and only if $\mathbb{E}(M_T) = \mathbb{E}(M_0)$ for every bounded stopping time T.

Proof. The "only if" part is the identity (3.2.5) in the stopping theorem. To prove the converse, choose $s < t$ and $A \in \mathcal{F}_s$. The random variable

$$T = s \cdot \mathbf{1}_{\Omega \setminus A} + t \cdot \mathbf{1}_A$$

is a bounded stopping time. By assumption

$$\begin{aligned} \mathbb{E}(M_s \mathbf{1}_{\Omega \setminus A}) + \mathbb{E}(M_t \mathbf{1}_A) &= \mathbb{E}(M_T) = \mathbb{E}(M_0) = \mathbb{E}(M_s) \\ &= \mathbb{E}(M_s \mathbf{1}_{\Omega \setminus A}) + \mathbb{E}(M_s \mathbf{1}_A) \end{aligned}$$

This implies $\mathbb{E}(M_t \mathbf{1}_A) = \mathbb{E}(M_s \mathbf{1}_A)$ for every $A \in \mathcal{F}_s$ and thus $\mathbb{E}(M_t \mid \mathcal{F}_s) = M_s$. Therefore M is a martingale. ■

The following is the continuous version of Doob's $\mathcal{L}^2$-inequality.

3.2.10 Theorem. Every separable $\mathcal{L}^2$-bounded martingale M satisfies

$$\|M_\infty\|_2 = \sup_{t \ge 0} \|M_t\|_2$$

and

$$\|M_\infty\|_2 \le \|\sup_{t \ge 0} |M_t|\,\|_2 \le 2\|M_\infty\|_2.$$

In particular, $\sup_{t \ge 0} |M_t|$ is a square-integrable random variable.

Proof. By (3.2.1) the function $t \longmapsto \mathbb{E}(M_t^2)$ is increasing . The $\mathcal{L}^2$-version of the increasing martingale convergence theorem 3.2.4(c) shows that the limit is $\mathbb{E}(M_\infty^2)$. This proves the identity. The first inequality then is obvious. For the second inequality use separability and the monotone convergence theorem to see that

$$\mathbb{E}((\sup_{t\geq 0} |M_t|)^2) = \mathbb{E}((\sup_{t\in U} |M_t|)^2) = \sup\left\{\mathbb{E}(\max_{t\in F} |M_t|^2) \ : F \subset U \text{ finite}\right\}$$

where U is the countable set of separability. Now Doob's inequality 3.1.3 implies for each finite F that

$$\mathbb{E}(\max_{t\in F} |M_t|^2) \leq 4 \max_{t\in F} \mathbb{E}(M_t^2) \leq 4\mathbb{E}(M_\infty^2)$$

Taking square roots completes the proof. ■

Since much of the later work is connected to the estimate of the last theorem we present an elegant elementary example (Dubins and Gilat (1978)) showing that the constant 2 in 3.2.10 (and hence the constant 4 in 3.1.4) is best possible.

3.2.11 Example. Let $(\Omega, \boldsymbol{F}, \mathbb{P})$ be the Lebesgue probability space over $]0,1[$. Let

$$\boldsymbol{F}_t = \left\{B \in \boldsymbol{F} : \text{either }]t,1[\subset B \text{ or } B \subset]0,t]\right\}$$

for $0 < t < 1$ and $\boldsymbol{F}_t = \boldsymbol{F}$ for $t \geq 1$. Then $(\boldsymbol{F}_t)_{t\geq 0}$ is a filtration and for every $Y \in \boldsymbol{L}^1(\mathbb{P})$ the martingale

$$(M_t)_{t\geq 0} = (\mathbb{E}(Y \mid \boldsymbol{F}_t))_{t\geq 0}$$

has the following representation

$$M_t(\omega) = \begin{cases} Y(\omega) & \text{for } 0 < \omega \leq t \\ \frac{1}{1-t}\int_t^1 Y(s)\,ds & \text{for } t < \omega < 1 \end{cases}.$$

Choose now $Y(\omega) = (1-\omega)^{-\alpha}$ for some $\alpha < ½$. Then $Y \in \boldsymbol{L}^2(\mathbb{P})$ and one easily computes

$$M_t(\omega) = \frac{1}{1-\alpha}(1-t)^{-\alpha} = \frac{1}{1-\alpha} Y(t) \text{ for } t < \omega < 1.$$

This (separable!) martingale thus satisfies

$$\sup_t M_t = \frac{1}{1-\alpha}\, Y$$

and hence

$$\|\sup_t M_t\|_2 = \frac{1}{1-\alpha}\, \|Y\|_2 = \frac{1}{1-\alpha}\, \|M_\infty\|_2.$$

As α approaches ½ the constant $\frac{1}{1-\alpha}$ approaches 2.

CHAPTER 4

LOCALIZATION AND APPROXIMATION

Stochastic integral processes will be constructed by approximation. One first defines the integral process in a straightforward manner for elementary integrands. Then one shows that the approximation of more general integrands by elementary ones yields a convergent sequence of the corresponding integral processes. The details depend on the choice of the respective spaces and concepts of convergence. In this chapter we discuss the part of the arguments which can be formulated without explicit reference to stochastic integrals. If you have ploughed through this somewhat technical chapter the way to the stochastic integral is open.

4.1 Localization and Stochastic Intervals

The technique of localization is an essential trick of today's theory of stochastic integration. Roughly speaking, one stops before things get nasty, i.e. one simply exhausts $\mathbb{R}_+ \times \Omega$ by an increasing sequence of "stochastic intervals" on which some given quantities do not become too large. To fix the ideas let us discuss the definition of a "locally bounded process". A function on $\mathbb{R}_+$ is locally bounded if $\sup_{s \le t}|f(s)| < \infty$ for every $t \ge 0$ or equivalently, if there is a sequence of intervals $[0,t_k]$ such that (i) $|f|$ is bounded by k on $[0,t_k]$ for eventually all k and (ii) $t_k \uparrow \infty$. If the function is random the intervals depend on ω and their endpoints should be given by stopping times. Here is the obvious definition of a stochastic interval.

4.1.1 Definition. Suppose that S and T are stopping times. The **stochastic interval** $]S,T]$ between S and T (open on the left and closed on the right) is defined by

$$]S,T] = \left\{ (t,\omega) \in \mathbb{R}_+ \times \Omega : S(\omega) < t \le T(\omega) \right\}.$$

The stochastic intervals $[S,T]$, $[S,T[$ are defined in an analogous way. Notice that even if T assumes the value ∞ the interval $]S,T]$ is a subset of $\mathbb{R}_+ \times \Omega$ and not of $(\mathbb{R}_+ \cup \{\infty\}) \times \Omega$.

One has to be more careful about the boundary of stochastic intervals than with ordinary intervals. For example in general one can not expect a half open stochastic interval $[0,T[$ to be the union of a sequence of closed stochastic intervals $[0,T_n]$, cf. the discussion of predictability in connection with proposition 2.3.5 and in chapter 6 . Therefore, it matters sometimes whether the "localizing" stochastic intervals are assumed to contain their right endpoint or not.

This distinction between the two sorts of stochastic intervals is the subject of difficult theorems in the general theory of processes. Of course one does not need to worry about these questions if one is considering only continuous processes. Only recently one has realized that also in the noncontinuous case basic stochastic integration can be developed with a minimum of reference to that theory.

4.1.2 Definition. Let $(\Omega, \boldsymbol{F}, \mathbb{P})$ be a probability space with the filtration $(\boldsymbol{F}_t)_{t \geq 0}$.

(a) A **localizing sequence** is a sequence $(T_k)_{k \geq 1}$ of stopping times such that $\mathbb{P}(T_k \uparrow \infty) = 1$.

(b) A process X is called **locally bounded** if there is a localizing sequence such that X is uniformly bounded on each of the stochastic intervals $]0,T_k]$.

Remarks. (a) The left endpoint 0 of the stochastic intervals is omitted since the random variable X_0 may be unbounded.

(b) An alternative way to express local boundedness is to say that for some localizing sequence $(T_k)_{k \geq 1}$ each of the processes $(\mathbf{1}_{\{T_k > 0\}} X_{t \wedge T_k})_{t \geq 0}$ is uniformly bounded: Under the latter condition X is bounded on $]0,T_k]$ and conversely if X is bounded on $]0,T_k]$ then $(\mathbf{1}_{\{S_k > 0\}} X_{t \wedge S_k})_{t \geq 0}$ is bounded where $S_k = \mathbf{1}_{\{|X_0| \leq k\}} T_k$. Obviously $(S_k)_{k \geq 1}$ is also a localizing sequence.

(c) If X is locally bounded then its paths are a.s. locally bounded real functions i.e. the process X satisfies

$$(4.1.1) \qquad \mathbb{P}\Big(\sup_{s \le t} |X_s| < \infty \Big) = 1 \ \text{ for every } t \ge 0.$$

This condition is e.g. fulfilled if X is a.s. continuous.
Concerning the converse, we shall encounter in the next section a simple right-continuous martingale M with bounded paths which is not locally bounded (cf. 4.2.6(a)). The difficulty is caused by the jumps whose size may depend on ω so that M_T is unbounded for every nontrivial stopping time T. However, for left-continuous processes this problem does not arise. Slightly more generally we have

4.1.3 Proposition. For an adapted process X the process X^- is locally bounded if (4.1.1) holds.

Proof. Suppose that X is adapted and that (4.4.1) holds. The process X^- is separable and adapted and therefore the random times

$$T_k = \inf\Big\{ t \ge 0 : |X_t^-| > k \Big\}$$

are stopping times according to proposition 2.3.4. Also (4.1.1) implies $\mathbb{P}(T_k \uparrow \infty) = 1$ since $T_k(\omega) \ge t$ if $\sup_{s \le t} |X_s(\omega)| \le k$. So $(T_k)_{k \ge 1}$ is a localizing sequence. Then for each ω the path $X_{.}(\omega)$ is bounded by k on $[0, T_k(\omega)[$ and hence the definition of X^- implies that $|X^-| \le k$ on $]0, T_k]$, i.e. X^- is locally bounded. ■

One nice and important fact about localization is expressed in the obvious

4.1.4 Lemma. If $(S_k)_{k \ge 1}$ and $(T_k)_{k \ge 1}$ are two localizing sequences then $(S_k \wedge T_k)_{k \ge 1}$ is also a localizing sequence.

From this one easily deduces e.g. that the sum of locally bounded processes is locally bounded: If X is bounded on $]0, S_k]$ and Y is bounded on $]0, T_k]$ then X+Y is bounded on $]0, S_k \wedge T_k]$. This kind of argument will be used without further references.

4.2 Local Martingales

In stochastic integration theory many important properties of martingales are shared by the larger class of local martingales. Since the integrability requirements are relaxed this class is much more flexible under a couple of useful transformations like random time-change and unbounded harmonic maps (cf. chapters 9-12). This is another illustration of the localization principle.

4.2.1 Definition. Let the filtration $(\boldsymbol{F}_t)_{t\geq 0}$ and a right-continuous stochastic process M be given. M is called a **local $\boldsymbol{L}^2$-martingale** (resp. **local martingale**) if there is a localizing sequence $(T_k)_{k\geq 1}$ such that for every $k \geq 1$ the stopped process $(M_{t\wedge T_k} - M_0)_{t\geq 0}$ is a $\boldsymbol{L}^2$-bounded (resp. uniformly integrable) martingale. The sequence $(T_k)_{k\geq 1}$ is called a **localizing sequence for the local ($\boldsymbol{L}^2$-) martingale** M.

Comment. (a) Note that M_t (even M_0) is not assumed to be integrable.
(b) We are mainly interested in the increments $M_t - M_s$ of a local martingale. Thus the initial variable M_0 often is irrelevant since $(M_t)_{t\geq 0}$ and $(M_t - M_0)_{t\geq 0}$ have the same increments.
(c) The condition of right-continuity is almost enforced by the fact that local martingales are closely linked to the application of the stopping theorem 3.2.7 which works only for right-continuous martingales. Theorem 3.2.6(a) implies that every local martingale is a.s. regular.

4.2.2 Examples. (a) Every martingale is a local martingale. In fact, the constant stopping times $T_k \equiv k$ define a localizing sequence for M since

$$\left(M_{t\wedge k}\right)_{t\geq 0} = \left(\mathbb{E}\left(M_k | \boldsymbol{F}_t\right)\right)_{t\geq 0}$$

is a uniformly integrable martingale by lemma 3.2.2.
(b) Plainly, a local $\boldsymbol{L}^2$-martingale is a local martingale. The converse does not hold in general (cf. example 4.2.6(a)). But every a.s. *continuous* local martingale is a local $\boldsymbol{L}^2$-martingale with localizing sequence

$$T_k = \inf\left\{t \geq 0 : |M_t - M_0| > k\right\}.$$

The T_k are stopping times by proposition 2.3.4 and increase to infinity a.s. since M has a.s. continuous paths. Let $(S_k)_{k \geq 1}$ be a localizing sequence of M. Then the processes

$$\left(M_{t \wedge T_k \wedge S_k} - M_0\right)_{t \geq 0}$$

are martingales since by right-continuity corollary 3.2.8(b) applies. They are uniformly bounded and hence bounded in $\boldsymbol{L}^2$.

The following result is a criterion for a stopped local martingale to be a martingale.

4.2.3 Lemma. Suppose that M is a local martingale and let T be a stopping time. If the set

$$\left\{M_S : S \leq T \text{ is a bounded stopping time}\right\}$$

is uniformly integrable then $(M_{t \wedge T})_{t \geq 0}$ is a martingale closed by M_T.

Proof. First note that the condition implies the integrability of M_0. Therefore replacing M_t by $M_t - M_0$ we may assume $M_0 \equiv 0$.
We use the criterion 3.2.9. To this end, choose a localizing sequence $(T_k)_{k \geq 1}$ for M and a bounded stopping time S. Then $S \wedge T \wedge T_k \leq T$ for each k. Hence the sequence $(M_{S \wedge T \wedge T_k})_{k \geq 1}$ is uniformly integrable. Plainly, it converges a.s. to $M_{S \wedge T}$ and application of the stopping theorem to the martingales $(M_{t \wedge T_k})_{t \geq 0}$ yields

$$\mathbb{E}(M_{S \wedge T}) = \lim_{k \to \infty} \mathbb{E}(M_{S \wedge T \wedge T_k}) = \mathbb{E}(M_0).$$

So $(M_{t \wedge T})_{t \geq 0}$ is a uniformly integrable martingale and hence closed by some limit-variable. Since $M_T = \lim_{k \to \infty} M_{T \wedge T_k}$ this limit-variable is M_T. ■

Remark. In the preceding lemma it was important to allow general stopping times S. Example 4.2.6(b) below implies that the uniform integrability of the set $\{ M_{t \wedge T} : t \geq 0 \}$ is not sufficient.

An analogue of lemma 4.1.4 holds for localizing sequences of local martingales.

4.2.4 Lemma. Suppose that $(T_k)_{k\geq 1}$ is a localizing sequence for the local $(\boldsymbol{L}^2)$-martingale M. If $(S_k)_{k\geq 1}$ is any other localizing sequence then $(S_k \wedge T_k)_{k\geq 1}$ is also a localizing sequence for M.

Proof. This follows from the stopping theorem. ∎

4.2.5 Corollary. The local $(\boldsymbol{L}^2$-)martingales form a linear space under pointwise operations.

4.2.6 Counterexamples. We construct
(a) a uniformly integrable (right-continuous) martingale which is not a local $\boldsymbol{L}^2$-martingale,
(b) a $\boldsymbol{L}^2$-bounded local martingale which is not a martingale.

(a) This example is a modification of example 3.2.11. Let $(\Omega, \boldsymbol{F}, \mathbb{P})$ be the Lebesgue probability space over $]0,1[$ and let $(\boldsymbol{F}_t)$ be the filtration introduced in 3.2.11. We consider the martingale $\left(\mathbb{E}(Y|\boldsymbol{F}_t)\right)_{t\geq 0}$ for the function $Y(\omega) = \omega^{-1/2}$. Then according to the explicit representation in 3.2.11 we have $M_t(\omega) = \omega^{-1/2}$ for $0 < \omega \leq t$ and $M_t(\omega)$ does not depend on ω as long as $t < \omega$. (If you want you can check that $M_t(\omega) = 2/(1 + t^{1/2})$ for $t < \omega$.) The martingale is right-continuous. We shall show that every stopping time except $T \equiv 0$ satisfies $T(\omega) \geq \omega$ on some ω-interval $]0,\varepsilon]$ with $\varepsilon > 0$. But then $M_T(\omega) = \omega^{-1/2}$ on this interval and therefore

$$\mathbb{E}\left(M_T^2\right) \geq \int_0^\varepsilon s^{-1}\, ds = \infty.$$

Hence M cannot be a local $\boldsymbol{L}^2$-martingale. Moreover, this martingale has bounded paths but $\mathbf{1}_{\{T>0\}} M_T$ is unbounded for every stopping time $T \neq 0$ (this justifies remark (c) following 4.1.2).
Suppose now that T is a nonzero stopping time, i.e. $T(\omega_0) > 0$ for some ω_0. Let $\varepsilon = \omega_0 \wedge T(\omega_0)$ and choose $0 < s < \varepsilon$. Then the set $\{T \geq s\} \cap]s,1[$ is not empty. Since $\{T \geq s\} \in \boldsymbol{F}_s$ the definition of $\boldsymbol{F}_s$ implies $]s,1[\subset \{T \geq s\}$, i.e. $T(\omega) \geq s$ for all $\omega > s$. Now for $\omega \leq \varepsilon$ let s increase to ω to get $T(\omega) \geq \omega$.

(b) An example of a continuous $\boldsymbol{L}^2$-bounded local martingale which is not a martingale is reported in Dellacherie-Meyer (1980), chapter VI, no. 29. In more detail it is treated in Chung-Williams (1983), chapter 1.10.

The following noncontinuous but more elementary example has been pointed out to us by M. Scheutzow (and independently in a similar form by J. Pitman). We use freely a little Markov chain theory. The main idea is to embed a one-dimensional symmetric random walk into continuous time but let the particle run faster if it is far outside so that it spends only little time at large values. This process becomes positive recurrent and therefore can be bounded in $\mathcal{L}^2$. Since it does not converge as $t \to \infty$ it cannot be a martingale. But because it is constructed from a martingale by time-change it is a local martingale.

Let $(\pi_n)_{n \in \mathbb{Z}}$ be a probability distribution on $\mathbb{Z}$ such that $\pi_n > 0$ for all n and $\sum_{n=-\infty}^{+\infty} n^2 \pi_n < \infty$. Define the Markov process $M = (M_t)_{t \geq 0}$ as follows: It starts with probability π_n in n and if it is in n at time t then after an exponential waiting time of expectation π_n it moves either to n-1 or to n+1 with probability ½ each. Fig. 4.2.1 shows a "typical" path.

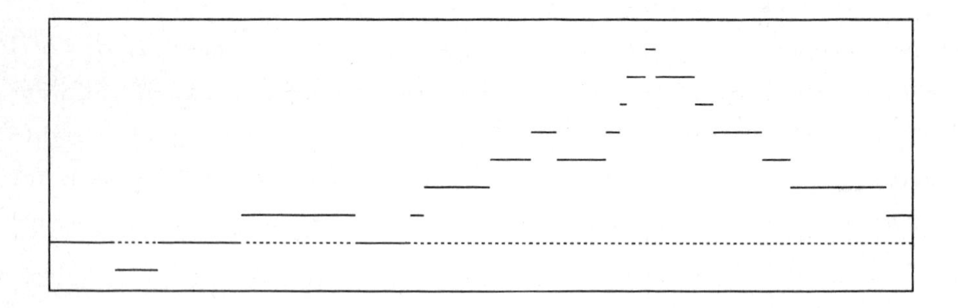

Fig. 4.2.1

This process has all required properties: If T_k denotes the time of the k-th jump then $(M_{T_k})_{k \geq 0}$ is a symmetric random walk. First of all this implies $T_k \uparrow \infty$ a.s. since after infinitely many steps the process has revisited its starting point infinitely often and by the strong law of large numbers the total duration of these visits is infinite. Secondly, M cannot be a $\mathcal{L}^2$-bounded martingale because the limit

$$\lim_{t \to \infty} M_t = \lim_{k \to \infty} M_{T_k}$$

does not exist contradicting the martingale convergence theorem. But for fixed k the stopped process $(M_{t \wedge T_k})_{t \geq 0}$ is a martingale. Let us be formal:

We have

$$(4.2.1) \qquad M_{t\wedge T_k} - M_0 = \sum_{i=1}^{k} \varepsilon_i \mathbf{1}_{\{T_i \le t\}}$$

where $\varepsilon_i \in \{-1,1\}$ is the i-th jump. For $s < t$ and every $F \in \boldsymbol{F}_s$ this jump is independent of the event $F \cap \{s < T_i \le t\}$. Thus

$$\mathbb{E}\left(\mathbf{1}_F(M_{t\wedge T_k} - M_{s\wedge T_k})\right) = \sum_{i=1}^{k} \mathbb{E}\left(\varepsilon_i \cdot \mathbf{1}_{\{s<T_i\le t\}\cap F}\right) = 0$$

proving the martingale property. This martingale is bounded in $\boldsymbol{L}^2$ since $|M_{t\wedge T_k} - M_0| \le k$. Hence M is a local $\boldsymbol{L}^2$-martingale.
Finally, in order to see that (M_t) is $\boldsymbol{L}^2$-bounded we prove that the process is stationary. The probabilities

$$p_n(t) = \mathbb{P}(M_t = n)$$

in a birth and death process are known to satisfy the forward equation

$$p_n'(t) = -(\mu_n + \lambda_n)p_n(t) + \mu_{n-1}p_{n-1}(t) + \lambda_{n+1}p_{n+1}(t)$$

where μ_n and λ_n are the intensities of the upward and downward jumping times, respectively, i.e. in our case

$$\mu_n = \lambda_n = \tfrac{1}{2}\frac{1}{\pi_n}.$$

Since the embedded random walk (M_{T_k}) is recurrent these equations have only one solution with initial condition (π_n) (see Feller (1966), vol. 2, p. 488, corollary 2). But $p_n(t) \equiv \pi_n$ is a (stationary) solution and hence

$$\mathbb{E}(M_t^2) = \sum_{n=-\infty}^{+\infty} n^2 \pi_n < \infty$$

for all t. ■

4.3 Local Approximation

Various concepts of local convergence of processes can be introduced. The following turns out to be the most flexible one. It can be expressed in several ways.

4.3.1 Lemma. Let X^n, $n \in \mathbb{N}$, and X be separable processes. For a random time T consider the condition

$$(4.3.1) \qquad \mathbb{P}\Big(\sup_{0 \le s \le T} |X_s^n - X_s| > \varepsilon \Big) \xrightarrow[n \to \infty]{} 0 \text{ for every } \varepsilon > 0 .$$

Then the following are equivalent:

(a) Condition (4.3.1) holds for every a.s. finite random time T.

(b) Condition (4.3.1) holds for every finite deterministic time $T \equiv t$.

(c) There is a localizing sequence $(T_k)_{k \ge 1}$ of stopping times such that (4.3.1) holds for each T_k.

Proof . The implications (a) → (b) and (b) → (c) are obvious. For (c) → (a) let T and $\varepsilon > 0$ be given and choose (T_k) as in (c). Then

$$\mathbb{P}\Big(\sup_{0 \le s \le T} |X_s^n - X_s| > \varepsilon \Big) \le \mathbb{P}\Big(\sup_{0 \le s \le T_k} |X_s^n - X_s| > \varepsilon \Big) + \mathbb{P}(T > T_k)$$

for all k and n. Since T is finite the second term on the right becomes small for large k. Choosing then n large enough we find that the left-hand side is arbitrarily small for sufficiently large n, i.e. (4.3.1) holds for T. ■

4.3.2 Definition. If one of the conditions in the preceding lemma holds then we say that (X^n) converges to X **locally uniformly in probability.** We will use the short-hand notation l.u.p..

Remark. This convergence is induced by a metric: For example $X^n \longrightarrow X$ if and only if $d(X^n,X) \longrightarrow 0$ where

$$d(X,Y) = \sum_{n=1}^{\infty} 2^{-n}\, \mathbb{E}\Big(1 \wedge \sup_{0 \le s \le n} |X_s - Y_s| \Big).$$

Rather than using a metric we formulate results in terms of convergence of sequences of processes.

A Cauchy-sequence of random variables with respect to convergence in probability has a limit and an a.s. convergent subsequence. An analogue holds for locally uniform convergence. This will be important for the construction of the integral processes.

4.3.3 Theorem. Let $(X^n)_{n\geq 1}$ be a sequence of separable adapted processes and let $(S_k)_{k\geq 1}$ a localizing sequence of (not necessarily finite !) stopping times. Suppose that for each $k \geq 1$ and every $\varepsilon > 0$

$$\lim_{n,m\to\infty} \mathbb{P}\Big(\sup_{0\leq s\leq S_k} |X^n_s - X^m_s| > \varepsilon\Big) = 0 . \tag{4.3.2}$$

Then the following holds:

(a) There is an adapted process X such that $X^n \longrightarrow X$ l.u.p., more precisely

$$\lim_{n,m\to\infty} \mathbb{P}\Big(\sup_{0\leq s\leq S_k} |X^n_s - X_s| > \varepsilon\Big) = 0 . \tag{4.3.3}$$

(b) There is a subsequence $(X^{n_r})_{r\geq 1}$ and a localizing sequence $(T_k)_{k\geq 1}$ such that $(X^{n_r})_{r\geq 1}$ converges uniformly on each of the right-open stochastic intervals $[0,T_k[$, i.e. that for every k,

$$\sup\Big\{\big|X^{n_r}(t,\omega) - X(t,\omega)\big| : (t,\omega) \in [0,T_k[\Big\} \xrightarrow[r\to\infty]{} 0.$$

(c) If the processes X^n are a.s. continuous then X is a.s. continuous. If the X^n are right-continuous then X *can be chosen* to be right-continuous.

Proof. First assume

$$\mathbb{P}\Big(\sup_{s\leq S_n} \big|X^{n+1}_s - X^n_s\big| > 2^{-n}\Big) < 2^{-n} \text{ for each } n \geq 1. \tag{4.3.5}$$

Consider

$$T_k = \inf\Big\{s \geq 0 : \big|X^{n+1}_s - X^n_s\big| > 2^{-n} \text{ for some } n\geq k\Big\}$$

and

$$X = \limsup_{n\to\infty} X^n .$$

We show that $(T_k)_{k\geq 1}$ is a localizing sequence and that for every k

$$\sup\Big\{\big|X^n_s(\omega) - X_s(\omega)\big| : 0 \leq s < T_k(\omega),\ \omega \in \Omega\Big\} \xrightarrow[n\to\infty]{} 0 .$$

Plainly $T_1 \leq T_2 \leq \ldots$ and each T_k is a stopping time because the processes X^n are separable (proposition 2.3.4). If $T_k < S_k$ then by (4.3.5) there are $s < S_k$ and $n\geq k$ such that $|X^{n+1}_s - X^n_s| > 2^{-n}$. Thus

$$\begin{aligned} \mathbb{P}(T_k < S_k) &\leq \sum_{n=k}^{\infty} \mathbb{P}\Big(\sup_{s<S_k} |X^{n+1}_s - X^n_s| > 2^{-n}\Big) \\ &\leq \sum_{n=k}^{\infty} \mathbb{P}\Big(\sup_{s<S_n} |X^{n+1}_s - X^n_s| > 2^{-n} \leq \sum_{n=k}^{\infty} 2^{-n} = 2^{-k+1} \end{aligned} \tag{4.3.6}$$

and hence $\mathbb{P}(T_k \uparrow \infty) = 1$. Thus $(T_k)_{k\geq 1}$ is a localizing sequence. By the very definition of these stopping times,

$$|X_s^{n+1} - X_s^n| \leq 2^{-n} \text{ for all } n \geq k \text{ on } [0,T_k[.$$

Hence for all $n > m \geq k$ on $[0,T_k[$,

$$\left|X_s^n - X_s^m\right| \leq \sum_{r=m}^{n} |X_s^{r+1} - X_s^r| \leq 2^{-m+1}.$$

Therefore (X^n) converges uniformly on $[0,T_k[$. A fortiori

$$X = \limsup_{n\to\infty} X^n = \lim_{n\to\infty} X^n \text{ on } [0,T_k[.$$

Now we are going to prove (a) and (b) under the assumptions of the theorem: If (4.3.2) holds then there *is* a subsequence (X^{n_r}) of (X^n) satisfying (4.3.5). Hence (b) follows from the above arguments. This subsequence satisfies (4.3.4). Now on the set $\{S_k < \infty\}$ one has $S_k < T_{k'}$ for eventually all k' since $(T_{k'})$ is a localizing sequence, and on the set $\{S_k = \infty\}$ one has $T_k = \infty$ a.s. for eventually all k because of (4.3.6). Thus (4.3.4) for the subsequence (X^{n_r}) implies

(4.3.7) $$\sup_{0\leq s\leq S_k} |X_s^{n_r} - X_s| \longrightarrow 0 \text{ a.s. for every } t \geq 0.$$

Hence (X^{n_r}) and X satisfy mutatis mutandis (4.3.3). Then (4.3.3) holds for the whole sequence by the Cauchy condition (4.3.2). This completes the proof of (a).

Finally, let us turn to (c). The relation (4.3.7) shows that X inherits a.s. continuity from the X^n. The main argument about right-continuity is formulated separately in the following lemma which together with part (b) implies the assertion. ■

Remark. It would be slightly more convenient if we could cancel the "a.s." in part (c) of the theorem. The approximation argument in the preceding proof definitely gives locally uniform convergence of the paths *only almost surely*. The question whether ω is in the exceptional set involves the whole time axis. Thus in general the exceptional set is not $\boldsymbol{F}_0$-measurable. The only obvious way to arrive at a strictly continuous adapted limit process defined *on the same space* Ω is to add all $\mathbb{P}$-nullsets to $\boldsymbol{F}_0$. The lemma which allows to choose a (strictly) right-continuous limit process uses an idea of Stroock and Varadhan (1979).

4.3.4 Lemma. Let $(X^n)_{n\geq 1}$ be a sequence of right-continuous adapted processes. Then there is a right-continuous adapted process X such that

$$X(t,\omega) = \lim_{n\to\infty} X^n(t,\omega) \text{ for all } (t,\omega) \text{ such that } t < T(\omega)$$

where

$$T(\omega) = \sup\left\{\tau : X^n(\cdot,\omega) \text{ converges uniformly on } [0,\tau]\right\}.$$

Proof. Clearly, for each $t > 0$

$$\left\{T < t\right\} = \left\{\limsup_{n,m\to\infty} \sup_{s\in\mathbb{Q}_+, s<t} \left|X_s^n - X_s^m\right| > 0\right\} \in \boldsymbol{F}_t$$

and hence T is a stopping time. The stopped processes $X^n_{\cdot\wedge T}$ are adapted according to proposition 2.3.13(b). Therefore the process

$$X = \limsup_{n\to\infty} X^n_{\cdot\wedge T}$$

is adapted. It is also right-continuous: on $[0,T[$ by pathwise uniform approximation and on $[T,\infty[$ it is constant. Since

$$X(t,\omega) = \lim X^n_{t\wedge T}(\omega) = \lim X^n_t(\omega)$$

for $t < T(\omega)$ the process X has all desired properties. ■

Here is another useful consequence of this idea. It shows that the requirement of right-continuity in the definition of a $(\boldsymbol{F}_t)$-Brownian motion is not an essential restriction.

4.3.5 Lemma. Each adapted and a.s. continuous process has an adapted version whose noncontinuous paths are right-continuous.

Proof. Let Y be adapted and a.s. continuous. Define the process X^n by

$$X^n = \sum_{k=0}^{\infty} Y_{k/n} \mathbf{1}_{[k/n,\,(k+1)/n[}\,.$$

Clearly X^n is adapted and right-continuous. Moreover $X^n(\cdot,\omega) \longrightarrow Y(\cdot,\omega)$ uniformly over compact intervals whenever $Y(\cdot,\omega)$ is continuous. Hence the stopping time T in the previous lemma is a.s. infinite and thus the right-continuous adapted process X associated with (X^n) is a.s. equal to Y. ■

4.4 Elementary Processes and their Limits

Elementary processes and their limits w.r.t. local uniform convergence in probability form the most important class of integrands in the stochastic integral. It is known that regular functions on a bounded interval are precisely the uniform limits of step functions. In proposition 4.4.4 below we prove a similar characterization of l.u.p. limits of elementary processes.

4.4.1 Definition. (a) We call a function $f : \mathbb{R}_+ \longrightarrow \mathbb{R}$ a **left-continuous step function** if it can be written in the form

$$f(t) = a_0 \, \mathbf{1}_{\{0\}}(t) + \sum_{i=1}^{n} a_i \cdot \mathbf{1}_{]t_i, t_{i+1}]}(t)$$

where $a_i \in \mathbb{R}$ and $t_i \in [0,\infty]$.

(b) Given the filtration $(\boldsymbol{F}_t)_{t \geq 0}$ a process H is called an **elementary process** if it is bounded and $(\boldsymbol{F}_t)$-adapted and if there is $n \in \mathbb{N}$ such that every path of H is a left-continuous step function with at most n steps.

The term $a_0 \mathbf{1}_{\{0\}}$ in (a) is introduced in order to allow arbitrary values at time 0 but for most purposes it is quite irrelevant.

We note some stability properties of elementary processes and prove a very useful representation.

4.4.2 Proposition. (a) The space of elementary processes is a linear space under pointwise operations and contains the pointwise product, minimum and maximum of any pair of its members.

(b) A bounded process H is elementary if and only if it can be written in the form

$$(4.4.1) \qquad H = H_0 \mathbf{1}_{\{0\}} + \sum_{i=1}^{n} H_i \mathbf{1}_{]T_i, T_{i+1}]}$$

where H_0 is $\boldsymbol{F}_0$-measurable, $(T_i)_{1 \leq i \leq n}$ is an increasing set of stopping times and H_i is $\boldsymbol{F}^+_{T_i}$-measurable for each i.

Proof. (a) This follows immediately from the corresponding statement for deterministic step functions.

(b) For both implications we may assume without loss of generality that $H_0 \equiv 0$.

Let H be given by the right-hand side of (4.4.1). Then every path is a left-continuous step function. We may write

$$H_t = \sum_{i=1}^{n} H_i\left(\mathbf{1}_{\{T_i<t\}} - \mathbf{1}_{\{T_{i+1}<t\}}\right).$$

Since H_i is $\boldsymbol{F}^+_{T_i}$-measurable and a fortiori $\boldsymbol{F}^+_{T_{i+1}}$-measurable (cf. lemma 2.2.5(a)) we conclude from lemma 2.2.4(b) that H is $(\boldsymbol{F}_t)$-adapted. Thus H is an elementary process.

Conversely, let H be an elementary process. Then H is $(\boldsymbol{F}_t)$-adapted. Let $T_1 = 0$ and for $i \geq 2$ let $T_i(\omega)$ be the time of the (i-1)-th jump of the step function $H.(\omega)$ and $T_i(\omega) = \infty$ if there is no jump after time $T_{i-1}(\omega)$. Let $H_i(\omega)$ be the value of $H.(\omega)$ in the interval $]T_i(\omega),T_{i+1}(\omega)]$. Since the number of jumps is bounded uniformly in ω the process H has the finite representation (4.4.1). The measurability part is a consequence of lemma 4.4.3 below: Choose there $\varepsilon = 0$. Then we see that $T_i = T_i^0$ and $H_i = H^+_{T_i^0}$ and the lemma implies that T_i is a stopping time and H_i is $\boldsymbol{F}^+_{T_i}$-measurable. ■

The following lemma is the key to the approximation mentioned above.

4.4.3 Lemma. Consider a separable process H and a number $\varepsilon \geq 0$. Using the process H^+ from (2.3.2) define inductively

$$T_1^\varepsilon \equiv 0,$$

$$T_{i+1}^\varepsilon = \begin{cases} \inf\{t > T_i^\varepsilon : |H_t - H^+_{T_i^\varepsilon}| > \varepsilon\} & \text{on } \{T_i^\varepsilon < \infty\} \\ +\infty & \text{on } \{T_i^\varepsilon = \infty\} \end{cases}.$$

Then the random times T_i^+ are stopping times and the random variable $H^+_{T_i^\varepsilon}$ is measurable w.r.t $\boldsymbol{F}^+_{T_i^\varepsilon}$ for each $i \geq 1$.

Proof. We prove both parts of the assertion simultaneously by induction. The initial step is obvious. For the induction step, suppose that T_i^ε is a stopping time. Then according to 2.3.13(a) the variable $H^+_{T_i^\varepsilon}$ is measurable w.r.t. $\boldsymbol{F}^+_{T_i^\varepsilon}$. We have to show that T_{i+1}^ε is a stopping time. The following argument is an adaption of the proof for proposition 2.3.4. Fix $\omega \in \Omega$ and $t \geq 0$. We see that $T_{i+1}^\varepsilon(\omega) < t$ if and only if $|H_s(\omega) - H^+_{T_i^\varepsilon}(\omega)| > \varepsilon$ for some

$s \in]T_i^\varepsilon(\omega), t[$ where in fact only points s in the separability set U of H need to be considered. Rewrite this as

$$\left\{T_{i+1}^\varepsilon < t\right\} = \bigcup_{s \in [0,t[\cap U} \left\{T_i^\varepsilon < s\right\} \cap \left\{|H_s - H_{T_i^\varepsilon}^+| > \varepsilon\right\}.$$

The variables H_s, $s < t$, and the restriction of $H_{T_i^\varepsilon}^+$ to the sets $\{T_i^\varepsilon < s\}$ are $\boldsymbol{F}_t$-measurable since $H_{T_i^\varepsilon}^+$ is $\boldsymbol{F}_{T_i^\varepsilon}^+$-measurable. Thus $\{T_{i+1}^\varepsilon < t\} \in \boldsymbol{F}_t$ for all t and hence T_{i+1}^ε is a stopping time. ■

4.4.4 Proposition. A separable $(\boldsymbol{F}_t)$-adapted process H is a.s. regular and a.s. left-continuous if and only if it can be approximated l.u.p. by elementary processes.

Note that the deterministic characterization of regular functions as uniform limits of step functions is contained in this result (except that for simplicity we have assumed left-continuous paths).

Proof. Suppose that $H^n \longrightarrow H$ l.u.p. for a sequence (H^n) of elementary processes. Then theorem 4.3.3 shows that almost every path of H can be approximated uniformly on every finite interval by the paths of a suitable subsequence of (H^n). Since on a bounded real interval a uniform limit of left-continuous step functions is regular and left-continuous the process H is a.s. regular and a.s. left continuous. This proves one implication.
For the converse suppose that H is separable a.s. regular and a.s. left-continuous. We consider the stopping times T_i^ε of lemma 4.4.3 for $\varepsilon > 0$. Choose a regular and left-continuous path ω. From the existence of the right limits it follows that $T_i^\varepsilon(\omega) < T_{i+1}^\varepsilon(\omega)$ for every i. The existence of the left limits implies that the increasing sequence $(T_i^\varepsilon(\omega))_{i\geq 1}$ cannot have a finite supremum t^* since in any left neighbourhood of the point t^* the path would have infinitely many oscillations of absolute value greater than ε. Thus $T_i^\varepsilon(\omega)\uparrow\infty$. On each interval $]T_i^\varepsilon(\omega), T_{i+1}^\varepsilon(\omega)]$ the path differs from $H_{T_i^\varepsilon}^+(\omega)$ at most by ε. Since the path is left-continuous we can include the right endpoint of the interval.

Now given $t \geq 0$ and $\varepsilon > 0$ choose n_ε large enough such that $\mathbb{P}(T^\varepsilon_{n_\varepsilon+1} \leq t) < \varepsilon$ and consider the elementary process

$$H^{\varepsilon,t} = H_0 1_{\{0\}} + \sum_{i=1}^{n_\varepsilon} H^+_{T^\varepsilon_i} 1_{]T^\varepsilon_i, T^\varepsilon_{i+1}]}.$$

Then

$$\sup_{0 \leq s \leq T^\varepsilon_{n_\varepsilon+1}} |H^{\varepsilon,t}_s - H_s| \leq \varepsilon$$

and hence

$$\mathbb{P}\Big(\sup_{0\leq s\leq t} |H^{\varepsilon,t}_s - H_s| > \varepsilon\Big) < \varepsilon.$$

Therefore H can be approximated l.u.p. by the processes $H^{\varepsilon,t}$. They may not be bounded but

$$H^{\varepsilon,t} = \lim_{n\to\infty} \big(-n \vee (H^{\varepsilon,t} \wedge n)\big) \text{ l.u.p.}$$

i.e. $H^{\varepsilon,t}$ can be approximated l.u.p. by elementary processes. Thus the same is true for H. This completes the proof. ■

4.4.5 Definition. We denote by ***H*** the space of all adapted a.s. regular and a.s. left-continuous processes .

Comment. If F_0 contains the $\mathbb{P}$-nullsets then you may omit the " a.s." from the definition of the space ***H*** . In fact, under this assumption on the filtration every element of ***H*** is a.s. equal to a regular left-continuous adapted process.

4.4.6 Remark. For every adapted a.s. regular process X the process X^- is a separable element of ***H***. If, moreover, X is a.s. right-continuous then $X = (X^-)^+$ a.s.. If $H \in$ ***H*** then $H = H^-$ a.s..

Proof. The separability of X^- was already noted in the remark after (2.3.2). Every regular path $X.(\omega)$ is bounded on $[0,t]$ for every t since it is a uniform limit of step functions according to the deterministic version of proposition 4.4.4. Hence $X^-_t(\omega)$ simply is the left-hand limit of the path at the point t. The values in an interval $]t,t+\varepsilon[$ of $X.(\omega)$ and hence also of $X^-.(\omega)$ do not differ much from the right-hand limit of $X.(\omega)$ at t. Thus $X^-.(\omega)$ is also regular and $(X^-.(\omega))^+_t = X^+_t(\omega)$. This implies the second and the third assertion. ■

CHAPTER 5

THE STOCHASTIC INTEGRAL

In this chapter, we define (stochastic) Itô- integrals $\int_0^t H\,dM$ for local $\mathcal{L}^2$-martingales M and a fairly large class of adapted processes H. The integral is a random variable. It will be constructed as a suitable limit of Riemann-Stieltjes type approximations like

$$\sum_{i=1}^{n} H_{s_i}\cdot(M_{s_{i+1}} - M_{s_i}).$$

A pathwise limit of these approximating sums does not exist in general since there are no nonconstant continuous martingales with bounded variation (section 5.3). K. Itô (1944), however, observed that in the case of Brownian motion M for many H the limit exists in $\mathcal{L}^2(\mathbb{P})$. We present an adaptation of his ideas which works also for general local $\mathcal{L}^2$-martingales.

With the strong fluctuation of the paths of M goes another peculiarity of the stochastic integral. Even if H is smooth in the time variable t, it matters at which point of the intervals $[s_i, s_{i+1}]$ the random function (H_s) is evaluated in the above sum. The main reason for choosing the left endpoints is that the integral then becomes a local martingale if the integrator M is a local martingale. This fact is basic for many applications. If one substitutes H_{s_i} by $(H_{s_{i+1}} + H_{s_i})/2$ then the result is the Stratonovich- Fisk integral which will be discussed shortly in section 8.5.

As in deterministic integration there are many ways to carry out the details of the construction. The key idea of the approach in this text is due to E. Lenglart (1983) and M.J. Sharpe: Introduce path by path an elementary integral for finite random step functions as a sum like above but allow the steps to be defined by stopping times. Then extend this integral to the class of locally uniform limits of these step functions. (This class was described in proposition 4.4.4. The extension is carried out in section 5.4). This integral corresponds to the Riemann integral in classical analysis. It is sufficient for many applications. In particular, it allows to construct the "increasing" or "quadratic variation" process $[M]$ of M which in Itô's situation of Brownian motion simply is given by $[M]_t = t$. Using $[M]_t$ instead of t the subsequent chapters follow fairly closely the lines of Itô's approach.

5.1 Stieltjes Integrals

In this section we introduce the Stieltjes integral in a very simple form. It represents the "easy part" of a stochastic integral. However integration with respect to (local) martingales (i.e. the "harder part") has many features in common with what we do here.

5.1.1 Definition. If

$$h = a_0 \, \mathbf{1}_{\{0\}} + \sum_{i=1}^{n} a_i \mathbf{1}_{]t_i, t_{i+1}]}$$

is a finite step function on $\mathbb{R}_+$ and if m is any function on $\mathbb{R}_+$ then the elementary integral $\int_0^\infty h \, dm$ is defined by

$$(5.1.1) \qquad \int_0^\infty \Big[a_0 \mathbf{1}_{(0)} + \sum_{i=1}^{n} a_i \, \mathbf{1}_{]t_i, t_{i+1}]} \Big] dm = \sum_{i=1}^{n} a_i \, \big(m(t_{i+1}) - m(t_i)\big).$$

For $0 \le a < b$ we write $\int_a^b h \, dm$ instead of $\int_0^\infty \mathbf{1}_{]a,b]} \, h \, dm$.

Obviously, the value $a_0 = h(0)$ does not play any role and therefore we will assume $a_0 = 0$ if convenient. Since

$$\mathbf{1}_{]0,t]} h = \sum_{i=1}^{n} a_i \mathbf{1}_{]t \wedge t_i, t \wedge t_{i+1}]}$$

we get in particular

$$(5.1.2) \qquad \int_0^t h \, dm = \sum_{i=1}^{n} a_i \big(m(t \wedge t_{i+1}) - m(t \wedge t_i)\big).$$

The interval function dm given by

$$(5.1.3) \qquad dm(]s,t]) = m(t) - m(s)$$

is additive and therefore the integral is well-defined, i.e. it does not depend on the particular representation of the step function h. That we have chosen the intervals to be open on the left and closed on the right corresponds to the fact that we will be interested in right-continuous integrating functions m.

The first aim of this chapter is to extend this definition to more general integrands. For this, the integral should be in some sense continuous in the integrated function. The most simple condition for such a continuity involves the variation of the "integrator" m.

5.1.2 Definition. Let $m : \mathbb{R}_+ \longrightarrow \mathbb{R}$ be a function. The **total variation** of m between a and b is the value

$$\int_a^b |dm| = \sup \left\{ \sum_{i=1}^n \left| m(s_{i+1}) - m(s_i) \right| : a = s_1 < \ldots < s_{n+1} = b,\ n \geq 1 \right\}$$

If $\int_0^t |dm|$ is finite then m is called a **function of finite** (or bounded) **variation on** $[0,t]$. If this holds for every $t > 0$ then m has **locally finite variation.**

In this case the function $\int_0^{\cdot} |dm|$ is itself of locally finite variation because it is nondecreasing. The integral of h over $]a,b]$ with respect to $\int_0^{\cdot} |dm|$ is denoted by $\int_a^b h\ |dm|$.

All functions m and all step functions h fulfill the estimate

$$\left| \int_0^t h\ dm \right| \leq \int_0^t |h|\ |dm| \leq \|h\|_\infty \int_0^t |dm| \tag{5.1.4}$$

where as usual $\|\cdot\|_\infty$ is the sup-norm. In fact, let $h = \sum_{i=1}^n a_i \mathbf{1}_{]t_i,t_{i+1}]}$. Then $\mathbf{1}_{]0,t]} \cdot h$ has the representation

$$\sum_{i=1}^k a_i \mathbf{1}_{]s_i,s_{i+1}]}$$

where $k = \max\{i : t_i < t\}$, $s_i = t_i$ for $i \leq k$ and $s_{k+1} = t_{k+1} \wedge t$. Therefore,

$$\left| \int_0^t h\ dm \right| \leq \sum_{i=1}^k |a_i|\ |m(s_{i+1}) - m(s_i)| \leq \sum_{i=1}^k |a_i| \int_{s_i}^{s_{i+1}} |dm|$$

$$\leq \left(\max_{1 \leq i \leq k} |a_i| \right) \sum_{i=1}^k \int_{s_i}^{s_{i+1}} |dm| \leq \|h\|_\infty \int_0^t |dm| .$$

Here the third expression can also be written as $\int_0^t |h|\ |dm|$. This proves (5.1.4).

This estimate allows the definition of $\int_0^t h\ dm$ for uniform limits of left-continuous step functions, i.e. for left-continuous regular functions h (cf. section 4.4).

5.1.3 Definition. Define the **Stieltjes integral** $\int_0^t h\,dm$ for every left-continuous regular function h (in particular every continuous h) and every function m of locally finite variation by

$$\int_0^t h\,dm = \lim_{n\to\infty} \int_0^t h_n\,dm \tag{5.1.5}$$

where $(h_n)_{n\geq 1}$ is any sequence of step functions converging uniformly to h.

Here we apply the general **extension principle**: if $\mathbf{E}_0$ is a dense subset of a metric space $\mathbf{E}$ and $\mathbf{F}$ is a complete metric space then every uniformly continuous map φ from $\mathbf{E}_0$ to $\mathbf{F}$ has a unique continuous extension to all of $\mathbf{E}$. Moreover, the extension has the same modulus of continuity which implies that the estimate (5.1.4) stays valid:

5.1.4 Proposition. Suppose that m has locally finite variation and h is left-continuous and regular. Then

$$\left|\int_0^t h\,dm\right| \leq \int_0^t |h|\,|dm| \leq \|h\|_\infty \int_0^t |dm|.$$

The following proposition is not needed in the logical development of the text. It shows that the functions of bounded variation are the only functions for which the preceding approach works.

5.1.5 Proposition. If m is such that

$$\left|\int_0^t h\,dm\right| \leq c\|h\|_\infty$$

for all step functions h and some $c < \infty$ then m is of finite variation on $[0,t]$ and $\int_0^t |dm| \leq c$.

Proof. For $0 = s_1 < \ldots < s_{n+1} = t$ set $h = \sum_{i=1}^n a_i \cdot \mathbf{1}_{]s_i,s_{i+1}]}$ where a_i is the sign of $m(s_{i+1}) - m(s_i)$. Then $\|h\|_\infty = 1$ and hence

$$\sum_{i=1}^n |m(s_{i+1}) - m(s_i)| = \sum_{i=1}^n a_i\big(m(s_{i+1}) - m(s_i)\big) = \int_0^t h\,dm \leq c.$$

By definition of $\int_0^t |dm|$ this implies $\int_0^t |dm| \leq c$. ■

What are examples of functions of (locally) finite variation? The identity $m(t) = t$ certainly is the most important one. It leads to the ordinary integral $\int_0^t h(s)\,ds$. If one has one function of finite variation m and $\varphi : \mathbb{R} \longrightarrow \mathbb{R}$ is continuously differentiable then $\varphi \circ m$ is another example. We show this under the simplifying assumption of continuity.

5.1.6 Proposition. Let $m : [0,t] \longrightarrow \mathbb{R}$ be continuous and of finite variation. Suppose that $\varphi : \mathbb{R} \longrightarrow \mathbb{R}$ is differentiable and has a continuous derivative φ'. Then $\varphi \circ m : [0,t] \longrightarrow \mathbb{R}$ is of finite variation and for every $t_0 < t$

$$\varphi(m(t)) - \varphi(m(t_0)) = \int_{t_0}^t \varphi'(m(s))\,dm(s)\,. \tag{5.1.6}$$

Proof. For every $n \in \mathbb{N}$ let $t_0 = s_0^n < \ldots < s_{n+1}^n = t$ be a partition of $[t_0,t]$ such that

$$\max_i (s_{i+1}^n - s_i^n) \xrightarrow[n\to\infty]{} 0,$$

e.g. $s_i^n = t_0 + \frac{i}{n+1}(t - t_0)$. Then by the mean value theorem of calculus

$$\begin{aligned} &\varphi\big(m(t)\big) - \varphi\big(m(t_0)\big) \\ &= \sum_{i=0}^n \varphi\big(m(s_{i+1}^n)\big) - \varphi\big(m(s_i^n)\big) \\ &= \sum_{i=0}^n \varphi'(\vartheta_i^n)\big(m(s_{i+1}^n) - m(s_i^n)\big) \\ &= \int_0^t h^n(s)\,dm(s) \end{aligned}$$

where ϑ_i^n is some point between $m(s_i^n)$ and $m(s_{i+1}^n)$ and

$$h^n = \sum_{i=0}^n \varphi'(\vartheta_i^n)\,\mathbf{1}_{]s_i^n, s_{i+1}^n]}\,.$$

Now let n tend to infinity. Since φ' is uniformly continuous on compact intervals the sequence $h^n(s)$ converges uniformly on $[0,t]$ to $h(s) = \varphi'(m(s))$. Thus

$$\varphi(m(t)) - \varphi(m(t_0)) = \lim_{n\to\infty} \int_0^t h^n(s)dm(s) = \int_{t_0}^t \varphi'(m(s))dm(s)\,. \blacksquare$$

Itô's formula which is the heart of stochastic calculus is the analogue of proposition 5.1.6 for the stochastic integral. Functions of finite variation are replaced there by semimartingales; instead of the mean value theorem second order Taylor approximation will be used and the limit arguments will be probabilistic. Otherwise the arguments are very similar. The special case $\varphi(x) = x^2$ is of particular importance (cf. also the relation (5.3.1) in section 5.3 and on the other hand theorems 5.6.2 and 8.1.1):

5.1.7 Corollary. If m is continuous and of locally finite variation then

$$m^2(t) - m^2(t_0) = 2 \int_{t_0}^{t} m(s)\, dm(s).$$

Every nondecreasing function and hence also every difference of non-decreasing functions is of locally finite variation. It is interesting to note that conversely every function m of locally finite variation can be written as a difference of two nondecreasing functions. For example $m = m_+ - m_-$ where

$$m_+(t) = \tfrac{1}{2}\left(\int_0^t |dm| + m(t)\right)$$

$$m_-(t) = \tfrac{1}{2}\left(\int_0^t |dm| - m(t)\right).$$

The functions m_+ and m_- are nondecreasing since the increments of m and $-m$ are smaller than those of $\int_0^{\cdot} |dm|$. These functions m_+ and m_- are minimal in the following sense: If n_+ and n_- are two other nondecreasing functions such that $m = n_+ - n_-$ then $m_+ \le n_+$ and $m_- \le n_-$. In fact continuing with the notation from (5.1.3) one has

$$|dm| = |dn_+ - dn_-| \le dn_+ + dn_-$$

and hence $\int_0^t |dm| \le n_+(t) + n_-(t)$. This implies

$$m_+ = \tfrac{1}{2}\left(\int_0^t |dm| + m(t)\right) \le \tfrac{1}{2}\left(n_+ + n_- + (n_+ - n_-)\right) = n_+$$

and similarly $m_- \le n_-$.

As a consequence of this decomposition one gets some continuity properties of functions of finite variation.

5.1.8 Proposition. Let m be a function of locally finite variation. Then

(a) m is regular.

(b) The total variation $\int_0^t |dm|$ is (right-)continuous in each t if m is (right-)continuous.

Proof. (a) Obviously, this statement is true for nondecreasing functions and it extends to the difference of such functions.

(b) Let m be continuous and let m_+ , m_- be defined as above. Suppose that m_+ is not continuous. Then m_+ jumps upwards by some $\varepsilon > 0$ at some point t_0. Since m is continuous and $m = m_+ - m_-$ the function m_- also jumps by the same amount ε at t_0. Letting

$$n_\pm(t) = \begin{cases} m_\pm(t) & \text{for} \quad t < t_0 \\ m_\pm(t) - \varepsilon & \text{for} \quad t \geq t_0 \end{cases}$$

one gets two increasing functions n_+, n_- such that $n_+ - n_- = m_+ - m_- = m$ which are smaller than m_+ and m_- respectively, contradicting the minimality of m_+ and m_-. Thus m_+ is continuous and hence also $\int_0^\cdot |dm| = 2m_+ - m$. The argument for right-continuity is similar. ■

Finally, we want to show how the Stieltjes integral of a regular function can also be computed as a limit of Riemann-Stieltjes sums. Note that in the following result the integrand h is evaluated at the left endpoints of the partition intervals in the Riemann-Stieltjes sums. This lack of symmetry is due to the fact that h is left continuous and m is right continuous. If either h or m is continuous then one can also take any other points in the corresponding interval. In the stochastic integral of section 5.4, however, it will turn out for completely different reasons that even in the continuous case the choice of the left endpoint is essential. Therefore, we content ourselves with the present formulation.

5.1.9 Proposition. For each n let $0 = t_0^n < \ldots < t_{k_{n+1}}^n$ be a finite sequence such that every given open interval eventually contains one of the t_i^n as $n \to \infty$. If either

(i) m is right-continuous and h is a left-continuous step function

or

(ii) m is right-continuous and of locally finite variation and h is left continuous regular then

$$\int_0^t h\,dm = \lim_{n\to\infty}\sum_{i=0}^{k_n} h(t_i^n)\big(m(t\wedge t_{i+1}^n) - m(t\wedge t_i^n)\big)$$

(5.1.7)

$$= \lim_{n\to\infty}\sum_{i=0}^{k_n} h(t_i^n+)\big(m(t\wedge t_{i+1}^n) - m(t\wedge t_i^n)\big)$$

uniformly in t on bounded intervals.

Case (i) is considered since we will be faced with functions m of infinite variation.

Proof. Case (i). By linearity it suffices to consider the case $h = \mathbf{1}_{]0,s]}$. Then for each t and n

$$\sum_{i=0}^{k_n} h(t_i^n)\big(m(t\wedge t_{i+1}^n) - m(t\wedge t_i^n)\big)$$

$$= \sum_{i=0}^{i_n}\big(m(t\wedge t_{i+1}^n) - m(t\wedge t_i^n)\big) = m(t\wedge t_{i+1}^n) - m(0)$$

where i_n is the last index such that $t_{i_n}^n \leq s$. If $h(t_i^n)$ is replaced by $h(t_i^n+)$ then i_n is the last index such that $t_{i_n}^n < s$. In both cases right-continuity implies $m(t_{i_n+1}^n) \longrightarrow m(s)$ and hence

$$m(t\wedge t_{i_n+1}^n) \longrightarrow m(t\wedge s)$$

uniformly in t. This proves (a).
Case (ii). Now assume that m is of bounded variation. Then the difference of the two sides in (5.1.7) can be estimated from above by

$$2\|h\|_\infty \int_0^t |dm|.$$

Thus by uniform approximation (5.1.7) carries over from step functions to general left-continuous regular h. ■

The reason why (at least in German university courses) the (Riemann-) Stieljes integral often is not discussed lies in the fact that integration with respect to functions of finite variation is a special case of integration on abstract measure spaces. We conclude this section by the standard trick which reduces Stieltjes integrals to Lebesgue integrals.

If m is increasing and right-continuous on $\mathbb{R}_+$ then the function

$$\sigma : \mathbb{R}_+ \longrightarrow [0,\infty], \quad \sigma(s) = \inf\{t : m(t) - m(0) \geq s\}$$

is also increasing and right-continuous. For a real number b let

$$\underline{b} = \inf\{t : m(t) = m(b)\}.$$

Then

$$\sigma^{-1}\big([0,b]\big) = \sigma^{-1}\big([0,\underline{b}]\big) = [0, m(b) - m(0)]$$

and therefore

$$\lambda\Big(\sigma^{-1}\big([0,b]\big)\Big) = m(b) - m(0) = dm\big(]0,b]\big).$$

This shows that the interval function dm is just the restriction to the system of intervals $]a,b]$ in $\mathbb{R}_+$ of the image under σ of Lebesgue measure λ on the interval $[0, \lim_{b\to\infty} m(b) - m(0)]$.

For every Borel-function h the integral $\int_0^t h\, dm$ can be defined by the transformation formula

$$\int_0^t h\, dm = \int_0^{m(t)-m(0)} h(\sigma(s))\, ds \tag{5.1.8}$$

whenever the right hand side exists.

If m is a general right-continuous function of locally finite variation and $\int_0^t |h||dm| < \infty$ then the integral $\int_0^t h\, dm$ is given by the the decomposition

$$\int_0^t h\, dm = \int_0^t h\, dm_+ - \int_0^t h\, dm_- . \tag{5.1.9}$$

This integral also satisfies the inequality

$$\left|\int_0^t h\, dm\right| \leq \|h\|_\infty \int_0^t |dm|$$

and therefore the definition using (5.1.8) and (5.1.9) coincides with definition 5.1.3 if the function h is left-continuous and regular.

The following proposition shows that in the present context an indefinite integral is again a potential integrator. This will be true also for the stochastic integrals.

5.1.10 Proposition. If m is (right)-continuous and of locally finite variation and if the Borel function h satisfies $\int_0^t |h|\, |dm| < \infty$ for every t then the indefinite integral $\int_0^{\cdot} h\, dm$ is also (right-)continuous and of locally finite variation.

Proof. That the integral is right-continuous and of locally finite variation is a direct consequence of the fact that the set function

$$\mu_h : A \longmapsto \int_0^t \mathbf{1}_A \, h \, dm$$

is a signed measure, i.e. a difference of finite nonnegative measures. If m is continuous then $\mu_h(\{t\}) = 0$ for every t and hence $\int_0^{\cdot} h \, dm$ is continuous. ■

5.2 The Elementary Integral

In this section we consider the stochastic integral process $\int_0^{\cdot} H \, dM$ for an elementary process H and any process M. The elementary integral is simply taken path by path using the definition 5.1.1. This is possible since by definition all paths of an elementary process are (left-continuous) step functions.

5.2.1 Definition. Let H be an elementary process and let M be any process. Then $\left(\int_0^t H \, dM\right)_{t\geq 0}$ or $\int_0^{\cdot} H \, dM$ denotes the process defined, path by path, by

$$\left(\int_0^t H \, dM\right)(\omega) = \int_0^t H(\cdot,\omega) \, dM(\cdot,\omega). \tag{5.2.1}$$

This process is called the **elementary stochastic integral of H with respect to M**. We also write $\int_0^t H_s \, dM_s$ to indicate the variable of integration.

Using the representation of an elementary process provided by proposition 4.4.2 one gets an explicit form of the integral.

5.2.2 Proposition. Let the elementary process H have the representation

$$H = H_0 \mathbf{1}_{\{0\}} + \sum_{i=1}^{n} H_i \mathbf{1}_{]T_i, T_{i+1}]} \tag{5.2.2}$$

with increasing stopping times T_i and $\boldsymbol{F}_{T_i}^{+}$-measurable random variables H_i. Then

$$\left(\int_0^t H \, dM\right)_{t\geq 0} = \left(\sum_{i=1}^{n} H_i\left(M_{t\wedge T_{i+1}} - M_{t\wedge T_i}\right)\right)_{t\geq 0}. \tag{5.2.3}$$

Let M be right-continuous and adapted and/or a.s. continuous. Then the process $\int_0^{\cdot} H \, dM$ has the same properties.

Proof. The definition of the integral of a finite step function (formula (5.1.2)) reads in the present case

$$\int_0^t H(s,\omega)\, dM(s,\omega) = \sum_{i=1}^{n} H_i(\omega)\left(M_{t\wedge T_{i+1}}(\omega) - M_{t\wedge T_i}(\omega)\right)$$

$$= \left[\sum_{i=1}^{n} H_i\left(M_{t\wedge T_{i+1}} - M_{t\wedge T_i}\right)\right](\omega).$$

In order to show that $\int_0^{\cdot} H\, dM$ is adapted if M is right-continuous and adapted it suffices to consider one term. We have

$$H_i\left(M_{t\wedge T_{i+1}} - M_{t\wedge T_i}\right) = H_i\, \mathbf{1}_{\{T_i<t\}}\left(M_{t\wedge T_{i+1}} - M_{t\wedge T_i}\right)$$

since

$$M_{t\wedge T_{i+1}} - M_{t\wedge T_i} = 0 \text{ for } T_i \geq t.$$

This difference is adapted by proposition 2.3.11(b) and the factor $(H_i\, \mathbf{1}_{\{T_i<t\}})_{t\geq 0}$ is adapted since H_i is $\mathcal{F}^+_{T_i}$-measurable. That the elementary integral process inherits the continuity properties from M is clear since the time variable t appears only in $M_{t\wedge T_{i+1}} - M_{t\wedge T_i}$. ■

Let us list some straightforward facts for future reference.

5.2.3 Proposition. Let G and H be elementary processes. Then the following holds:

(a) Let B be a subset of Ω and T be a random time such that $G(s,\omega) = H(s,\omega)$ for almost all $\omega \in B$ and all $s \leq T(\omega)$. Then

$$\left(\int_0^t G\, dM\right)(\omega) = \left(\int_0^t H\, dM\right)(\omega) \text{ for } \omega \in B \text{ and } 0 \leq t \leq T(\omega).$$

(If on B the paths of G and H coincide up to time T then the same is true for the paths of the corresponding stochastic integral processes.)

(b) For every stopping time T one has

$$\int_0^{t\wedge T} H_s\, dM_s = \int_0^t (\mathbf{1}_{[0,T]}H)_s\, dM_s = \int_0^t H_s\, dM_{s\wedge T}\, .$$

(c) The integral is linear, i.e. if $a, b \in \mathbb{R}$ then

$$\left(\int_0^t aG + bH\, dM\right)_{t\geq 0} = \left(a\int_0^t G\, dM\right)_{t\geq 0} + \left(b\int_0^t H\, dM\right)_{t\geq 0}.$$

Proof. (a) This property is a trivial consequence of the pathwise definition of the integral. (In the later parts where the integral will be constructed in a less direct manner one tends to forget this fact. Nevertheless, it stays valid and is quite useful - in particular, since no measurability of B is required.)
(b) First note that $\mathbf{1}_{[0,T]}H$ is an elementary processes according to proposition 4.4.2(a). Since

$$\mathbf{1}_{[0,t\wedge T]}H = \mathbf{1}_{[0,t]}\mathbf{1}_{[0,T]}H$$

we have by definition 5.1.1

$$\int_0^{t\wedge T} H_s\, dM_s = \int_0^{t\wedge T} \mathbf{1}_{[0,T]} H_s\, dM_s.$$

On the other hand, by the preceding proposition

$$\int_0^{t\wedge T} H_s\, dM_s = \sum_{i=1}^{n} H_i(M_{t\wedge T\wedge T_{i+1}} - M_{t\wedge T\wedge T_i}) = \int_0^t H_s\, dM_{s\wedge T}.$$

(c) This follows from the linearity of the deterministic integral of section 1. (One chooses joint partitions for the two step functions in question.) ■

5.3 Pathwise Stieltjes Integrals and why they are not Sufficient

Before we leave the classical approach let us convince ourselves that the stochastic integral of a process w.r.t. a (local) martingale really cannot be defined path by path as a Lebesgue-Stieltjes integral. The crucial point is that no nonconstant continuous (local) martingale has paths of locally finite variation. Besides its motivational value this result will be used to settle uniqueness questions. Finally, we construct integral processes in those cases where the pathwise Stieltjes integral makes sense.

We introduce the stochastic analogue of functions of locally finite variation.

5.3.1 Definition. We denote by $\mathcal{A}$ the space of all adapted right-continuous processes A with $A_0 \equiv 0$ which have **locally finite variation**, i.e. for which there is a localizing sequence $(T_k)_{k\geq 1}$ such that

$$\mathbb{P}\left(\int_0^{T_k} |dA| < \infty\right) = 1 \text{ for every } k$$

or equivalently

$$\mathbb{P}\left(\int_0^{t} |dA| < \infty\right) = 1 \text{ for every } t.$$

Remark. In the definition of $\left(\int_0^t |dA_s|\right)_{t\geq 0}$ (cf. definition 5.1.2) we have only to consider rational partitions because A is right-continuous; hence the process $\int_0^{\cdot} |dA|$ is adapted (and right-continuous where it is finite by proposition5.1.8(b)). Here the term locally *bounded* variation would be misleading since in general the process $\int_0^{\cdot} |dA|$ is not uniformly bounded on any stochastic interval $]0,T]$.

The announced result reads:

5.3.2 Theorem. Every a.s. continuous local martingale in $\boldsymbol{A}$ is a.s. equal to the zero process.

Note that the continuity assumption is important since e.g. the "Poisson martingale" $(N_t - t)_{t\geq 0}$ has paths of locally finite variation.

Proof. Denote the process in question by A. We start by reducing the result to the case where A is a continuous martingale of uniformly bounded variation. First we may restrict ourselves to the case where *all* paths are continuous and locally of finite variation. This is done by the null-set elimination argument mentioned at the end of section 2.1. Set

$$T_n = \inf\left\{t \geq 0 : \int_0^t |dA_s| > n\right\};$$

this defines a sequence of stopping times by proposition 2.3.4. Whenever $0 \leq t \leq T_n(\omega)$ one has $|A_t(\omega)| \leq n$. Therefore $(A_t^n)_{t\geq 0} = (A_{t\wedge T_n})_{t\geq 0}$ defines a continuous martingale by proposition 4.2.3 which in addition is of finite variation. If the assertion is proved for A^n instead of A then

$$\mathbb{P}(A_{t\wedge T_n} \neq 0 \text{ for some } t \geq 0) = 0 \text{ for every } n \geq 0.$$

Since $T_n \uparrow \infty$ this implies

$$\mathbb{P}(A_t \neq 0 \text{ for some } t \geq 0) = 0$$

and we are done.

Therefore we may assume that A is a continuous martingale of uniformly bounded variation starting at zero. The transformation rule (cf. corollary 5.1.7) and the Riemann approximation for Stieltjes integrals 5.1.9 yield

$$(5.3.1)\quad \begin{aligned} A_t^2(\omega) &= A_t^2(\omega) - A_0^2(\omega) = 2\int_0^t A_s(\omega)\, dA_s(\omega) \\ &= 2 \lim_{n\to\infty} \sum_{k=0}^{n-1} A_{\frac{kt}{n}} \left(A_{\frac{(k+1)t}{n}} - A_{\frac{kt}{n}}\right)(\omega). \end{aligned}$$

The summands have expectation zero since the increments of a martingale are orthogonal (cf. lemma 3.1.1). Because of

$$\sum_{k=0}^{n-1} \left| A_{\frac{kt}{n}} \left(A_{\frac{(k+1)t}{n}} - A_{\frac{kt}{n}}\right)\right| \le \sup\left\{|A_s| : s \le t\right\} \int_0^t |dA_s| < \infty,$$

we may use dominated convergence in (5.3.1) to get $\mathbb{E}(A_t^2) = 0$. Consequently $\mathbb{P}(A_t \neq 0, t \in \mathbb{Q}_+) = 0$ and finally by path continuity

$$\mathbb{P}(A_t \neq 0 \text{ for some } t \ge 0) = 0.$$

The proof is now complete. ■

Nevertheless, pathwise Stieltjes integration sometimes is useful. Let A be a process of locally finite variation. Then the expression

$$(5.3.2)\qquad \int_0^t H(s,\omega)dA_s(\omega)$$

makes sense for all ω for which the path $H(\cdot,\omega)$ is a Borel function such that

$$\int_0^t |H(s,\omega)|\,|dA_s(\omega)| < \infty.$$

(This Stieltjes integral is understood in the sense of (5.1.9).) Even if this holds on a set of full $\mathbb{P}$-measure it is not yet clear whether the collection of pathwise integrals defines an adapted process. In order to get a nice process we must possibly (re-)define the pathwise integrals on a nullset.

5.3.3 Definition. We say that a process $\int_0^\cdot H\, dA$ is a **version of the pathwise integral of** H **w.r.t.** A if for almost all paths ω the whole path $\left(\int_0^\cdot H\, dA\right)(\omega)$ is given by the Stieltjes integrals (5.3.2).

The versions in ***A*** will be constructed by approximation. The lemma will be used again for a similar purpose in section 6.1 .

5.3.4 Lemma. Let A be a process of locally finite variation and let H^n, $n \ge 1$, H be processes such that

$$(5.3.3) \qquad \int_0^t |H^n(s,\omega) - H(s,\omega)|\ |dA_s(\omega)| \xrightarrow[n\to\infty]{} 0 \text{ for all } t \geq 0$$

for all ω outside a $\mathbb{P}$-nullset. If for each n the pathwise integral of H^n w.r.t. A has a version in $\boldsymbol{A}$ then the same holds for H.

Proof. Let N be the exceptional set outside which (5.3.3) holds. Let $\int_0^{\cdot} H^n\, dA$ be right-continuous adapted versions of the respective pathwise integrals. For $\omega \notin N$ and each t^* we have by (5.3.3) that

$$\sup_{t\leq t^*} \left|\int_0^t H^n(s,\omega)\, dA_s(\omega) - \int_0^t H(s,\omega)\, dA_s(\omega)\right| \longrightarrow 0.$$

Therefore, according to lemma 4.3.4 there is a right-continuous adapted process $\int_0^{\cdot} H\, dA$ such that

$$\int_0^t H^n\, dA \longrightarrow \int_0^t H\, dA \text{ for all } t \text{ and } \omega \notin N.$$

Then $\int_0^{\cdot} H\, dA$ is a version of the pathwise integral of H. It has locally finite variation in view of proposition 5.1.10. Hence $\int_0^{\cdot} H\, dA \in \boldsymbol{A}$. ■

We conclude that there are appropriate versions of the integral processes.

5.3.5 Propositon. For $H \in \boldsymbol{H}$ and $A \in \boldsymbol{A}$ the pathwise integral of H w.r.t. A has a version in $\boldsymbol{A}$.

Proof. By remark 4.4.6 there is a separable version of H. For this version there is according to proposition 4.4.4 a sequence $(H^n)_{n\geq 1}$ of elementary processes which converges l.u.p. to H. According to theorem 4.3.3(b) and lemma 4.3.1 we may assume that almost all paths of H^n converge uniformly on bounded t-intervals to the corresponding paths of H. Since the H^n are bounded the assumption of the last lemma is satisfied by dominated convergence. Since the pathwise integrals of the H^n are right-continuous and adapted by proposition 5.2.2 the lemma implies the assertion. ■

We may now close this section with

5.3.6 Definition. For $H \in \boldsymbol{H}$ and $A \in \boldsymbol{A}$ the symbol $\int_0^{\cdot} H\, dA$ always denotes a right-continuous adapted version of the pathwise integral.

5.4 The Martingale Integral for Regular Integrands

In this section we define the stochastic integral $\int_0^{\cdot} H\,dM$ for regular left-continuous integrands H and right-continuous local L^2-martingales M as integrators.

As we know from the Stieltjes integral, in one-dimensional analysis the simplest class of functions for integration are the regular functions. Since this class contains the continuous functions on bounded intervals it is rich enough to establish the connection to differentiation via the fundamental theorem of calculus and thus the introduction of the full Lebesgue (-Stieltjes) integral can be postponed.

The key to and the first step of a similar approach to the stochastic integral is the estimate (5.4.2) in theorem 5.4.1 for the elementary integral with respect to L^2-bounded martingales. It corresponds to the estimate (5.1.4) for the Stieltjes integral. The extension to left-continuous regular processes as integrands and local L^2-martingales as integrators was sketched already in section 1.3(B). Here we combine step 2 and step 3 in that description to a single approximation in theorem 5.4.4.

5.4.1 Theorem. Consider an elementary process

$$H = H_0 \mathbf{1}_{\{0\}} + \sum_{i=1}^{n} H_i \mathbf{1}_{]T_i, T_{i+1}]}$$

and let M be a right-continuous martingale. Then $\int_0^{\cdot} H\,dM$ is also a martingale. If M is L^2-bounded then

$$\mathbb{E}\Big(\Big(\int_0^T H\,dM\Big)^2\Big) = \mathbb{E}\Big(\sum_{i=1}^{n} H_i^2 \big(M_{T\wedge T_{i+1}} - M_{T\wedge T_i}\big)^2\Big) \tag{5.4.1}$$

holds for every stopping time T. Further

$$\Big\| \sup_{t\ge 0} \Big| \int_0^t H\,dM \Big| \Big\|_2 \le 2h\, \|M_\infty\|_2 \tag{5.4.2}$$

where $h = \sup\big\{ |H(t,\omega)| : t > 0,\ \omega \in \Omega \big\}$.

Proof. Let us recall from proposition 5.2.2 that

$$\Big(\int_0^t H\,dM\Big)_{t\ge 0} = \Big(\sum_{i=1}^{n} H_i\big(M_{t\wedge T_{i+1}} - M_{t\wedge T_i}\big)\Big)_{t\ge 0} \tag{5.4.3}$$

and that the integral process is right-continuous. To check the martingale property we use the criterion of proposition 3.2.9. We first consider a single term. For every bounded stopping time S the stopping theorem implies

$$\mathbb{E}\left(M_{S\wedge T_{i+1}} \middle| \boldsymbol{F}^{+}_{T_i}\right) = M_{S\wedge T}$$

and from the $\boldsymbol{F}^{+}_{T_i}$-measurability of H_i follows

$$\mathbb{E}\left(\int_0^S H_i \mathbf{1}_{]T_i,T_{i+1}]}\, dM \,\middle|\, \boldsymbol{F}^{+}_{T_i}\right)$$

$$= \mathbb{E}\left(H_i M_{S\wedge T_{i+1}} - H_i M_{S\wedge T_i} \,\middle|\, \boldsymbol{F}^{+}_{T_i}\right)$$

$$= H_i M_{S\wedge T_i} - H_i M_{S\wedge T_i} = 0.$$

It is not overstressing to say that this identity makes the whole theory of stochastic integration tick. The point is that the elementary integrand H is adapted - which implies that H_i is $\boldsymbol{F}^{+}_{T_i}$-measurable. In particular, the ordinary expectation vanishes. Hence the criterion 3.2.9 applies and the process

$$\left(\int_0^t H_i \mathbf{1}_{]T_i,T_{i+1}]}\, dM\right)_{t\geq 0}$$

is a martingale. Therefore $\left(\int_0^t H\, dM\right)_{t\geq 0}$ itself is a martingale by linearity 5.2.3(c).

Assume now that M is $\boldsymbol{L}^2$-bounded. Since H is bounded the representation 5.4.3 shows that this martingale is $\boldsymbol{L}^2$-bounded and hence closed. Given the stopping time T set

$$N_m = \int_0^{T\wedge T_{m+1}} H\, dM = \sum_{i=1}^{m} H_i\left(M_{T\wedge T_{i+1}} - M_{T\wedge T_i}\right).$$

The process $(N_m)_{1\leq m\leq n}$ is a discrete parameter martingale by corollary 3.2.8(a). In particular, the terms in the sum are orthogonal. Thus applying the "Parseval-identity" of lemma 3.1.1(b) we get

$$\mathbb{E}\left(\left(\int_0^T H\, dM\right)^2\right) = \mathbb{E}(N_n^2) = \mathbb{E}\left(\sum_{i=1}^{n} (N_i - N_{i-1})^2\right)$$

$$= \mathbb{E}\left(\sum_{i=1}^{n} H_i^2\left(M_{T\wedge T_{i+1}} - M_{T\wedge T_i}\right)^2\right).$$

This proves the identity (5.4.1). This expectation is estimated from above by

$$h^2\,\mathbb{E}\Big(\sum_{i=1}^{n}\big(M_{T\wedge T_{i+1}} - M_{T\wedge T_i}\big)^2\Big)$$

$$= h^2\,\mathbb{E}\Big(M^2_{T\wedge T_{n+1}} - M^2_{T\wedge T_1}\Big) \le h^2\,\mathbb{E}(M^2_{T\wedge T_{n+1}})$$

$$\le h^2\,\mathbb{E}(M^2_\infty)\,.$$

We conclude that the process $\int_0^{\cdot} H\,dM$ is a L^2-martingale. The inequality (5.4.2) is now a consequence of Doob's L^2-inequality 3.2.10. ■

The stochastic integral will be defined as a limit of elementary integrals in the space of local L^2-martingales.

5.4.2 Definition. Denote by $\boldsymbol{M}$ the space of all local L^2-martingales and by $\boldsymbol{M}_c$ the space of all a.s. continuous local martingales.

According to example 4.2.2(b) we have $\boldsymbol{M}_c \subset \boldsymbol{M}$. If $\boldsymbol{F}_0$ contains all $\mathbb{P}$-nullsets then every element of $\boldsymbol{M}_c$ is a.s. equal to a strictly continuous local martingale and in this case you can suppress the "right-continuous and a.s." in the definition of $\boldsymbol{M}_c$. The reason why this strange condition appears was discussed at the end of section 4.3.

In order to extend the definition of the integral we use the following completeness of the space $\boldsymbol{M}$:

5.4.3 Propositon. Let a sequence $(N^n)_{n\ge1}$ of right-continuous a.s. regular processes be given and suppose that $(T_k)_{k\ge1}$ is a localizing sequence such that the processes $(N^n_{t\wedge T_k})_{t\ge0}$ are L^2-bounded and

$$\lim_{n,m\to\infty} \mathbb{E}\Big(\sup_{0\le s\le T_k}\big(N^n_s - N^m_s\big)^2\Big) = 0 \text{ for each } k.$$

Then there is a right-continuous a.s. regular process N such that

$$\lim_{n\to\infty} \mathbb{E}\Big(\sup_{0\le s\le T_k}\big(N^n_s - N_s\big)^2\Big) = 0 \quad \text{for each } k.$$

If each N^n is in $\boldsymbol{M}$ (resp. $\boldsymbol{M}_c$) then N is in $\boldsymbol{M}$ (resp. $\boldsymbol{M}_c$).

Proof. Since $\mathcal{L}^2$-convergence implies convergence in probability the sequence $(N^n)_{n\geq 1}$ is a Cauchy sequence in the sense of theorem 4.3.3 (just replace S_k there by T_k). Thus by 4.3.3(c) there is a right-continuous process N such that for every k

$$\lim_{n\to\infty} \sup_{s\leq T_k} |N_s^n - N_s| = 0$$

in probability. This convergence is also in $\mathcal{L}^2$. In fact, passing to a subsequence we may assume a.s. convergence and then Fatou's lemma implies

$$\lim_{n\to\infty} \mathbb{E}\Big(\sup_{s\leq T_k} \big(N_s^n - N_s\big)^2\Big) \leq \lim_{n\to\infty} \liminf_{m\to\infty} \mathbb{E}\Big(\sup_{s\leq T_k} \big(N_s^n - N_s^m\big)^2\Big) = 0.$$

By this argument, every subsequence has a $\mathcal{L}^2$-convergent subsequence which implies $\mathcal{L}^2$-convergence. The a.s. regularity carries over to the limit. Suppose now that every N^n is a local $\mathcal{L}^2$-martingale. Since by lemma 4.2.3 the processes $(N^n_{t\wedge T_k})_{t\geq 0}$ are martingales we get by the $\mathcal{L}^2$-convergence

$$\mathbb{E}\big(N_{t\wedge T_k} \mid \mathcal{F}_s^+\big) = \lim_{n\to\infty} \mathbb{E}\big(N^n_{t\wedge T_k} \mid \mathcal{F}_s^+\big) = \lim_{n\to\infty} N^n_{s\wedge T_k} = N_{s\wedge T_k},$$

i.e. $\big(N_{t\wedge T_k}\big)_{t\geq 0}$ is a ($\mathcal{L}^2$-bounded) martingale for each k and hence $N \in \mathcal{M}$. That N inherits a.s. continuity from the N^n follows from theorem 4.3.3(c). This completes the proof. ■

We are now ready for the main step in the construction of the integral. Recall that H^- is a separable version of H for all $H \in \mathcal{H}$ (cf. remark 4.4.6).

5.4.4 Theorem. Suppose $H \in \mathcal{H}$ and $M \in \mathcal{M}$. Then the following hold:

(a) There are a localizing sequence $(T_k)_{k\geq 1}$ and a sequence $(H^n)_{n\geq 1}$ of elementary processes such that for every $k \geq 1$

(i) the sequence $(H^n)_{n\geq 1}$ converges uniformly to H^- on $[0,T_k[$,

(ii) the process $(M_{t\wedge T_k})_{t\geq 0}$ is a $\mathcal{L}^2$-bounded martingale.

(b) There is an element $\int_0^\cdot H\,dM$ of $\mathcal{M}$ such that for all sequences $(H^n)_{n\geq 1}$ and $(T_k)_{k\geq 1}$ as in (a)

$$\mathbb{E}\Big(\sup_{0\leq t\leq T_k} \Big(\int_0^t H^n\,dM - \int_0^t H\,dM\Big)^2\Big) \xrightarrow[n\to\infty]{} 0 \text{ for every k.} \qquad (5.4.4)$$

In particular, $\int_0^\cdot H^n\,dM$ converges l.u.p. to $\int_0^\cdot H\,dM$.

5.4.5 Definition. Let $M \in \boldsymbol{M}$ and $H \in \boldsymbol{H}$. Then the process $\int_0^{\cdot} H\, dM \in \boldsymbol{M}$ of theorem 5.4.4 is called the **stochastic integral of** H **with respect to** M.

Proof of theorem 5.4.4. (a) According to proposition 4.4.4 there is a sequence $(H^n)_{n\geq 1}$ of elementary processes which converges l.u.p. to H^-. There is a localizing sequence such that - passing to a subsequence if necessary - (i) holds (theorem 4.3.3(b)). Combining this localizing sequence with any localizing sequence for which (ii) is true yields a localizing sequence $(T_k)_{k\geq 1}$ for which both (i) and (ii) hold.

(b1) Consider sequences $(H^n)_{n\geq 1}$ and $(T_k)_{k\geq 1}$ as in (a). Then (i) yields

$$h_k^{n,m} = \sup\left\{\left|H^n(\omega,s) - H^m(\omega,s)\right| \;:\; \omega \in \Omega,\ 0 \leq s < T_k(\omega)\right\} \xrightarrow[n,m\to\infty]{} 0.$$

The left-continuity of the elementary processes implies that

$$\mathbf{1}_{]0,T_k]}\, |H^n - H^m| \leq h_k^{n,m} \qquad \text{uniformly.}$$

Since $\left(M_{s\wedge T_k}\right)_{s\geq 0}$ is an $\boldsymbol{L}^2$-bounded martingale closed by M_{T_k} we can apply the estimate (5.4.2) to the elementary process $\mathbf{1}_{]0,T_k]}(H^n - H^m)$. Hence for every k

$$\begin{aligned} &\left\| \sup_{0\leq t\leq T_k} \left|\int_0^t H^n\, dM - \int_0^t H^m\, dM\right| \right\|_2 \\ &= \left\| \sup_{0\leq t} \int_0^t \mathbf{1}_{]0,T_k]}(H^n - H^m)_s\, dM_{s\wedge T_k} \right\|_2 \qquad (5.4.5)\\ &\leq 2h_k^{n,m} \cdot \|M_{T_k}\|_2 \xrightarrow[n,m\to\infty]{} 0. \end{aligned}$$

(b2) According to theorem 5.4.1 and the stopping formula 5.2.3(b) for elementary integrals the processes

$$N^n = \int_0^{\cdot} H^n\, dM$$

are local $\boldsymbol{L}^2$-martingales and they are right-continuous by proposition 5.2.2. Hence they form a sequence in $\boldsymbol{M}$ which by (5.4.5) satisfies the Cauchy condition of proposition 5.4.3. Therefore, there is an element $\left(\int_0^t H\, dM\right)_{t\geq 0}$ of $\boldsymbol{M}$ for which (5.4.4) holds.

Finally, we have to show that (5.4.4) holds for all other sequences $(\tilde{H}^n)$ and $(\tilde{T}_k)_{k\geq 1}$ as in (a). Starting from these new sequences we get a process $\left(\int_0^t H\, dM\right)^{\sim}_{t\geq 0}$ for which (5.4.4) holds with twiddles. For every k,

$$\Big\| \sup_{t\geq 0} \Big| \int_0^{t\wedge \tilde{T}_k\wedge T_k} H^n \, dM - \int_0^{t\wedge \tilde{T}_k\wedge T_k} \tilde{H}^n \, dM \Big| \Big\|_2$$

$$\leq 2 \sup_{0<t\leq T_k\wedge \tilde{T}_k} |H^n - \tilde{H}^n| \ \mathbb{E}\big(M^2_{T_k\wedge \tilde{T}_k}\big) \xrightarrow[n\to\infty]{} 0.$$

This implies

$$\Big(\int_0^t H \, dM\Big)_{t\geq 0} = \Big(\int_0^t H \, dM\Big)^{\sim}_{t\geq 0} \text{ a.s.}$$

which completes the proof. ■

These are first properties of the integral:

5.4.7 Proposition. Suppose that $M \in \boldsymbol{M}$. Then:

(a) if $M \in \boldsymbol{M}_c$ then $\int_0^{\cdot} H \, dM \in \boldsymbol{M}_c$ for every $H \in \boldsymbol{H}$,

(b) if $H, H' \in \boldsymbol{H}$ and $a, b \in \mathbb{R}$ then

$$\int_0^{\cdot} aH + bH' \, dM = a \int_0^{\cdot} H \, dM + b \int_0^{\cdot} H' \, dM \text{ a.s.},$$

(c) if R is a random time such that $H = H'$ a.s. on $]0,R]$ then

$$\Big(\int_0^{t\wedge R} H \, dM\Big)_{t\geq 0} = \Big(\int_0^{t\wedge R} H' \, dM\Big)_{t\geq 0} \text{ a.s.},$$

(d) if T is a stopping time then

$$\Big(\int_0^{t\wedge T} H \, dM\Big)_{t\geq 0} = \Big(\int_0^t 1_{[0,T]} H \, dM\Big)_{t\geq 0} = \Big(\int_0^t H_s \, dM_{s\wedge T}\Big)_{t\geq 0} \text{ a.s.}.$$

Proof. (a) The a.s. continuity of the stochastic integral has been shown for elementary integrands in proposition 5.2.2. For $H \in \boldsymbol{H}$ it follows by approximation and by the last part of proposition 5.4.3.

(b) If H^n converges uniformly to H on $[0,T_k[$ and H'^n converges uniformly to H' on $[0,T'_k[$ then $aH^n + bH'^n$ converges uniformly to $aH + bH'$ on $[0,T_k\wedge T'_k[$. By (5.4.4) the linearity of the elementary integral carries over to the limit. The parts (c) and (d) follow similarly. Use the approximation (5.4.4) and the corresponding parts of proposition 5.2.3 for the elementary integral. ■

5.5 Continuity of the Integral

In this section we collect a couple of properties of the stochastic integral most of which express in one way or another the continuity of the integral. First we extend the inequality (5.4.2).

5.5.1 Theorem. Suppose that $H \in \boldsymbol{H}$ and $M \in \boldsymbol{M}$. Let T be a stopping time such that $(M_{t \wedge T})_{t \geq 0}$ is a $\boldsymbol{L}^2$-bounded martingale and $\sup_{0 \leq t < T} |H_t| \leq h$. Then

$$\Big\| \sup_{0 \leq t \leq T} \Big| \int_0^t H \, dM \Big| \, \Big\|_2 \leq 2 \, h \, \big\| M_T \big\|_2 . \tag{5.5.1}$$

If the left-hand side of (5.5.1) is finite then $\left(\int_0^{t \wedge T} H \, dM \right)_{t \geq 0}$ is a $\boldsymbol{L}^2$-bounded martingale. In particular $\mathbb{E}\left(\int_0^S H \, dM \right) = 0$ for every stopping time $S \leq T$.

Proof. We may assume that H is separable. Let (T_k) be a localizing sequence for M and let (H^n) be a sequence of elementary processes converging uniformly to H on each $[0,T_k[$. Then the elementary integrals satisfy the inequality

$$\Big\| \sup_{t \leq T \wedge T_k} \Big| \int_0^t H^n \, dM \Big| \, \Big\|_2 \leq 2 \Big(\sup_{0 < t \leq T \wedge T_k} \big| H_t^n \big| \Big) \big\| M_{T \wedge T_k} \big\|_2 . \tag{5.5.2}$$

This follows by (5.4.2) applied to the elementary process $\mathbf{1}_{]0,T \wedge T_k]} H^n$ and the martingale $(M_{t \wedge T \wedge T_k})_{t \geq 0}$. Now we can replace H^n by H in (5.4.2): The left-hand side of (5.5.2) converges to the corresponding term for H by (5.4.4). The bracket on the right is asymptotically at most h by uniform convergence. Thus

$$\Big\| \sup_{t \leq T \wedge T_k} \Big| \int_0^t H \, dM \Big| \, \Big\|_2 \leq 2 \, h \, \big\| M_{T \wedge T_k} \big\|_2$$

Then we let k tend to ∞ to get (5.5.1). Note that

$$\lim_{k \to \infty} \big\| M_{T \wedge T_k} \big\|_2 = \big\| M_T \big\|_2$$

by our assumption on M and T and the $\boldsymbol{L}^2$-convergence of martingales 3.2.2(c). The rest follows from the criterion 4.2.3 and the stopping theorem since the stochastic integral is a right-continuous local $\boldsymbol{L}^2$-martingale. ■

Up to now we have reduced all results on the stochastic integral to the

$\mathcal{L}^2$-bounded situation by suitable localization. Often it is convenient to replace $\mathcal{L}^2$-convergence by convergence in probability. Then one is less restricted in the choice of the stopping times. A first step is the following estimate in which H does not need to be bounded on $[0,T[$ anymore.

5.5.2 Corollary. Suppose that $M \in \boldsymbol{M}$ and let T be a stopping time such that $(M_{t\wedge T})_{t\geq 0}$ is an $\mathcal{L}^2$-bounded martingale. Then for every $H \in \boldsymbol{H}$ and all $\varepsilon > 0$, $h > 0$

$$\mathbb{P}\Big(\sup_{0\leq t\leq T} \Big|\int_0^t H\, dM\Big| \geq \varepsilon\Big) \leq \frac{4h^2\mathbb{E}(M_T^2)}{\varepsilon^2} + \mathbb{P}\Big(\sup_{0\leq t<T} |H_t| > h\Big) \tag{5.5.3}$$

Proof. As usual, we may and will assume that H is separable. Then $S = \inf\{t : |H_t| > h\}$ is a stopping time. Plainly,

$$\sup_{0\leq t< S\wedge T} |H_t| \leq h$$

and hence by (5.5.1) and Tschebychev's inequality

$$\mathbb{P}\Big(\sup_{t\leq T}\Big|\int_0^t H\, dM\Big| \geq \varepsilon\Big) \leq \mathbb{P}\Big(\sup_{t\leq S\wedge T}\Big|\int_0^t H_t\, dM\Big| \geq \varepsilon\Big) + \mathbb{P}\big\{S > T\big\}$$

$$\leq \Big(\frac{4h^2\,\mathbb{E}(M_{S\wedge T}^2)}{\varepsilon^2}\Big) + \mathbb{P}\Big(\sup_{0\leq t<T}|H_t| > h\Big). \blacksquare$$

The most important consequence of this inequality is the following continuity of the integral.

5.5.3 Proposition. Let $M \in \boldsymbol{M}$. If a sequence (H^n) in $\boldsymbol{H}$ converges l.u.p. to $H \in \boldsymbol{H}$ then the same is true for the corresponding integral processes with respect to M.

Proof. By linearity it suffices to consider the case $H \equiv 0$. Let the finite stopping times T_k form a localizing sequence for M and let $(H^n)_{n\geq 1}$ be a sequence in $\boldsymbol{H}$ converging l.u.p. to 0. Then for all $\varepsilon > 0$, $h > 0$ and $k \in \mathbb{N}$ the inequality (5.5.3) yields

$$\mathbb{P}\Big(\sup_{t\leq T_k}\Big|\int_0^t H^n\, dM\Big| \geq \varepsilon\Big) \leq \frac{4h^2\,\mathbb{E}(M_{T_k}^2)}{\varepsilon^2} + \mathbb{P}\Big(\sup_{0\leq t<T_k} |H_t^n| > h\Big).$$

Given $\varepsilon > 0$ and k, we may choose first $h > 0$ small enough to get the first term on the right small and then n large enough to get the second term also small. Thus the left-hand side converges to 0 for every k and $\varepsilon > 0$. This completes the proof. ■

As an illustration and a first application we extend the explicit formula of the elementary integral to infinite sums.

5.5.4 Corollary. Let the process H be of the form $H_0 \mathbf{1}_{\{0\}} + \sum_{i=1}^{\infty} H_i \mathbf{1}_{]T_i, T_{i+1}]}$ where each H_i is $\boldsymbol{F}^+_{T_i}$-measurable and $\mathbb{P}(T_i \uparrow \infty) = 1$. Then for every $M \in \boldsymbol{M}$ the integral can be written as

$$\int_0^t H \, dM = \sum_{i=1}^{\infty} H_i \left(M_{t \wedge T_{i+1}} - M_{t \wedge T_i} \right) \text{ a.s. .}$$

Proof. The process H is adapted and a.s. regular and a.s. left-continuous, i.e. $H \in \boldsymbol{H}$. The corresponding partial sums are elementary processes which converge to H locally uniformly. Thus the preceding proposition applies and yields the result. ■

Also for the approximation of the stochastic integral by "Riemann-sums" locally uniform convergence in probability is the appropriate concept. The following stochastic version of interval partitions is convenient.

5.5.5 Definition. For each $n \in \mathbb{N}$ let $\rho_n = \left(T_1^n \le T_2^n \le \dots \right)$ be a **partition** given by a localizing sequence. We write $|\rho_n| \longrightarrow 0$ if

$$\sup_i T_{i+1}^n - T_i^n \xrightarrow[n \to \infty]{} 0 \text{ a.s.}$$

With this convention one has

5.5.6 Proposition. Let $H \in \boldsymbol{H}$ and $M \in \boldsymbol{M}$ be given. Then

$$\begin{aligned} \int_0^t H \, dM &= \lim_{n \to \infty} \sum_{i=1}^{\infty} H_{T_i^n}^+ \left(M_{t \wedge T_{i+1}^n} - M_{t \wedge T_i^n} \right) \\ &= \lim_{n \to \infty} \sum_{i=1}^{\infty} H_{T_i^n} \left(M_{t \wedge T_{i+1}^n} - M_{t \wedge T_i^n} \right) \end{aligned} \tag{5.5.4}$$

l.u.p. for every sequence $(\rho_n)_{n\geq 1}$ of partitions such that $|\rho_n| \longrightarrow 0$.

Proof. First assume that H is an elementary process. Then (5.5.4) holds pathwise uniformly on bounded t-intervals by proposition 5.1.9. For general $H \in \boldsymbol{H}$ note that the first sum on the right-hand side of (5.5.4) can be written as $\left(\int_0^t H^n \, dM\right)_{t\geq 0}$ where

$$H^n = \sum_{i=1}^{\infty} H^+_{T_i^n} \mathbf{1}_{]T_i^n, T_{i+1}^n]}.$$

For every elementary process $\tilde{H}$ and every stopping time T we have

$$\mathbb{P}\left(\sup_{t\leq T} \left| \int_0^t H - H^n \, dM\right| > \varepsilon\right)$$

(5.5.5)

$$\leq \mathbb{P}\left(\sup_{t\leq T} \left|\int_0^t \tilde{H}^n - \tilde{H} dM\right| > \frac{\varepsilon}{2}\right) + \mathbb{P}\left(\sup_{t\leq T} \left|\int_0^t H - \tilde{H} + (H^n - \tilde{H}^n) dM\right| > \frac{\varepsilon}{2}\right).$$

If $|H - \tilde{H}| < \delta$ then also $|H^n - \tilde{H}^n| < \delta$ on $]0,T]$. So the second probability on the right-hand side can be made arbitrarily small (independent of n) by choosing $\tilde{H}$ close to H (corollary 5.5.3). The first probability converges to 0 since $\tilde{H}$ is elementary. Therefore the left hand side converges to 0 which proves the first identity in (5.5.4). Exactly the same argument yields the second identity if one works with $H_{T_i^n}$ instead of $H^+_{T_i^n}$. ■

5.6 The Increasing Process of a Local L^2-Martingale

Intuitively, the increasing process of a local martingale M gives at each instant the total quadratic fluctuation (or variation) which M went through up to that time. The quadratic variation vanishes for continuous functions of finite (absolute) variation. Therefore in classical analysis this concept does not show up.

In the present approach the increasing process [M] is constructed indirectly by means of the stochastic integral. The process [M] will play two roles in the further development of the theory. On the one hand, in Itô's formula it measures the deviation from the classical calculus; on the other hand it is the main tool in the construction of the stochastic integral for nonregular integrands.

The point of the following definition is to integrate a local $\mathcal{L}^2$-martingale with respect to itself. Since the integrand should be a.s. regular left-continuous one takes the a.s. left continuous version M^- of M, cf. remark 4.4.6.

5.6.1 Definition. Let M be in $\mathcal{M}$. The **increasing process** or **quadratic variation process** [M] of M is defined by

$$[M]_t = M_t^2 - M_0^2 - 2\int_0^t M^- \, dM. \tag{5.6.1}$$

If M is a.s. continuous then the quadratic variation process is also denoted by $\langle M \rangle$ instead of [M].

Remark. For general $M \in \mathcal{M}_c$ the symbol $\langle M \rangle$ denotes the "predictable increasing process" (cf. section 6.6). For $M \in \mathcal{M}_c$ these two processes coincide and this is the reason why in this case often the symbol [M] is replaced by $\langle M \rangle$.

The central result about the increasing process is the following one.

5.6.2 Theorem. Suppose that M is a right-continuous local $\mathcal{L}^2$-martingale. Then
(a) The process [M] starts at 0 and has right-continuous a.s. nondecreasing paths. Moreover $M^2 - [M]$ is a local $\mathcal{L}^2$-martingale.
(b) If M is a.s. continuous then [M] is also a.s. continuous. In this case it is (up to a.s. equality) the only a.s. continuous process with the properties in (a).
(c) The increasing process has the representation

$$[M]_t = \lim_{n \to \infty} \sum_{i=1}^{\infty} \left(M_{T_{i+1}^n \wedge t} - M_{T_i^n \wedge t} \right)^2 \tag{5.6.2}$$

as a l.u.p. limit for every sequence $(\rho_n = (T_1^n \le T_2^n \le , \ldots) : n \in \mathbb{N})$ of partitions with $|\rho_n| \longrightarrow 0$ (for this concept cf. definition 5.5.5).

Remark. (a) For $M \in \mathcal{M}_c$ the characterization (b) can be used as a definition of the increasing process. The representation in (c) shows that [M] measures the fluctuation or variation of M as mentioned in the introductory remarks. We will learn more about this aspect in chapter 7.
(b) The theorem provides a decomposition

$$M^2 = (M^2 - [M]) + [M]$$

of the "local submartingale" M^2 into a local $\boldsymbol{L}^2$-martingale and a nondecreasing process. So it is a version of the "Doob-Meyer decomposition" of a submartingale into a martingale and an increasing process. This and related decompositions will be discussed in section 6.6.
(c) There is a modification of the uniqueness statement in part (b) of the theorem which carries over to the noncontinuous case. (cf. proposition 7.3.18).

Proof of the theorem. Let us start with (b). Since a stochastic integral with respect to an a.s. continuous local martingale is a.s. continuous according to proposition 5.4.7(a) we read off from (5.6.1) that $[M]$ is a.s. continuous. As for uniqueness, suppose that M is an a.s. continuous local martingale and that A is an a.s. continuous nondecreasing process such that $A_o \equiv 0$ and $M^2 - A$ is a local martingale. Then the process $A' = [M] - A$ starts at 0 and is a.s. continuous. Being the difference of a.s. nondecreasing processes A' is locally of finite variation. Finally, A' is a local martingale because it is the difference of the local martingales $M^2 - A$ and $M^2 - [M]$. Therefore by theorem 5.3.1 the process A' vanishes, i.e. $A = [M]$ a.s. .
For (c) let $\rho_n = (T_1^n \le T_2^n \le \ldots)$, $n \in \mathbb{N}$, be a sequence of partitions such that $|\rho_n| \longrightarrow 0$. Then $M_t^2 - M_o^2$ may be expressed as

$$(5.6.3)\quad \begin{aligned} M_t^2 - M_o^2 &= \sum_{i=1}^{\infty} \left(M^2_{t\wedge T_{i+1}^n} - M^2_{t\wedge T_i^n}\right) \\ &= \sum_{i=1}^{\infty} \left(M_{t\wedge T_{i+1}^n} - M_{t\wedge T_i^n}\right)^2 + 2\sum_{i=1}^{\infty} M_{t\wedge T_i^n}\left(M_{t\wedge T_{i+1}^n} - M_{t\wedge T_i^n}\right). \end{aligned}$$

Denote by V_t^n (quadratic variation along the partition) the infinite sum of squares in the preceding line. Right-continuity and a.s. regularity of M implies $(M^-)^+ = M$ a.s. by remark 4.4.6. Thus the second infinite sum is a "Riemann-Stieltjes sum" for $2\int_o^{\cdot} M^- \, dM$ in the sense of propositon 5.5.6. Hence it converges l.u.p. to $2\int_o^{\cdot} M^- \, dM$ as $n \longrightarrow \infty$ and (V_t^n) converges to $[M]$ l.u.p. by the definition of $[M]$.
Finally, let us turn to (a). The process $M^2 - M_o^2 - [M]$ is a local $\boldsymbol{L}^2$-martingale because it is a stochastic integral with respect to M. To see that $[M]$ is nondecreasing, note first that for every n and j

$$V^n_{T_j^n} = \sum_{i=1}^{j-1} \left(M_{T_{i+1}^n} - M_{T_i^n}\right)^2 \le \sum_{i=1}^{j} \left(M_{T_{i+1}^n} - M_{T_i^n}\right)^2 = V^n_{T_{j+1}^n}$$

i.e. V^n is nondecreasing along the partition $T_1^n \le T_2^n \le \dots$. Given two points s, t such that $s < t$ we may choose the partitions such that both s and t belong to each partition. Thus $V_s^n \le V_t^n$ for this particular choice of the partition sequence. Passing to the limit gives $[M]_s \le [M]_t$ a.s.. Right continuity then implies that almost all paths are nondecreasing. ■

Let M be in $\boldsymbol{M}$. From the very definition of local ($\boldsymbol{L}^2$-)martingales we know that suitable stopping produces a $\boldsymbol{L}^2$-bounded martingale from $(M - M_0)$ and, similarly, a uniformly integrable martingale from $(M - M_0)^2 - [M]$. The corresponding stopping times can be characterized by the integrability of the stopped increasing process.

5.6.3 Proposition. Suppose $M \in \boldsymbol{M}$ and let T be a stopping time. Then the following are equivalent:

(a) $\mathbb{E}([M]_T) < \infty$

(b) $(M_{t\wedge T} - M_0)_{t\ge 0}$ is a $\boldsymbol{L}^2$-bounded martingale

(c) $(M_{t\wedge T}^2 - M_0^2 - [M]_{t\wedge T})_{t\ge 0}$ is a uniformly integrable martingale and $\mathbb{E}((M_T - M_0)^2) < \infty$.

In particular, for a $\boldsymbol{L}^2$-martingale M the process $M^2 - [M]$ is a martingale.

Proof. We may and shall assume $M_0 \equiv 0$. We first note that (c) implies (a): If (c) holds then $M_T^2 - [M]_T$ closes the martingale in (c) and hence it is in $\boldsymbol{L}^1$. Together with $\mathbb{E}(M_T^2) < \infty$ this implies (a). For the other two implications we choose a joint localizing sequence $(T_k)_{k\ge 1}$ for the local $\boldsymbol{L}^2$-martingale M and the local martingale $M^2 - [M]$ (cf. lemma 4.2.4). Thus if T is replaced by T_k then (b), (c) and hence (a) are valid.

(a) → (b) Since $\boldsymbol{L}^2$-boundedness implies uniform integrability, lemma 4.2.3 shows that it is sufficient to prove that the family $\{M_S : S \text{ stopping time} \le T\}$ is bounded in $\boldsymbol{L}^2$. For the latter we apply Doob's $\boldsymbol{L}^2$-inequality to the $\boldsymbol{L}^2$-bounded martingales $(M_{t\wedge T\wedge T_k})_{t\ge 0}$ and use monotone convergence:

$$\mathbb{E}(M_S^2) \le \mathbb{E}(\sup_{t\le T} M_t^2) = \sup_{k\ge 1} \mathbb{E}(\sup_{t\le T\wedge T_k} M_t^2) \le \sup_{k\ge 1} 4\,\mathbb{E}(M_{T\wedge T_k}^2).$$

Now recall that each $((M_{t\wedge T_k}^2 - [M]_{t\wedge T_k})_{t\ge 0}$ is a martingale starting at 0 and hence the last expression may be rewritten as

$$\sup_{k\geq 1} 4\Big[\ \mathbb{E}([M]_{T\wedge T_k})\Big].$$

This quantity is bounded by $\mathbb{E}([M]_T)$ since $[M]$ is nondecreasing and according to (a) we have a bound for $\|M_S\|_2$.

(b) → (c) Doob's inequality shows that $\sup_{t\leq T} M_t^2$ is integrable. In particular $\mathbb{E}(M_T^2) < \infty$. Because of the estimate

$$\mathbb{E}\Big(\sup_{t\leq T} [M]_T\Big) = \sup_k \mathbb{E}\Big([M]_{T\wedge T_k}\Big) = \sup_k \mathbb{E}\Big(M^2_{T\wedge T_k}\Big) \leq \mathbb{E}\Big(\sup_{t\leq T} M_t^2\Big)$$

we have also $\sup_{t\leq T}[M]_t \in \mathcal{L}^1$. Therefore $\{M_S^2 - [M]_S : S \leq T\}$ is uniformly integrable and hence $((M^2 - [M])_{t\wedge T})_{t\geq 0}$ is a uniformly integrable martingale by lemma 4.2.3. This concludes the proof of the equivalences.

The last assertion in the proposition is just the implication (b)→(c) for $T \equiv \infty$ and $M_0 \equiv 0$. ■

Moreover, the squared increments of M and the increments of [M] have the same (conditional) expectations as long as they are stopped according to the previous proposition.

5.6.4 Corollary. Let $M \in \boldsymbol{M}$ and assume that T satisfies one of the conditions (a), (b) or (c) of proposition 5.6.3. Then for every stopping time $S \leq T$ we have the identities

$$\mathbb{E}\Big((M_T - M_S)^2|\boldsymbol{F}_S^+\Big) = \mathbb{E}\big(M_T^2 - M_S^2|\boldsymbol{F}_S^+\big) = \mathbb{E}\Big([M]_T - [M]_S|\boldsymbol{F}_S^+\Big), \tag{5.6.4}$$

$$\mathbb{E}\Big(M_T^2 - M_0^2\Big) = \mathbb{E}\Big([M]_T\Big). \tag{5.6.5}$$

Proof. Follows from the stopping theorem, part (c) of the last proposition and lemma 3.1.1. ■

Another useful consequence is

5.6.5 Corollary. (**Doob's inequality for local $\mathcal{L}^2$-martingales**) For every $M \in \boldsymbol{M}$ and every pair S,T of stopping times such that $S \leq T$ one has

$$\mathbb{E}\Big(\sup_{S\leq t\leq T}\big(M_t - M_S\big)^2\Big) \leq 4\,\mathbb{E}\Big([M]_T - [M]_S\Big). \tag{5.6.6}$$

Proof. First suppose $S \equiv 0$. We may assume $\mathbb{E}([M]_T) < \infty$. Then by proposition 5.6.3 we can apply Doob's inequality in the form 3.2.10 to the L^2-bounded martingale $(M_{t\wedge T} - M_0)_{t\geq 0}$ to get

$$\mathbb{E}\Big(\sup_{0\leq t\leq T}\big(M_t - M_0\big)^2\Big) \leq 4\mathbb{E}\big((M_T - M_0)^2\big) = 4\mathbb{E}\big([M]_T\big). \tag{5.6.7}$$

For general S we consider the filtration $(\mathcal{G}_r)_{r\geq 0} = (\mathcal{F}_{S+r})_{r\geq 0}$ and the process $(M_{S+r} - M_S)_{r\geq 0}$ which is a local L^2-martingale with respect to $(\mathcal{G}_r)$. Its increasing process is $([M]_{S+r} - [M]_S)_{r\geq 0}$ as can be seen for example by the approximation in theorem 5.6.2(c). Also $T - S$ is a $(\mathcal{G}_r)$-stopping time. Then (5.6.6) is nothing but (5.6.7) written down in this new situation. ∎

The classical and most important increasing process is that of Brownian motion.

5.6.6 Example. (**The increasing process of Brownian motion**) Let B be an $(\mathcal{F}_t)$-Brownian motion. Then

$$\langle B\rangle_t = t \text{ a.s..}$$

In fact, the deterministic process $(t, \omega) \longmapsto t$ is increasing and

$$\mathbb{E}\big(B_t^2 - B_s^2 \mid \mathcal{F}_s\big) = \mathbb{E}\big((B_t - B_s)^2 \mid \mathcal{F}_s\big) = t - s$$

shows that $(B_t^2 - t)_{t\geq 0}$ is a martingale. The uniqueness statement in theorem 5.6.2(b) implies $[B]_t = t$.

As an illustration we note **Wald's identity**: Let T be a stopping time. Then corollary 5.6.4 implies:

$$\text{If } \mathbb{E}(T) < \infty \text{ then } \mathbb{E}(B_T^2) = \mathbb{E}(T).$$

From theorem 5.6.2(c) it follows in particular that for a sequence $t_1^n \leq \ldots \leq t_i^n \leq \ldots$, $n \in \mathbb{N}$, of partitions such that $\max t_{i+1}^n - t_i^n \xrightarrow[n\to\infty]{} 0$ one has

$$t = \lim_{n\to\infty} \sum_{t_i^n \leq t} \Big(B_{t_{i+1}^n} - B_{t_i^n}\Big)^2 \tag{5.6.8}$$

in probability. Passing to a subsequence we get a.s. convergence in (5.6.8). Actually, for Brownian motion (5.6.8) holds almost surely for *every increasing* sequence (t_i^n) of deterministic partitions i.e. if each t_i^m is also some t_j^n for all $n > m$. (cf. Freedman (1971), 1.4(61))). However, the exceptional set depends on the sequence of partitions. The so-called "true quadratic variation"

$$\sup\left\{ \sum_{i=1}^{k} \left(B(t_{i+1},\omega) - B(t_i,\omega)\right)^2 \ : \ 0 \le t_i < \ldots < t_k = t \right\}$$

is infinite a.s. (cf. Freedman, p.48). This means that the partitions in (5.6.8) cannot be allowed to depend in an arbitrary fashion on ω. It is essential in theorem 5.6.2(c) that the partitions are defined by stopping times.

Finally, let us consider the process $M_t = CB_t$ where C is a random variable independent of B (cf. example 2.1.4). From (5.6.8) we get for an appropriate sequence of partitions

$$\frac{1}{t} \lim_{n\to\infty} \sum_{t_i^n \le t} \left(M_{t_{i+1}^n} - M_{t_i^n}\right)^2 = \frac{C^2}{t} \lim_{n\to\infty} \sum_{t_i^n \le t} \left(B_{t_{i+1}^n} - B_{t_i^n}\right)^2 = C^2$$

with probability 1 *for every* $t > 0$. If $(\boldsymbol{F}_t)$ is the natural filtration of (M_t) then it follows that C^2 (and hence C if $C \ge 0$) is $\mathbb{P}$-a.s. equal to a $\boldsymbol{F}_0^+$-measurable random variable as was claimed in example 2.1.4.(c). If $\mathbb{E}(C^2) < \infty$ then it is easily checked that M is a square integrable $(\boldsymbol{F}_t)$-martingale. Its increasing process is C^2t.

Let us compare this continuous situation with its natural noncontinuous counterpart.

5.6.7 Example. **(The increasing process of the Poisson martingale)** The Poisson case is not of the same fundamental importance for us but it is nevertheless quite instructive. Let N be a (right-continuous) Poisson process and $M_t = N_t - t$. Then M is a $\boldsymbol{L}^2$-martingale and by the same calculation as for Brownian motion using independence of the increments and $\mathrm{var}(N_t) = t$ one gets that $M_t^2 - t$ is a $\boldsymbol{L}^2$-martingale. However, since M unlike Brownian motion is not continuous we cannot conclude that $[M]_t = t$. In order to compute $[M]$ we fix ω such that $N_.(\omega)$ has countably many jumps of size 1 (cf. the description of the Poisson process in section 1.1) and a (deterministic) sequence $(t_1^n \le t_2^n \le \ldots)$, $n \in \mathbb{N}$, of partitions such that $\max_i t_{i+1}^n - t_i^n \xrightarrow[n\to\infty]{} 0$. The path $M_.(\omega)$ can be written in the form

$$M_t(\omega) = k - t \text{ for } T_k(\omega) \le t < T_{k+1}(\omega)$$

where T_k is the time of the k-th jump of the Poisson process N. For sufficiently large n each interval $]t_i^n, t_{i+1}^n] \cap [0,t]$ contains at most one jump.

Compute now the increment of M in these intervals

$$M_{t^n_{i+1}}(\omega) - M_{t^n_i}(\omega) = \begin{cases} -\left(t^n_{i+1} - t^n_i\right) & \text{if }]t^n_i, t^n_{i+1}] \text{ contains no jump} \\ 1 - \left(t^n_{i+1} - t^n_i\right) & \text{if }]t^n_i, t^n_{i+1}] \text{ contains a jump} \end{cases}$$

Since

$$\sum_{t^n_i \le t} \left(t^n_{i+1} - t^n_{i+1}\right)^2 \le \left(\max_i t^n_{i+1} - t^n_i\right) \sum_{t^n_i \le t} t^n_{i+1} - t^n_i \xrightarrow[n \to \infty]{} 0$$

we conclude from theorem 5.6.2(c) that $[M]_t(\omega) = N_t(\omega)$.

In summary, the increasing process of the Poisson martingale $(N_t - t)_{t \ge 0}$ *is the Poisson process* N.

In chapter 9 (lemma 9.2.2) we shall need the fact that a continuous local martingale is constant where quadratic variation is constant. This is proved below. It illustrates again how quadratic variation "controls" a martingale. The result does not extend to the noncontinuous case since e.g. in the previous example we have $[M] = N = 0$ but $M_t = -t$ on $[0, T_1[$.

5.6.8 Proposition. Suppose that $M \in \boldsymbol{M}_c$. Then

$$\mathbb{P}\left(\left\{ [M]_s = [M]_t \text{ but } M_s \neq M_t \text{ for some } s < t \right\}\right) = 0.$$

Proof. Fix s and consider the stopping time

$$S = \inf\left\{ t \ge s : [M]_t > [M]_s \right\}.$$

Then the a.s. continuity of [M] implies $[M]_S = [M]_s$ a.s. and hence

$$\mathbb{E}\left(\sup_{s \le t \le S} \left(M_t - M_s\right)^2\right) \le 4\, \mathbb{E}\left([M]_S - [M]_s\right) = 0$$

by Doob's inequality 5.6.5. Moreover $t > s$ and $[M]_t = [M]_s$ implies $t \le S$. Thus for $s < t$ the event that $[M]_t = [M]_s$ and $M_t \neq M_s$ has probability 0. Therefore

$$\mathbb{P}\left(\bigcup_{s,t \ge 0} \left\{ [M]_t = [M]_s \text{ and } M_t \neq M_s \right\}\right) = 0$$

since by the (right-)continuity it is sufficient to form the union over rational s and t. ■

Since a stochastic integral process $\int_0^\cdot H\, dM$ is itself in $\boldsymbol{M}$ one might wonder how its increasing process looks like. The proof of the following proposition is postponed to chapter 7 where it is established in a more general form (corollary 7.4.3) .

5.6.9 Proposition. The increasing process of a stochastic integral with respect to a local $\boldsymbol{L}^2$- martingale is given by the formula

$$[\int_0^{\cdot} H\, dM]_t = \int_0^t H^2\, d[M].$$

5.7 PASTA: An Application of the Poisson Integral

During the construction of the integral we have been careful to use only arguments which work also for noncontinuous martingales. However due to limitation of energy and space, in most of the later parts of this text we shall restrict ourselves to applications of continuous martingales. By way of example we show that the stochastic integral is a useful tool even in the simple case of a Poisson process. We prove the following result which sometimes is called the PASTA-theorem: **P**oisson **A**rrivals **S**ee **T**ime **A**verages. For background material consult S. Asmussen's monograph (1987).

5.7.1 Theorem. Let N be a $(\boldsymbol{F}_t)$-Poisson process with intensity α and let $(T_n)_{n\geq 1}$ be the sequence of its jumping times. Then for every bounded process $H \in \boldsymbol{H}$ the limit

$$\lim_{n\to\infty} \frac{1}{n} \sum_{i=1}^{n} H_{T_i} \tag{5.7.1}$$

exists a.s. if and only if the limit

$$\lim_{t\to\infty} \frac{1}{t} \int_0^t H_s\, ds \tag{5.7.2}$$

exists a.s.. In this case the two limits are a.s. equal.

Proof. We first reduce the theorem to the case $\alpha = 1$. Introduce the new time variable $t' = \alpha t$. Then $(N_{t'})_{t'\geq 0}$ is a $(\boldsymbol{F}_{t'})$-Poisson process with intensity 1. The limit (5.7.1) is not affected by this time change and neither is the limit in (5.7.2) since

$$\lim_{t\to\infty} \frac{1}{t} \int_0^t H_s\, ds = \lim_{t'\to\infty} \frac{1}{t'} \int_0^{t'} H_{s'}\, ds'.$$

Thus we may assume $\alpha = 1$. Consider the integral of the process

$$H' = \sum_{i=1}^{n} \frac{1}{i} H \mathbf{1}_{]T_{i-1},T_i]}$$

with respect to the martingale $M_t = N_t - t$. Recall from example 5.6.7 that $[M] = N$ and hence for every n

$$\mathbb{E}\left([M]_{T_n} - [M]_{T_{n-1}}\right) = \mathbb{E}\left(N_{T_n} - N_{T_{n-1}}\right) = 1 .$$

The sequence $\left(\int_0^{T_n} H'\, dM\right)_{n\geq 1}$ is a discrete-time $\boldsymbol{L}^2$- martingale by theorem 5.5.1 and the stopping theorem 3.2.8(a) since H' is bounded .

Let $h = \sup |H|$. The orthogonality of the increments and theorem 5.5.1 imply

$$\mathbb{E}\left(\left(\int_0^{T_n} H'\, dM\right)^2\right) = \sum_{i=1}^{n} \mathbb{E}\left[\left(\int_{T_{i-1}}^{T_i} H'\, dM\right)^2\right]$$

$$= \sum_{i=1}^{n} \mathbb{E}\left(\frac{1}{i^2}\left(\int_{T_{i-1}}^{T_i} H\, dM\right)^2\right) \leq 4\cdot h^2 \sum_{i=1}^{\infty} \frac{1}{i^2} .$$

Thus this martingale is $\boldsymbol{L}^2$-bounded. Set

$$Y_i = \int_{T_{i-1}}^{T_i} H_s\, dM_s = H_{T_i} - \int_{T_{i-1}}^{T_i} H_s\, ds.$$

Here the second identity is obvious for elementary integrands and extends to general H by l.u.p. approximation. Then

$$\lim_{n\to\infty} \int_0^{T_n} H'\, dM = \lim_{n\to\infty} \sum_{i=1}^{n} \frac{Y_i}{i}$$

exists a.s. and in $\boldsymbol{L}^2$.

Kronecker's lemma (Breiman (1968), 3.28) implies that $\frac{1}{n} \sum_{i=1}^{n} Y_i$ converges a.s. to 0. Therefore $\lim \frac{1}{n} \sum_{i=1}^{n} H_{T_i}$ exists a.s. if and only if $\lim \frac{1}{n} \int_0^{T_n} H_s\, ds$ exists a.s. and in this case the two limits are equal.

The sequence $(T_n - T_{n-1})_{n\geq 1}$ is i.i.d. with expectation 1 and by the strong law of large numbers

$$\lim_{n\to\infty} \frac{T_n}{n} = 1 \text{ a.s.} .$$

This implies that

$$\frac{1}{T_n} \int_{T_n}^{T_{n+1}} |H_s|\, ds \leq \frac{h(T_{n+1} - T_n)}{T_n} \longrightarrow 0 \text{ as } n \longrightarrow \infty$$

and hence

$$\lim_{n\to\infty} \frac{1}{n} \sum_{i=1}^{n} H_{T_i} = \lim_{n\to\infty} \frac{1}{n} \int_0^{T_n} H_s \, ds$$

$$= \lim_{n\to\infty} \frac{1}{T_n} \int_0^{T_n} H_s \, ds = \lim_{t\to\infty} \frac{1}{t} \int_0^{t} H_s \, ds$$

provided one of these limits exists almost surely. ■

Comment. Using the extension of the stochastic integral in section 6.2, the same proof yields this theorem also for bounded *predictable* processes H.

The theorem is e.g. applied in the following context: Imagine a single server queue such that the average service time is smaller than the average time between two customer arrivals. Then by a renewal theorem the system converges to a stationary state. Let W_n be the total time which the n-th customer waits until his service starts. Let Q_t be the time which the last member of the queue at time t is going to stay in the queue (so that $t + Q_t$ will be his time of departure). If T_n is the n-th arrival time then W_n equals $Q_{T_n}^-$, i.e. the residual time of the last (or (n-1)-th) customer at time T_n. Then the equilibrium distributions $\mathbb{P}_Q$ of Q_t as $t \longrightarrow \infty$ and $\mathbb{P}_W$ of W_n as $n \longrightarrow \infty$ are exhibited by a.s. averaging

$$\int f \, d\mathbb{P}_W = \lim_{n\to\infty} \frac{1}{n} \sum_{i=1}^{n} f(W_i) \text{ a.s.}$$

and

$$\int f \, d\mathbb{P}_Q = \lim_{t\to\infty} \frac{1}{t} \int_0^t f(Q_s) ds = \lim_{t\to\infty} \frac{1}{t} \int_0^t f(Q_s^-) ds \text{ a.s.}$$

where f is any bounded continuous function. The theorem applied to $H_t = f(Q_t^-)$ shows that $\mathbb{P}_Q = \mathbb{P}_W$ if the arrivals form a Poisson process. Such queues are called M/G/1 queues.

Thus for a "M/G/1 queue" the waiting variables W_n and Q_t asymptotically have the same distribution.

CHAPTER 6

PREDICTABILITY

In chapter 5 we got as an analogue of the Riemann integral the stochastic integral for regular left-continuous integrands. The integrand was approximated (locally) uniformly by elementary processes and then, the integral was defined as a limit in $\boldsymbol{M}$ of the respective elementary integrals (theorem 5.4.4). In order to define a more general Itô integral uniform approximation has to be replaced by a weaker concept of convergence. One uses convergence in a (local) $\boldsymbol{L}^2$-sense. The measure which governs this $\boldsymbol{L}^2$-convergence is the "Föllmer-Doleans measure" on $\mathbb{R}_+ \times \Omega$ given by $d[M]_t \, d\mathbb{P}(\omega)$ where $[M]$ is the increasing (quadratic variation) process of the integrator M.

Section 1 below provides basic results on predictable sets and processes. The latter are the candidates for the new integrands. In section 2 the program sketched above is carried out. For special integrators the integral is further extended in section 3. Sections 4 - 6 aim at a deeper understanding of non-continuous semimartingales which are studied in the (optional) section 7.2. The discussion of predictable stopping times and the predictable stopping theorem in section 4 and the predictable projection and desintegration in section 5 are of independent interest. But they also serve as a preparation for the decomposition of a (uniformly integrable) submartingale into a martingale and a predictable increasing process in section 6.

6.1 Predictable Sets and Processes

In the further extension of the stochastic integral measurability questions come up. More precisely, one looks out for a σ-field $\boldsymbol{P}$ over $\mathbb{R}_+ \times \Omega$ such that the stochastic integral has a meaning for all $\boldsymbol{P}$-measurable processes satisfying some boundedness condition. If one recalls the representation

$H = \sum_{i=1}^{n} H_i \mathbf{1}_{]T_i,T_{i+1}]}$ of elementary processes one is quite naturally led to the σ-field generated by the stochastic intervals $]S,T]$.

Technical note. It does not really matter which subsets of $\{0\} \times \Omega$ belong to $\boldsymbol{P}$. One possibility is to take only $\emptyset$ and $\{0\} \times \Omega$, another possibility is to allow all sets of the form $\{0\} \times F$, $F \in \boldsymbol{F}_0$. Both choices have their drawbacks. So we follow Rogers-Williams (1987) and remove $\{0\} \times \Omega$ from the underlying set: $\boldsymbol{P}$ is a σ-field over $]0,\infty[\times \Omega$.

6.1.1 Definition. Denote by $\boldsymbol{P}$ the σ-field over $]0,\infty[\times \Omega$ generated by the stochastic intervals of the form $]S,T]$. A set in $\boldsymbol{P}$ is called a **predictable** set. A process $X = (X_t)_{t>0}$ is **predictable** if it is $\boldsymbol{P}$-measurable as the function $(t,\omega) \longmapsto X_t(\omega)$ on $]0,\infty[\times \Omega$. Also, we call a process on $\mathbb{R}_+ \times \Omega$ **predictable** if X_0 is $\boldsymbol{F}_0$-measurable and its restriction to $]0,\infty[\times \Omega$ is $\boldsymbol{P}$-measurable.

There are several equivalent descriptions of the σ-field $\boldsymbol{P}$:

6.1.2 Proposition. The σ-field $\boldsymbol{P}$ is generated by each of the following systems of sets or processes:

(a) the stochastic intervals $]S,T]$ where S and T are stopping times,

(b) the stochastic intervals $]0,T]$ where T is a strict stopping time,

(c) the **predictable rectangles** i.e. the sets $]s,t] \times F$ where $s < t$ and $F \in \boldsymbol{F}_s$,

(d) the adapted continuous processes $(X_t)_{t>0}$,

(e) the adapted left-continuous processes $(X_t)_{t>0}$.

Proof. For the proof denote the σ-fields generated by the sets or processes above by $\sigma_a, \ldots, \sigma_e$, respectively. So by definition, $\sigma_a = \boldsymbol{P}$.

$\sigma_b \subset \sigma_a$. This is obvious.

$\sigma_c \subset \sigma_b$. If $F \in \boldsymbol{F}_s$ then the rectangle $]s,t] \times F$ can be written as $]0,T] \setminus]0,S]$ where the strict stopping times S and T are defined by $S = s \cdot \mathbf{1}_F + \infty \cdot \mathbf{1}_{F^c}$ and $T = t \cdot \mathbf{1}_F + \infty \cdot \mathbf{1}_{F^c}$. Thus $\sigma_c \subset \sigma_b$.

$\sigma_e \subset \sigma_c$. We have another look at the argument in proposition 2.3.11. Let $((t_n^k)_{1 \le k \le k_n})_{n \ge 1}$ be an enumeration of all increasing finite subfamilies of $\mathbb{Q}_+ \backslash \{0\}$

and let X be an adapted process. Then the process $X^{(n)}$ defined by

$$X^{(n)}(t,\omega) = \sum_{k=1}^{k_n} X_{t_n^k} \mathbf{1}_{]t_n^k, t_n^{k+1}]}$$

are easily seen to be measurable with respect to σ_c. If, moreover, X is left-continuous then $X = \lim_{n\to\infty} X^{(n)}$ and hence X is σ_c-measurable. This proves $\sigma_e \subset \sigma_c$.

$\sigma_d \subset \sigma_e$ is trivial.

$\sigma_c \subset \sigma_d$. For a predictable rectangle $]s,s'] \times F$, $F \in \boldsymbol{F}_s$, choose continuous functions f_n on $\mathbb{R}_+$ such that $f_n \longrightarrow \mathbf{1}_{]s,s']}$ pointwise and $f_n|[0,s] \equiv 0$. Then $f_n(t)\cdot\mathbf{1}_F(\omega)$ defines a sequence of continuous and adapted processes converging pointwise to $\mathbf{1}_{]s,s']\times F}$ which proves the inclusion.

$\sigma_a \subset \sigma_e$. If S and T are stopping times then the process $\mathbf{1}_{]S,T]}$ is left-continuous and adapted (cf. proposition 4.4.2(b)) . ■

It is easy to modify the argument for $\sigma_e \subset \sigma_c$ to get

6.1.3 Lemma. If X is adapted then X^- is predictable.

Proof. Define the predictable process $X^{(n)}$ as in the proof of $\sigma_e \subset \sigma_c$ above and let $T_k = \inf\{t : \sup_{s<t, s\in\mathbb{Q}} |X_s| > k\}$. Then the T_k are stopping times and by definition (2.3.3)

$$X^- = \begin{cases} \limsup_{n\to\infty} X^n & \text{on } \bigcup_{k=1}^{\infty}]0,T_k] \\ 0 & \text{otherwise} \end{cases}$$

Thus X^- is predictable. ■

As a preparation for the next section we show that in analogy to proposition 5.3.5 pathwise Stieltjes integrals of predictable processes have nice versions. In the proof we use the monotone class theorem. Here is one version (cf. Dellacherie-Meyer, (1978), thm. I 21).

6.1.4 Lemma. (**Monotone class theorem**) Let $\boldsymbol{V}$ be a linear space of bounded real-valued functions which contains the constants, is closed under uniform convergence and has the following property: for every uniformly bounded increasing sequence of positive functions in $\boldsymbol{V}$ the pointwise

limit-function belongs to $\boldsymbol{V}$. Suppose further that $\boldsymbol{E}$ is a subset of $\boldsymbol{V}$ which is closed under multiplication.
Then $\boldsymbol{V}$ contains all bounded functions which are measurable w.r.t. the σ-field generated by $\boldsymbol{E}$.

6.1.5 Proposition. Let $A \in \boldsymbol{A}$ and let H be a predictable process such that for almost every ω

$$\int_0^t |H(s,\omega)| \; |dA_s(\omega)| < \infty \text{ for all t.}$$

Then the pathwise integral of H w.r.t. A has a version $\int_0^{\cdot} H \, dA$ in $\boldsymbol{A}$.

Proof. Let $\boldsymbol{V}$ denote the class of all bounded predictable processes H for which the conclusion holds. The class $\boldsymbol{V}$ contains all elementary processes by the explicit representation in proposition 5.2.2. By dominated convergence and lemma 5.3.4 it satisfies the assumption of the monotone class theorem 6.1.4. Therefore $\boldsymbol{V}$ contains all bounded predictable processes. Applying lemma 5.3.4 once more the boundedness can be removed. ∎

The general theory of stochastic processes studies also other, larger σ-fields on $\mathbb{R}_+ \times \Omega$ resp. $]0,\infty[\times \Omega$. We already mentioned the (fairly wide) concept of progressively measurable processes in section 2.3.

6.1.6 Lemma. Every predictable set is a progressively measurable subset of $\mathbb{R}_+ \times \Omega$. In particular, each predictable process is adapted and for every ω the path $X_{\cdot}(\omega)$ is a Borel function on $]0,\infty[$.

Proof. Plainly, each predictable rectangle is progressively measurable. Since the predictable σ-field is generated by the predictable rectangles the assertion is proved. ∎

Even larger than the σ-field of progressively measurable sets is the σ-field generated by all $\mathcal{B}(\mathbb{R}_+) \otimes \boldsymbol{F}$-measurable $(\boldsymbol{F}_t)$-adapted processes. Another intermediate measurability concept is given by the "optional" **σ-field** which may be introduced as follows. In our definition of the elementary processes the stochastic intervals $]S,T]$ are open to the left, making the elementary processes continuous from the left. If intervals $[S,T[$ are chosen instead - leading to right-continuous "elementary" processes - then the corresponding σ-field on $\mathbb{R}_+ \times \Omega$ becomes larger (which is somewhat surprising from the deterministic point of view). The stochastic integral can also be defined for optional processes provided one sticks to *continuous integrating* processes M. In that case, however, one can pass directly to the progressively measurable processes (see section 6.3).

The following result shows that for special filtrations all these different measurability concepts are equivalent. A more abstract version may be found in (Dellacherie-Meyer (1978), IV 97). We will not use theorem 6.1.7 explicitely but it helps to clarify the assumptions in chapter 12.

6.1.7 Theorem. Let Ω be the space either of all continuous or of all left-continuous functions on $\mathbb{R}_+$ with values in $\mathbb{R}^n$. Let $(\boldsymbol{F}_t)_{t\geq 0}$ be the natural filtration of the canonical projection process $(\Pi_t)_{t\geq 0}$ and let $\boldsymbol{F} = \sigma\Big(\bigcup_{t\geq 0} \boldsymbol{F}_t\Big)$. If a process X is $\mathcal{B}(\mathbb{R}_+) \otimes \boldsymbol{F}$-measurable and $(\boldsymbol{F}_t)$-adapted then it is predictable w.r.t. $(\boldsymbol{F}_t)$.

Proof. The idea is to write $X = X \circ a$ with a transformation $a : \mathbb{R}_+ \times \Omega \longrightarrow \mathbb{R}_+ \times \Omega$ which is measurable w.r.t. $\boldsymbol{P}$ and $\mathcal{B}(\mathbb{R}_+) \otimes \boldsymbol{F}$.
We consider for each $t > 0$ the map $a_t : \Omega \longrightarrow \Omega$ defined by $(a_t(\omega))_s = \omega_{s\wedge t}$ for $s \geq 0$. Then for every fixed $s \geq 0$ the process $(t,\omega) \longmapsto \Pi_s(a_t(\omega)) = \omega_{s\wedge t}$ is $\boldsymbol{P}$-measurable. In fact, it is adapted and has (left-)continuous paths by the choice of Ω. Since the functions Π_s, $s \geq 0$, generate $\boldsymbol{F}$ this implies the $\boldsymbol{P}$-$\boldsymbol{F}$-measurability of $(t,\omega) \longmapsto a_t(\omega)$ and hence the $\boldsymbol{P}$-$\mathcal{B}(\mathbb{R}_+) \otimes \boldsymbol{F}$-measurability of $a : (t,\omega) \longmapsto (t, a_t(\omega))$.
Now let $(X_t)_{t\geq 0}$ be $\mathcal{B}(\mathbb{R}_+) \otimes \boldsymbol{F}$-measurable and $(\boldsymbol{F}_t)$-adapted. Since X_t is $\boldsymbol{F}_t$-measurable and $\boldsymbol{F}_t$ is generated by Π_s, $s \leq t$, the value $X_t(\omega)$ depends only on $\omega|_{]0,t]}$. Therefore

$$X(t,\omega) = X_t(\omega) = X_t\big(a_t(\omega)\big) = X\big(t,a_t(\omega)\big) = X\big(a(t,\omega)\big).$$

Thus $X = X \circ a$ is a predictable process. ■

6.2 Predictable Integrands

The stochastic integral will be introduced for a fairly large class of predictable processes. We follow closely the lines of section 5.4.

The integrands will be in the following space:

6.2.1 Definition. Let M be a right-continuous local $\boldsymbol{L}^2$-martingale. Denote by $\boldsymbol{L}^{2,loc}(M)$ the space of those predictable processes H for which there is a localizing sequence $(T_k)_{k\geq 1}$ such that

$$\mathbb{E}\left(\int_0^{T_k} H^2 \, d[M]\right) < \infty \text{ for each } k \geq 1.$$

According to proposition 6.1.5 the process $\int_0^{\cdot} H^2 \, d[M]$ has a version in $\boldsymbol{A}$. For this version each $\int_0^{T_k} H^2 \, d[M]$ is a nonnegative random variable and therefore the expectation makes sense. The key is the following analogue of theorem 5.4.4.

6.2.2 Theorem. Suppose that $M \in \boldsymbol{M}$ and $H \in \boldsymbol{L}^{2,loc}(M)$. Then the the following holds:

(a) There are a localizing sequence $(T_k)_{k\geq 1}$ and a sequence $(H^n)_{n\geq 1}$ of elementary processes such that for every $k \geq 1$ the following holds

(i) $\mathbb{E}\left(\int_0^{T_k} (H - H^n)^2 \, d[M]\right) \xrightarrow[n\to\infty]{} 0$,

(ii) the process $(M_{t\wedge T_k} - M_0)_{t\geq 0}$ is a $\boldsymbol{L}^2$-bounded martingale.

(b) There is an element $\int_0^{\cdot} H \, dM$ of $\boldsymbol{M}$ such that for all sequences $(H^n)_{n\geq 1}$ and $(T_k)_{k\geq 1}$ as in (a) and each $k \geq 1$

$$\mathbb{E}\left(\sup_{0\leq t\leq T_k} \left(\int_0^t H^n \, dM - \int_0^t H \, dM\right)^2\right) \xrightarrow[n\to\infty]{} 0. \tag{6.2.1}$$

In particular, $\int_0^{\cdot} H^n \, dM$ converges l.u.p. to $\int_0^{\cdot} H \, dM$.

(c) If $M \in \boldsymbol{M}_c$ then $\int_0^{\cdot} H \, dM \in \boldsymbol{M}_c$.

This justifies

6.2.3 Definition. For $M \in \boldsymbol{M}$ and $H \in \boldsymbol{L}^{2,loc}(M)$ the process $\int_0^{\cdot} H\,dM$ of the preceding theorem is called the **stochastic integral of** H **with respect to** M.

The analogy between theorems 6.2.2 and 5.4.4 shows that for a predictable $H \in \boldsymbol{H}$ this new definition is compatible with the one in 5.4.5.

That $\boldsymbol{L}^2$-space techniques are appropriate in this context is underlined by the following property of the integral:

6.2.4 Proposition. Suppose that $M \in \boldsymbol{M}$ and $H \in \boldsymbol{L}^{2,loc}(M)$. Let T be a stopping time such that

$$\mathbb{E}\Big(\int_0^T H^2\,d[M]\Big) < \infty .$$

Then $\int_0^{\cdot\wedge T} H\,dM$ is a $\boldsymbol{L}^2$-bounded martingale. If $\int_0^T H\,dM$ denotes the limit variable then

$$\mathbb{E}\Big(\Big(\int_0^T H\,dM\Big)^2\Big) = \mathbb{E}\Big(\int_0^T H^2\,d[M]\Big). \tag{6.2.2}$$

Moreover

$$\mathbb{E}\Big(\sup_{0\le t\le T}\Big(\int_0^t H\,dM\Big)^2\Big) \le 4\,\mathbb{E}\Big(\int_0^T H^2\,d[M]\Big). \tag{6.2.3}$$

Before we go into the details of the proofs let us introduce a measure associated to the increasing process of the integrator M.

6.2.5 Definition. Suppose that $M \in \boldsymbol{M}$. Define a measure μ_M on $\mathscr{B}(\mathbb{R}_+) \otimes \boldsymbol{F}$ by

$$\int_{\mathbb{R}_+\times\Omega} Z\,d\mu_M = \mathbb{E}\Big(\int_0^\infty Z\,d[M]\Big) \tag{6.2.4}$$

where $Z : \mathbb{R}_+ \times \Omega \longrightarrow \mathbb{R}_+$ is product-measurable. This measure is called the **Föllmer-Doleans measure**. The $\boldsymbol{L}^2$-norm associated the Föllmer-Doleans measure is denoted by $\|\ \|_{2,\mu_M}$.

Here $\int_0^\infty Z\,d[M]$ is a.s. defined as a pathwise Stieltjes integral. The scrupulous reader may apply proposition 6.1.5 to the constant filtration to make sure that $\int_0^\infty Z\,d[M]$ can be chosen to be $\boldsymbol{F}$-measurable.

Let us shortly comment on the role of this measure.

Remark. (a) Set $T \equiv \infty$ in proposition 6.2.4. Then the integrability condition reads $\mathbb{E}\left(\int_0^\infty H^2 \, d[M]\right) < \infty$. It simply means that $H \in \boldsymbol{L}^2(\mathbb{R}_+ \times \Omega, \boldsymbol{P}, \mu_M)$. The identity (6.2.2) expresses that the linear operator

$$I : \boldsymbol{L}^2(\mathbb{R}_+ \times \Omega, \boldsymbol{P}, \mu_M) \longrightarrow \boldsymbol{L}^2(\Omega, \boldsymbol{F}, \mathbb{P}), \; H \longmapsto \int_0^\infty H \, dM$$

preserves the $\boldsymbol{L}^2$-norms.

(b) Suppose that M is a $\boldsymbol{L}^2$-martingale and that $]s,t] \times F$ is a predictable rectangle i.e. $F \in \boldsymbol{F}_s$. Then

$$\mu_M\left(]s,t] \times F\right) = \mathbb{E}\left(\mathbf{1}_F([M]_t - [M]_s)\right)$$

and hence by corollary 5.6.4

$$\mu_M\left(]s,t] \times F\right) = \mathbb{E}\left(\mathbf{1}_F(M_t^2 - M_s^2)\right).$$

This relation is for some authors the starting point to *define* the Föllmer-Doleans measure. Then one has to prove that this set function extends to a (σ-additive) measure (we had a similar but simpler construction in chapter 1). We do not have this problem since the increasing process is already constructed.

(c) In passing, we mention that one cannot define the integral in such a way that for fixed ω the value $\left(\int_0^t H_s \, dM_s\right)(\omega)$ is linear in H on $\boldsymbol{L}^2(\mu_M)$ (cf. S. Mohammed (1987)).

For convenient reference we note a reformulation of proposition 5.6.3 in terms of the Föllmer-Doleans measure.

6.2.6 Lemma. Suppose that $M \in \boldsymbol{M}$ and that T is a stopping time. Then the following are equivalent:

(i) $\mu_M(]0,T]) < \infty$, (ii) $\mathbb{E}([M]_T) < \infty$, (iii) $M_{\cdot\wedge T} - M_0$ is a $\boldsymbol{L}^2$-bounded martingale.

We will show in chapter 7 that $[M_{\cdot\wedge T}]_t = [M]_{t\wedge T}$ and thus the general integrability condition in proposition 6.2.4 is equivalent to the condition $H \in \boldsymbol{L}^2(\mathbb{R}_+ \times \Omega, \boldsymbol{P}, \mu_{M_{\cdot\wedge T}})$. We continue with some further illustrations.

6.2.7 Examples. (a) The Föllmer-Doleans measure of Brownian motion is the product $\lambda \otimes \mathbb{P}$ of Lebesgue measure λ and $\mathbb{P}$. In fact, in (6.2.4) the term $d[M]$ can simply be replaced by ds (cf. example 5.6.6).

(b) The second standard example is of course the Poisson martingale $M_t = N_t - t$ where N_t is a (right-continuous) Poisson process with intensity $\alpha = 1$. In this case $[M]_t = N_t$ as we have seen in example 5.6.7. Thus

$$\int_{\mathbb{R}_+\times\Omega} Z\, d\mu_M = \int_\Omega \sum_{k=1}^{\infty} Z\big(T_k(\omega),\omega\big)\, d\mathbb{P}(\omega)$$

where $(T_k)_{k\geq 1}$ is the sequence of jumping times of the Poisson process N. If Z is considered as a process we may simply write $\mathbb{E}\big(\sum_{k=1}^{\infty} Z_{T_k}\big)$ for the right-hand side. This is strictly different from the Brownian motion case, isn't it?

But now let us compute this expectation for a predictable rectangle $]s,t] \times F$, $F \in \boldsymbol{F}_s$. Since $[M] = N$ and since $N_t - t$ is a martingale we get

$$\mu_M\big(]s,t] \times F\big) = \mathbb{E}\big(\mathbf{1}_F(N_t - N_s)\big) = \mathbb{E}\big(\mathbf{1}_F(t - s)\big) = \lambda \otimes \mathbb{P}\big(]s,t] \times F\big).$$

The predictable rectangles generate the σ-field $\boldsymbol{P}$ and therefore we conclude that

μ_M and $\lambda \otimes \mathbb{P}$ coincide on all predictable sets!

But for instance

$$\mu_M(N = 0) = \mu_M(]0,T_1[) = 0$$

whereas

$$\lambda \otimes \mathbb{P}(N = 0) = \lambda \otimes \mathbb{P}(]0,T_1[) = \mathbb{E}(T_1) = 1.$$

This shows in particular that the Poisson process N cannot be predictable. On the other hand it is progressively measurable being right-continuous. So for a filtration $(\boldsymbol{F}_t)$ the predictable and progressively measurable sets are different if there is a $(\boldsymbol{F}_t)$-adapted right-continuous Poisson process.

Let us start now with the proof of the theorem. The next two results will be useful in the construction. For later reference they are stated in a slightly more general form than actually needed in this section. An a.s. increasing process A will be called **locally integrable** if there is a localizing sequence $(R_k)_{k\geq 1}$ such that $\mathbb{E}(A_{R_k}) < \infty$ for each k. Note that the increasing process of every right-continuous local L^2-martingale is locally integrable by lemma 6.2.6.

6.2.8 Lemma. Let A be an adapted right-continuous a.s. increasing and locally integrable process with $A_0 \equiv 0$. Suppose further that H is a predictable process and T a stopping time with $\mathbb{E}\left(\int_0^T H^2 \, dA\right) < \infty$.
Then for each $\varepsilon > 0$ there is an elementary process H^ε such that

$$\mathbb{E}\left(\int_0^T (H - H^\varepsilon)^2 \, dA\right) < \varepsilon.$$

Proof. Define like in (6.2.4) the measure μ on $\boldsymbol{P}$ by

$$\int f \, d\mu = \mathbb{E}\left(\int_0^T f \, dA\right)$$

where f is a measurable nonnegative function. Let $\boldsymbol{E}$ denote the space of all elementary processes

$$H' = \sum_{i=1}^{n} H_i \mathbf{1}_{]T_i, T_{i+1}]}$$

such that $\mu(]0, T_{n+1}]) < \infty$. Then $\boldsymbol{E} \subset \boldsymbol{L}^2(\mu)$. We shall use the monotone class theorem to show that $\boldsymbol{E}$ is dense in $\boldsymbol{L}^2(\mu)$. Since the integrability assumption on H simply means that $H \in \boldsymbol{L}^2(\mu)$ this will prove the lemma.
Choose a localizing sequence $(R_k)_{k \geq 1}$ with

$$\mu(]0, R_k]) = \mathbb{E}(A_{R_k}) < \infty.$$

Consider the space $\boldsymbol{V}$ of all bounded $\boldsymbol{P}$-measurable functions H such that $H \cdot \mathbf{1}_{]0,R_k]}$ belongs to the closure $\bar{\boldsymbol{E}}$ of $\boldsymbol{E}$ in $\boldsymbol{L}^2(\mu)$. Then $\boldsymbol{V}$ satisfies the assumptions of the monotone class theorem 6.1.4. Since stochastic intervals of finite μ-measure are stable under finite unions $\boldsymbol{E}$ is a linear space which is closed under multiplication. A minute of reflection shows that $\boldsymbol{E}$ generates the σ-field $\boldsymbol{P}$. Thus $H \cdot \mathbf{1}_{]0,R_k]}$ is in $\bar{\boldsymbol{E}}$ for every bounded $\boldsymbol{P}$-measurable process H. Since every $H \in \boldsymbol{L}^2(\mu_M)$ can be approximated by such processes, more precisely

$$H = \lim_{k \to \infty} \left(-k \vee (H \wedge k)\right) \mathbf{1}_{]0,R_k]} \text{ in } \boldsymbol{L}^2(\mu),$$

the proof is complete. ■

The next lemma is an immediate consequence.

6.2.9 Lemma. Let H be a predictable process and A as in the preceding lemma. Suppose that $(T_k)_{k\geq 1}$ is a localizing sequence such that

$$\mathbb{E}\Big(\int_0^{T_k} H^2\, dA\Big) < \infty \text{ for every } k \geq 1.$$

Then there is a sequence $(H^n)_{n\geq 1}$ of elementary processes such that

$$\mathbb{E}\Big(\int_0^{T_k} (H - H^n)^2\, dA\Big) \xrightarrow[n\to\infty]{} 0 \text{ for every } k \geq 1.$$

The $\boldsymbol{L}^2$-isometry (6.2.2) will now be established for elementary processes.

6.2.10 Lemma. Let M be in $\boldsymbol{M}$ and let H be an elementary process

$$H = \sum_{i=1}^{n} H_i \mathbf{1}_{]T_i,T_{i+1}]}$$

such that $\mu_M\big(]0,T_{n+1}]\big) < \infty$. Then for every stopping time T the identity

$$\mathbb{E}\Big(\Big(\int_0^T H\, dM\Big)^2\Big) = \mathbb{E}\Big(\int_0^T H^2\, d[M]\Big)$$

holds true.

Proof. The elementary integral and the increasing process do not change if we replace M by $M - M_0$. Hence we may and shall assume that $M_0 \equiv 0$. We noted in lemma 6.2.6 that $\big(M_{t\wedge T_{n+1}}\big)_{t\geq 0}$ is a $\boldsymbol{L}^2$-bounded martingale. Proposition 5.2.3(b), theorem 5.4.1 and corollary 5.6.4 yield

$$\mathbb{E}\Big(\Big(\int_0^T H_s\, dM_s\Big)^2\Big)$$

$$= \mathbb{E}\Big(\Big(\int_0^T H_s\, dM_{s\wedge T_{n+1}}\Big)^2\Big) = \mathbb{E}\Big(\sum_{i=1}^{n} H_i^2 (M_{T\wedge T_{i+1}} - M_{T\wedge T_i})^2\Big)$$

$$= \mathbb{E}\Big(\sum_{i=1}^{n} H_i^2 ([M]_{T\wedge T_{i+1}} - [M]_{T\wedge T_i})^2\Big) = \mathbb{E}\Big(\int_0^T H_s^2\, d[M]_s\Big). \blacksquare$$

Proof of theorem 6.2.2. We may assume $M_0 \equiv 0$.

(a) Suppose $H \in \boldsymbol{L}^{2,loc}(M)$. Choose a localizing sequence $(T_k)_{k\geq 1}$ such that for every $k \geq 1$

(i) the process $(M_{t\wedge T_k})_{t\geq 0}$ is a $\boldsymbol{L}^2$-bounded martingale,

(ii) $\mathbb{E}\Big(\int_0^{T_k} H^2\, d[M]\Big) < \infty$.

Since the increasing process of M is locally integrable by lemma 6.2.6 and because (ii) holds lemma 6.2.9 provides the desired sequence.

(b) Let the sequences $(H^n)_{n\geq 1}$ and $(T_k)_{k\geq 1}$ be given as in (a). Then proposition 5.2.3(b) yields

$$\int_0^{\cdot \wedge T_k} (H^n - H^m)\, dM = \int_0^{\cdot} (H^n - H^m)_s \, dM_{s\wedge T_k} .$$

These processes are right-continuous by proposition 5.2.2 and $\mathcal{L}^2$-bounded martingales by theorem 5.4.1. Hence Doob's $\mathcal{L}^2$-inequality applies. Using then lemma 6.2.10 we get

$$\mathbb{E}\Big(\sup_{0\leq s\leq T_k} \Big(\int_0^s (H^n - H^m)\, dM\Big)^2\Big) \leq 4\; \mathbb{E}\Big(\Big(\int_0^{T_k} (H^n - H^m)\, dM\Big)^2\Big)$$
$$= 4\; \mathbb{E}\Big(\int_0^{T_k} (H^n - H^m)^2\, d[M]\Big).$$

Since $(H^n \mathbf{1}_{]0,T_k]})_{n\geq 1}$ converges in $\mathcal{L}^2(\mu_M)$ it is also a Cauchy sequence in this space and the last expectations tend to zero as m and n tend to infinity. The rest of the proof of (b) is the same as in the proof of theorem 5.4.5, part (b2).

(c) If $M \in \mathcal{M}_c$ then the elementary integrals $\int_0^{\cdot} H^n\, dM$ are in $\mathcal{M}_c$ and therefore (6.2.1) implies that also the limit process $\int_0^{\cdot} H\, dM$ is in $\mathcal{M}_c$ by proposition 5.4.3. ■

Proof of proposition 6.2.4. Assume now $\mathbb{E}\big(\int_0^T H^2 d[M]\big) < \infty$. Let $S \leq T$ be a bounded stopping time. Then the identity

$$\mathbb{E}\Big(\Big(\int_0^{S\wedge T_k} H^n\, dM\Big)^2\Big) = \mathbb{E}\Big(\int_0^{S\wedge T_k} (H^n)^2\, d[M]\Big)$$

holds according to lemma 6.2.10 for the elementary processes in the proof of (b) above. Then we may replace H^n by H for fixed k and then let k tend to ∞ to arrive at

$$\mathbb{E}\Big(\Big(\int_0^S H\, dM\Big)^2\Big) = \mathbb{E}\Big(\int_0^S H^2\, d[M]\Big) \leq \mathbb{E}\Big(\int_0^T H^2 d[M]\Big) < \infty .$$

By the criterion 4.2.3 the process $\int_0^{\cdot\wedge T} H\, dM$ is a $\mathcal{L}^2$-bounded martingale closed by a limit variable $\int_0^T H\, dM$. Hence the left-hand side of the identity makes sense and the identity holds. The rest reduces now to Doob's $\mathcal{L}^2$-inequality 3.2.10. ■

The following criteria help to decide whether a predictable process is in $\mathcal{L}^{2,loc}(M)$.

6.2.11 Proposition. Suppose that $M \in \mathcal{M}$ and let H be a predictable process. Then the following holds:

(a) If H is locally bounded then H belongs to $\mathcal{L}^{2,loc}(M)$.

(b) Let M be continuous. Then H belongs to $\mathcal{L}^{2,loc}(M)$ if and only if

$$\mathbb{P}\left(\int_0^t H^2 \, d[M] < \infty\right) = 1 \text{ for each } t. \tag{6.2.5}$$

Proof. To prove (a) just take the minimum of the localizing sequences for H and M.

For (b) assume that (6.2.5) holds. Since [M] is a.s. continuous the Lebesgue-Stieltjes integral $\int_0^t H^2 \, d[M]$ is a.s. continuous in t (proposition 5.1.10). Moreover

$$T_k = \inf\left\{t : \int_0^t H^2 \, d[M] > k\right\}$$

defines a localizing sequence and continuity implies $\int_0^{T_k} H^2 \, d[M] \le k$ a.s. and hence

$$\mathbb{E}\left(\int_0^{T_k} H^2 \, d[M]\right) \le k < \infty.$$

So $H \in \mathcal{L}^{2,loc}(M)$. The converse implication is obvious. ■

If M is not continuous then there is no reason why (6.2.5) should imply that $\mathbb{E}\left(\int_0^{T_k} H_s^2 \, d[M]_s\right)$ be finite since $\int_{[T_k]} H_s^2 \, d[M]$ may be large.

Proposition 6.2.4 implies that $\int_0^{\cdot} H \, dM$ is a $\mathcal{L}^2$-bounded martingale whenever $H \in \mathcal{L}^2(\mu_M)$. There is also a converse to this.

6.2.12 Proposition. For $H \in \mathcal{L}^{2,loc}(\mu_M)$ assume that $\int_0^{\cdot} H \, dM$ is a $\mathcal{L}^2$-bounded martingale. Then $H \in \mathcal{L}^2(\mu_M)$.

Proof. For a $\mathcal{L}^2$-bounded martingale N closed by the random variable N_∞ one has the estimate $\|N_T\|_2 \le \|N_\infty\|_2$ for every stopping time T. Let (T_k) be a localizing sequence as in definition 6.2.1. Then

$$\|H\|_{2,\mu_M}^2 = \mathbb{E}\left(\int_0^\infty H^2 \, d[M]\right)$$

$$= \sup_k \mathbb{E}\left(\int_0^{T_k} H^2 \, d[M]\right) = \sup \left\|\int_0^{T_k} H \, dM\right\|_2^2 \le \left\|\int_0^\infty H \, dM\right\|_2^2 < \infty . \quad \blacksquare$$

6.3 Relaxing Predictability of the Integrand

Sometimes the predictability of an integrand is too strong an assumption. In this section we look for processes which agree with a suitable predictable process μ_M-almost everywhere on $\mathbb{R}_+ \times \Omega$. For these processes the definition of the stochastic integral can be extended in an obvious manner. The situation is particularly pleasant if μ_M equals $\lambda \otimes \mathbb{P}$ or if μ_M is at least absolutely continuous with respect to this measure. In fact, we have the following result:

6.3.1 Theorem. Let the process H be product-measurable, i.e. $\mathscr{B}(\mathbb{R}_+) \otimes \boldsymbol{F}$-measurable and $(\boldsymbol{F}_t^+)$-adapted. Then there is a predictable process H^* such that

$$\lambda \otimes \mathbb{P}\left(H \neq H^*\right) = 0. \tag{6.3.1}$$

Proof. We first remark that the theorem may easily be reduced to the case of a bounded process H. The obvious difficulty in the proof is that there is only little relation between the σ-field $\mathscr{B}(\mathbb{R}_+) \otimes \boldsymbol{F}$ and the filtration $(\boldsymbol{F}_t)$. The missing link is established by conditional exceptions. So we prove a slightly different statement:

For every bounded $\mathscr{B}(\mathbb{R}_+) \otimes \boldsymbol{F}$-measurable H there is a predictable process H^* such that

$$\lambda \otimes \mathbb{P}\left(\left\{(t,\omega) : H^*(t,\omega) \neq \mathbb{E}(H_t \mid \boldsymbol{F}_t^+)(\omega)\right\}\right) \tag{6.3.2}$$

where $\mathbb{E}(H_t \mid \boldsymbol{F}_t^+)$ denotes a conditional expectation of the random variable $H_t = H(t,\cdot)$ with respect to $\boldsymbol{F}_t^+$. Fubini's theorem implies that in (6.3.2) the choice of the version of the conditional expectation does not matter. Thus (6.3.2) implies the theorem since $H_t = \mathbb{E}(H_t \mid \boldsymbol{F}_t^+)$ if H is $(\boldsymbol{F}_t^+)$-adapted.

Concerning (6.3.2) let $\boldsymbol{D}$ be the class of all bounded $\mathscr{B}(\mathbb{R}_+) \otimes \boldsymbol{F}$-measurable H for which a predictable H^* with (6.3.2) exists. Clearly, $\boldsymbol{D}$ is closed under monotone uniformly bounded limits of sequences. By the monotone class theorem it is sufficient to show that $\boldsymbol{D}$ contains all processes H of the form $H(t,\omega) = f(t)g(\omega)$ where f is continuous on $\mathbb{R}_+$ and g is $\boldsymbol{F}$-measurable. Let (Y_t) be a a.s. right-continuous regular version of the martingale $\mathbb{E}(g \mid \boldsymbol{F}_t^+)$. Such a version exists according to theorem 3.2.6.

Then the process H^* given by $H^*(t,\omega) = f(t)Y_t^-(\omega)$ is predictable (see lemma 6.1.3) and for every regular path ω the number of those t for which $Y_t(\omega) \neq Y_t^-(\omega)$ is countable. (For every n the set of jumps of size $\geq \frac{1}{n}$ is locally finite since it has no accumulation point.) Thus by Fubini's theorem we have $\lambda \otimes \mathbb{P}$-a.s.

$$H^*(t,\omega) = f(t)Y_t^-(\omega) = f(t)Y_t(\omega) = f(t)\,\mathbb{E}(g \mid \boldsymbol{F}_t^+)(\omega)$$

$$= \mathbb{E}\left(f(t)g \mid \boldsymbol{F}_t^+\right)(\omega) = \mathbb{E}(H_t \mid \boldsymbol{F}_t^+)(\omega).$$

This implies (6.3.2). ■

We want to prove a similar statement for the Föllmer-Doleans measures of continuous local martingales. The following lemma is of independent interest.

6.3.2 Lemma. Suppose that A is an a.s. increasing right-continuous process starting with $A_0 \equiv 0$. Set

$$S_t = \inf\left\{s \geq 0 : A_s > t\right\} \text{ and } \varphi(t,\omega) = \left(S_t(\omega),\omega\right).$$

Then

(a) The map $\varphi : \mathbb{R}_+ \times \Omega \longrightarrow (\mathbb{R}_+ \cup \{\infty\}) \times \Omega$ is measurable w.r.t. $\mathscr{B}(\mathbb{R}_+) \otimes \boldsymbol{F}$ and $\mathscr{B}(\mathbb{R}_+ \cup \{\infty\}) \otimes \boldsymbol{F}$.

(b) The measure on $\mathscr{B}(\mathbb{R}_+) \otimes \boldsymbol{F}$ associated with A is the image of $\lambda \otimes \mathbb{P}$ under φ. More precisely, for every $\mathscr{B}(\mathbb{R}_+) \otimes \boldsymbol{F}$-measurable $f \geq 0$ one has

$$\int_\Omega \int_0^\infty f(s,\omega)\,dA_s(\omega)\,d\mathbb{P}(\omega) = \int_\Omega \int_0^{A_\infty} f(S_t(\omega),\omega)\,dt\,d\mathbb{P}(\omega) \tag{6.3.3}$$

$$= \int_{[0,A_\infty]} f \circ \varphi \; d\lambda \otimes \mathbb{P}.$$

Proof. Choose $B = [0,s[\times F$, where $0 \leq s < \infty$ and $F \in \boldsymbol{F}$. The right-continuity of A implies that $S_t < s$ if and only if $S_r < s$ for some rational $r > t$. Thus

$$\varphi^{-1}[B] = \left\{(t,\omega) : \omega \in F \text{ and } t < r,\ S_r(\omega) < s \text{ for some } r \in \mathbb{Q}_+\right\}$$

$$= \bigcup_{r \in \mathbb{Q}_+} [0,r[\times \left(\left\{A_s < r\right\} \cap F\right)$$

proving the measurability of φ claimed in (a).

That $t \longmapsto S_t$ transforms Lebesgue measure into dA_s is a well-known one-dimensional fact: For the nondecreasing paths of A one gets

$$\lambda\left(\left\{t : S_t \le s\right\}\right) = \lambda\left(\left\{t : A_s \ge t\right\}\right) = A_s = \int_0^s dA.$$

So in (6.3.3) the inner integrals are a.s. equal which proves part (b). Note that $\varphi^{-1}(\mathbb{R}_+ \times \Omega)$ is the stochastic interval $[0,A_\infty[$ since $A_\infty = \sup_{t\ge 0} A_t$. ■

Reflect a minute on part (b) of the lemma. Assume that A is the increasing process of some continuous local martingale M. Recall that the increasing process of Brownian motion generates Lebesgue measure for every path. So φ transforms the Föllmer-Doleans measure of Brownian motion into the Föllmer-Doleans measure of M. Considering the form of φ, the thought might occur that every continuous martingale arises from Brownian motion simply by rescaling the time axis by means of the increasing process. This turns out to be true; it is made precise in the remarkable theorem of Dambis, Dubins and Schwarz which will be discussed in detail in section 9.2.

6.3.3 Corollary. Let H be progressively measurable and assume that $M \in \boldsymbol{M}$ has a continuous increasing process. Then there is a predictable process H^* such that

$$\mu_M\left(H \ne H^*\right) = 0.$$

Proof. We use the nullset elimination argument of section 2.1 and assume that $\langle M\rangle$ has only continuous paths. We apply the preceding lemma to $A = \langle M\rangle$ and consider the filtration $(\boldsymbol{G}_t) = (\boldsymbol{F}^+_{S_t})$. The process $(X_t) = \left(H_{S_t} 1_{\{S_t<\infty\}}\right)$ is $(\boldsymbol{G}_t)$-adapted by proposition 2.2.11. Thus there is a $(\boldsymbol{G}_t)$-predictable process X^* such that $\lambda \otimes \mathbb{P}(X \ne X^*) = 0$ according to theorem 6.3.1. Now we claim that for every $(\boldsymbol{G}_t)$-predictable process (Y_t) the process $(Z_t) = \left(Y_{\langle M\rangle_t}\right)_{t\ge 0}$ is $(\boldsymbol{F}_t)$-predictable: This is certainly true if Y is continuous because then Z is also continuous; the rest is a simple argument with monotone classes. Therefore we get an $(\boldsymbol{F}_t)$-predictable process H^* by setting $H^*_t = X^*_{\langle M\rangle_t}$. Then $X = H \circ \varphi$, $X^* = H^* \circ \varphi$ and by lemma 6.3.2,

$$\mu_M\left(H \ne H^*\right) = \mu_M\left(\varphi^{-1}\left[\left\{H \ne H^*\right\}\right]\right) = \lambda \otimes \mathbb{P}\left(X \ne X^*\right) = 0.$$

This proves the assertion. ■

It is now obvious how to extend the integral to the more general integrands considered in theorem 6.3.1 and corollary 6.3.3.

6.3.4 Definition. Suppose that $M \in \boldsymbol{M}_c$. Assume that H is a progressively measurable process (or $\mathcal{B}(\mathbb{R}_+)\otimes\boldsymbol{F}$-measurable and adapted if M is a Brownian motion) such that

$$\mathbb{P}\Big(\int_0^t H_s^2 \, d\langle M\rangle_s < \infty\Big) = 1 \text{ for every } t \geq 0.$$

Then the stochastic integral $\int_0^{\cdot} H \, dM$ is defined by

$$\int_0^t H_s \, dM_s = \int_0^t H_s^* \, dM_s ,$$

where H^* is predictable with $\mu_M(H \neq H^*) = 0$.

6.4 Predictable Stopping

Predictable stopping is a refinement of the previous stopping techniques. It is very useful in the study of martingales which are not a priori continuous. In particular, it provides the "predictable" version 6.4.10 of the martingale stopping theorem. This is needed for the next section and it also yields the continuity criterion 6.4.11 for predictable martingales.

6.4.1 Definition. A stopping time T is called **predictable** if $T > 0$ and the graph [T] is a predictable set.

The set]0,T] is always predictable. Therefore a strictly positive stopping time T is predictable if and only if]0,T[is predictable. Every predictable stopping time is a strict stopping time since for every t the indicator

$$\mathbf{1}_{\{T\leq t\}} = \mathbf{1}_{\{T<t\}} + \mathbf{1}_{[T]}(t)$$

is $\boldsymbol{F}_t$-measurable.

The following example shows that predictable stopping times really are a new concept.

6.4.2 Examples. (a) As noted in example 6.2.7(b) for the first jumping time T_1 of a Poisson process the stochastic interval $]0,T_1[$ is not a predictable set, i.e. T_1 is not predictable.

(b) Suppose X is a continuous adapted process and C is a closed set such that $X_0(\omega) \notin C$ for all ω. Then the entrance time of X into C is predictable: In proposition 2.3.5 we proved that there is a sequence (T_n) of stopping times such that $T_n < T$ and $T_n \uparrow T$. Then

$$]0,T[\; = \bigcup_{n \in \mathbb{N}}]0,T_n[\; \in \boldsymbol{P}.$$

In Theorem 6.4.5 we shall show that up to $\mathbb{P}$-nullsets the existence of such an announcing sequence is also necessary for the predictability of a stopping time T.

In this and the next section the nullset augmentation of the filtration is a very useful tool. With the filtration $(\boldsymbol{F}_t)$ one can associate the filtration $({}^\circ\boldsymbol{F}_t^+)_{t \geq 0}$ which satisfies conditions (i), (ii) of the "usual conditions".

6.4.3 Remark. A process X is $({}^\circ\boldsymbol{F}_t^+)$-predictable if and only if it is $\mathscr{B}(\mathbb{R}_+) \otimes \boldsymbol{F}$-measurable and a.s. equal to a $(\boldsymbol{F}_t)$-predictable process.

Proof. Since predictability implies progressive measurability a $({}^\circ\boldsymbol{F}_t^+)$-predictable process is $\mathscr{B}(\mathbb{R}_+) \otimes {}^\circ\boldsymbol{F}$-measurable. Clearly a $({}^\circ\boldsymbol{F}_t^+)$-predicable rectangle $]s,t] \times F$, $F \in {}^\circ\boldsymbol{F}_s^+$ is a.s. equal to a rectangle $]s,t] \times F^*$, $F^* \in \boldsymbol{F}_s^+$ whose indicator is left-continuous and $(\boldsymbol{F}_t)$-adapted. Hence $]s,t] \times F^* \in \boldsymbol{P}$. Then a monotone class argument shows that every $({}^\circ\boldsymbol{F}_t^+)$-predicable process is a.s. equal to a $(\boldsymbol{F}_t)$-predictable process.

Conversely, assume that Y is $(\boldsymbol{F}_t)$-predictable and that X is $\mathscr{B}(\mathbb{R}_+) \otimes \boldsymbol{F}$-measurable with $X_.(\omega) = Y_.(\omega)$ for all $\omega \notin N$ where $N \in \boldsymbol{F}$ is a $\mathbb{P}$-nullset. Let ${}^\circ\boldsymbol{P}$ be the σ-field of $({}^\circ\boldsymbol{F}_t)$-predictable sets. Then

$$(]0,\infty[\times N) \cap (]s,t] \times F) =]s,t] \times (F \cap N) \in {}^\circ\boldsymbol{P}$$

for every $F \in \boldsymbol{F}$ and hence the σ-fields $\mathscr{B}(\mathbb{R}_+) \otimes \boldsymbol{F}$ and ${}^\circ\boldsymbol{P}$ have the same trace on $]0,\infty[\times N$. Thus $\mathbf{1}_{]0,\infty[\times N}(X - Y)$ and

$$X = Y + \mathbf{1}_{]0,\infty[\times N}(X - Y)$$

are ${}^\circ\boldsymbol{P}$-measurable. ■

As for stopping times we shall see as part of theorem 6.4.5 that the filtrations $(\boldsymbol{F}_t)$ and $({}^\circ\boldsymbol{F}_t^+)$ (and hence also $(\boldsymbol{F}_t^+)$ and $({}^\circ\boldsymbol{F}_t)$) have "almost the same" predictable stopping times. For not necessarily *predictable* stopping times this is straightforward (see lemma 6.4.4 below). But for strict stopping times the corresponding result is not valid.

6.4.4 Lemma. A $\boldsymbol{F}$-measurable map $T : \Omega \longrightarrow [0,\infty]$ is a $({}^\circ\boldsymbol{F}_t^+)$-stopping time if and only if there is a $(\boldsymbol{F}_t)$-stopping time T^* such that $T = T^*$ a.s..

Proof. The "if"-part is obvious. For the converse let T be a $({}^\circ\boldsymbol{F}_t^+)$-stopping time. For every $t \geq 0$ there is a set $C_t \in \boldsymbol{F}_t^+$ and a $\mathbb{P}$-nullset N_t such that

$$C_t \,\Delta\, \{T < t\} \subset N_t$$

where the symbol Δ denotes symmetric difference: $A \,\Delta\, B = (A \cap B^c) \cup (A^c \cap B)$. Define T^* by

$$T^*(\omega) = \inf\{t : \omega \in C_s \text{ for some rational } s < t\}.$$

Then we have for every $t \geq 0$

$$\{T^* < t\} = \bigcup_{s<t,\, s\in\mathbb{Q}} C_s \in \sigma\Big(\bigcup_{s<t} \boldsymbol{F}_s^+\Big) \subset \boldsymbol{F}_t$$

and $\{T^* < t\} \,\Delta\, \{T < t\}$ is contained in the nullset $\bigcup_{s<t,\, s\in\mathbb{Q}} N_s$. This shows that T^* is a $(\boldsymbol{F}_t)$-stopping time (cf. lemma 2.2.2) and $\mathbb{P}(T^* = T) = 1$. ■

The next theorem gives us several descriptions of "a.s. predictable stopping times".

6.4.5 Theorem. Suppose that $T : \Omega \longrightarrow]0,\infty]$ is $\boldsymbol{F}$-measurable. Then the following are equivalent:

(a) T is a predictable stopping time for $({}^\circ\boldsymbol{F}_t^+)_{t\geq 0}$.

(b) There is a $(\boldsymbol{F}_t)$-predictable stopping time S such that $T = S$ a.s..

(c) There is a sequence (T_n) of $(\boldsymbol{F}_t)$-stopping times such that $T_n < T$ a.s. and $T_n \uparrow T$ a.s. .

Remark. The proof will show that the stopping times T_n in (c) can be chosen to be themselves almost predictable in the sense of this theorem.

Proof. The main implication is (b) → (c) . The implication (a) → (c) then is an easy consequence. The converse implications (c) → (a) and (c) → (b) are straightforward. Let us first do the easy parts.

(c) → (a). If (T_n) is given by (c) and N is the union of the exceptional sets in (c) then define

$$T_n^* = \left(\left(T - \tfrac{1}{n}\right) \vee 0\right) \cdot \mathbf{1}_N + T_n \cdot \mathbf{1}_{N^c}$$

Since T is measurable and N is a $\boldsymbol{F}$-measurable nullset each T_n^* is $\boldsymbol{F}$-measurable and a.s. equal to the $(\boldsymbol{F}_t)$-stopping time T_n. So each T_n^* is a $({}^o\boldsymbol{F}_t^+)$-stopping time by the preceding lemma. Moreover $T_n^* < T$ and $T_n^* \uparrow T$ everywhere. Therefore T is $({}^o\boldsymbol{F}_t^+)$-predictable since

$$]0,T[\;=\; \bigcup_{n\geq 1} \;]0,T_n^*]$$

is a $({}^o\boldsymbol{F}_t^+)$-predictable set.

(c) → (b). Let again be (T_n) as in (c). Define S by $S = \sup R_n$ where

$$R_n = \begin{cases} T_n & \text{on } \bigcap_{k\geq 1} \bigcup_{m\geq k} \{T_m < T_{m+1}\} \\ +\infty & \text{otherwise} . \end{cases}$$

Each R_n is a stopping time and

$$]0,S[\;=\; \bigcup_{m\geq 1} \;]0,R_n] \in \boldsymbol{P} .$$

Therefore S is a predictable stopping time. Since $T_m < T_{m+1}$ a.s. infinitely often we have $S = \sup R_n = \sup T_n = T$ a.s.. This proves (b).

(a) → (c). We argue that it is sufficient to prove (b) → (c): Assume (a). If (b) → (c) holds then this implication can be applied also to the filtration $({}^o\boldsymbol{F}_t^+)$. Therefore (a) implies (c) except that now the T_n are only $({}^o\boldsymbol{F}_t^+)$-stopping times. But the preceding lemma 6.4.4 allows then to change them into $(\boldsymbol{F}_t)$-stopping times without changing the a.s. statements in (c).

(b) → (c). Now let us turn to this nontrivial implication. It relies on the fact that predictable sets can be approximated from inside by zero sets of continuous adapted processes, see lemma 6.4.6 below. We may replace T by S or equivalently, we may assume that T itself is a predictable $(\boldsymbol{F}_t)$-stopping time. Define a finite measure μ_T on $\mathcal{B}([0,\infty]) \otimes \boldsymbol{F}$ by

$$\mu_T(D) = \mathbb{P}\left(\left\{\omega : (T(\omega),\omega) \in D\right\}\right).$$

This makes sense since T is $\boldsymbol{F}$-$\mathscr{B}([0,\infty])$-measurable and hence $\omega \longmapsto (T(\omega),\omega)$ is $\boldsymbol{F}$-$\mathscr{B}([0,\infty]) \otimes \boldsymbol{F}$-measurable. Fix $\varepsilon > 0$. According to lemma 6.4.6 below there is a continuous adapted process Y such that

$$\{Y = 0\} \subset [T] \text{ and } \mu_T\big([T] \setminus \{Y = 0\}\big) < \varepsilon.$$

Let R be the entrance time of Y into $\{0\}$. Because $\{Y = 0\}$ is contained in the graph $[T]$ which meets a set $\mathbb{R}_+ \times \{\omega\}$ at most in a point $(T(\omega),\omega)$, a path $Y_.(\omega)$ vanishes only at time $t = R(\omega) = T(\omega)$. Thus

$$[R] = \{Y = 0\} \subset [T]$$

and on the set $\{R < \infty\}$ the stopping times R and T coincide. By proposition 2.3.5 there is a sequence (R_m) of finite stopping times such that $R_m \uparrow R$ and $R_m < R$.

(Actually these R_m can be chosen to be predictable themselves: Since $\{Y = 0\} \subset [T]$ and $T > 0$ we have $Y_0 \neq 0$ and thus the process $(Y_t/Y_0)_{t \geq 0}$ is adapted and continuous with the same zero set as Y. Therefore we may assume $|Y_0| = 1$ and choose R_m to be the entrance time of Y into the closed set $[-\frac{1}{2m}, \frac{1}{2m}]$. Thus each R_m is predictable.)

With these stopping times,

$$\mathbb{P}\big(R_m \geq T > 0\big) \leq \mathbb{P}\big(R \neq T)\big) = \mu_T\big([T] \setminus [R]\big) < \varepsilon$$

and for m sufficiently large,

$$\mathbb{P}\big(R_m + \varepsilon < T < \infty\big) \leq \mathbb{P}\big(R \neq T\big) + \mathbb{P}\big(R_m + \varepsilon < R < \infty\big) < \varepsilon .$$

We have shown that for every $\varepsilon > 0$ there is a finite predictable stopping time $S(\varepsilon)$ - namely the previous R_m - such that

$$\mathbb{P}\big(S(\varepsilon) \geq T\big) < \varepsilon \text{ and } \mathbb{P}\big(S(\varepsilon) + \varepsilon < T < \infty\big) < \varepsilon .$$

Then according to lemma 2.2.5(b)

$$T_n = \inf_{k \geq n} S(2^{-k})$$

defines a sequence of finite stopping times such that $T_n < T$ a.s. and $T_n \uparrow T$ a.s. which proves (c). This completes the proof of the theorem. ∎

Proof of the remark. We show that T_n is $({}^\circ\boldsymbol{F}_t^+)$-predictable. One has a.s. $T_n < T - 2^{-k} \leq S(2^{-k})$ for eventually all k. Hence the infimum in the definition of T_n is a.s. attained. The indicators of the two sets

$$]0,T_n[\text{ and } \bigcap_{k\geq 1}]0,S(2^{-k})[$$

are a.s. equal. Since the second set is in $\boldsymbol{P}$ the set $]0,T_n[$ and hence the stopping time T_n is $({}^{\circ}\boldsymbol{F}_t^{+})$-predictable by remark 6.4.3. ■

In the proof we used that finite measures on the predictable σ-field are inner regular in the following sense:

6.4.6 Lemma. For every finite measure μ on the σ-field $\boldsymbol{P}$ over $\mathbb{R}_+ \times \Omega$ and for every predictable set $P \in \boldsymbol{P}$ one has

$$\text{(6.4.1)} \quad \mu(P) = \sup\{\mu(Y = 0) : Y \text{ is continuous and adapted and } \{Y = 0\} \subset P\}.$$

Proof. For the proof denote by $\boldsymbol{P}_0$ the system of all sets P in $\boldsymbol{P}$ fulfilling (6.4.1). First we show that $\boldsymbol{P}_0$ is stable under countable intersections and countable unions.

Given the sets $P_n \in \boldsymbol{P}_0$, $n \in \mathbb{N}$, and $\varepsilon > 0$, choose continuous adapted processes Y^n such that

$$\{Y^n = 0\} \subset P_n \text{ and } \mu\left(P_n \setminus \{Y^n = 0\}\right) < \varepsilon \cdot 2^{-n} \text{ for each n.}$$

In order to prove that the intersection of the P_n belongs to $\boldsymbol{P}_0$ consider the process

$$Y = \sum_{n=1}^{\infty} 2^{-n}\left(1 \wedge |Y^n|\right)$$

which is continuous and adapted and satisfies

$$\{Y = 0\} = \bigcap_{n=1}^{\infty} \{Y^n = 0\} \subset \bigcap_{n=1}^{\infty} P_n$$

and

$$\mu\Big(\bigcap_{n=1}^{\infty} P_n \setminus \{Y = 0\}\Big) \leq \sum_{n=1}^{\infty} \varepsilon \cdot 2^{-n} = \varepsilon.$$

Thus $\bigcap_{n=1}^{\infty} P_n \in \boldsymbol{P}_0$. For the case of countable unions consider for every $N \in \mathbb{N}$ the (continuous adapted) process

$$Y_N = \prod_{n=1}^{N} Y^n.$$

Then

$$\{Y_N = 0\} = \bigcup_{n=1}^{N} \{Y^n = 0\} \subset \bigcup_{n=1}^{\infty} P_n$$

and

$$\mu\Big(\bigcup_{n=1}^{\infty} P_n \setminus \{Y_N = 0\}\Big) \leq \mu\Big(\bigcup_{n=1}^{\infty} P_n \setminus \bigcup_{n=1}^{N} P_n\Big) + \sum_{n=1}^{N} \varepsilon \cdot 2^{-n} < 2\varepsilon$$

for sufficiently large N, i.e. $\bigcup_{n=1}^{\infty} P_n \in \boldsymbol{P}_O$.

Therefore the system

$$\{P \in \boldsymbol{P} : P \text{ and } P^c \text{ belong to } \boldsymbol{P}_O\}$$

is a σ-field. It is actually equal to $\boldsymbol{P}$ since for every continuous and adapted process X and every $a \in \mathbb{R}$ the sets

$$\{X \geq a\} = \{X - a \geq 0\} = \{\min(X - a, 0) = 0\}$$

and

$$\{X < a\} = \bigcup_{n=1}^{\infty} \{-X \geq -a + \tfrac{1}{n}\}$$

are in $\boldsymbol{P}_O$. This proves the lemma and completes the proof of theorem 6.4.5. ■

The proof of the lemma yields a more abstract result: if $\boldsymbol{C}$ is a vector lattice of real functions on some set such that $\boldsymbol{C}$ contains the constants and is closed under uniform limits then every finite measure on the σ-field generated by $\boldsymbol{C}$ is regular with respect to the zero sets of elements of $\boldsymbol{C}$.

The following corollary shows that even for non predictable stopping times a "predictable part" can also be announced as in theorem 6.4.5(c). This result will be used only once in section 11.1. So the reader may skip it. In the formulation we write A = B a.s. if A and B are two subsets of $\mathbb{R}_+ \times \Omega$ such that $\mathbf{1}_A = \mathbf{1}_B$ a.s., i.e. $A \triangle B \subset \mathbb{R}_+ \times N$ for some nullset N.

6.4.7 Corollary. Let T be a stopping time. Suppose that $G \subset]0,\infty[\times \Omega$ is a predictable set such that $G \cap [0,T[= \emptyset$ a.s.. Then there is a sequence (T_n) of stopping times such that $T_n \uparrow T$ a.s. and $G \cap]0,T_n] = \emptyset$ a.s. for all n.

Proof. Define S by

$$S(\omega) = \begin{cases} T(\omega) & \text{if } (\omega,T(\omega)) \in G \\ +\infty & \text{otherwise} . \end{cases}$$

Then $[S] = [0,T] \cap G$ a.s.. Thus S is a $({}^\circ\boldsymbol{F}_t)$-predictable stopping time. According to the theorem there is a sequence (S_n) of stopping times such that $S_n \uparrow S$ a.s. and $S_n < S$ a.s. on $\{S > 0\}$. Thus $T_n = S_n \wedge T$ satisfies $T_n \uparrow T$ a.s.. Moreover $(\omega,T_n(\omega)) \notin G$ a.s. on $\{T_n > 0\}$: If $T_n < T$ then $(\omega,T_n(\omega)) \notin G$ a.s. by the assumption on G. If $T_n = T$ then $T \neq S$ a.s. and thus $(\omega, T(\omega)) \notin G$ by definition of S. This completes the proof. ■

The main significance of predictable stopping times lies in the predictable stopping theorem. For its formulation we associate to every stopping time a new σ-field.

6.4.8 Definition. For every stopping time T let $\boldsymbol{F}_T^-$ be the σ-field

$$\sigma\{F \cap \{T > t\} : t \geq 0, F \in \boldsymbol{F}_t\}.$$

$\boldsymbol{F}_T^-$ is called **the σ-field of events strictly before T.**

6.4.9 Remark. (a) If T is a constant stopping time t then it is predictable and $\boldsymbol{F}_t^- = \sigma\Big(\bigcup_{s<t} \boldsymbol{F}_s\Big)$. The family $(\boldsymbol{F}_t^-)_{t\geq 0}$ is itself a filtration. The reader should be warned that $\boldsymbol{F}_T^-$ is in general *not* equal to the σ-field which one gets by applying the definition 2.2.3 to this filtration $(\boldsymbol{F}_t^-)_{t\geq 0}$.
(b) If X is a predictable process then $\mathbf{1}_{\{T<\infty\}} X_T$ is $\boldsymbol{F}_T^-$-measurable. It suffices to consider the indicator $X = \mathbf{1}_{]s,t]\times F}$, $F \in \boldsymbol{F}_s$ of a predictable rectangle and in this case

$$\mathbf{1}_{\{T<\infty\}} X_T = \mathbf{1}_{\{T>s\}\cap F \setminus \{T>t\}\cap F}$$

is $\boldsymbol{F}_T^-$-measurable.
(c) Every stopping time T is measurable w.r.t. its own σ-field $\boldsymbol{F}_T^-$. In fact, the process $X : (t,\omega) \longmapsto t\mathbf{1}_{[0,T]}(t,\omega)$ is predictable and hence

$$\{0 < T < s\} = \{0 < \mathbf{1}_{\{T<\infty\}} X_T < s\} \in \boldsymbol{F}_T^-$$

for all $s < \infty$ by part (b). This implies the assertion.

Now we can formulate the announced version of the stopping theorem.

6.4.10 Theorem (Predictable stopping theorem). Let $(M_t)_{t\geq 0}$ be a martingale closed by M_∞. Then for every predictable stopping time T,

$$M_T^- = \mathbb{E}\Big(M_\infty \;\Big|\; \boldsymbol{F}_T^-\Big). \qquad (6.4.2)$$

Proof. Let T be predictable and let (T_n) an announcing sequence as in theorem 6.4.5(c). Using the nullset elimination argument we may assume that $T_n < T$ on $\{T > 0\}$ and $T_n \uparrow T$ everywhere. Also we may asume that every path

of M^- is regular (note that M^- is separable and hence theorem 3.2.6(a) can be applied). Thus M^- is left-continuous and admits right-hand limits.
Now $M_t^+ = \lim_{t'\downarrow t} M_{t'}^-$ defines a right-continuous process which is a $(\boldsymbol{F}_t^+)$-martingale by decreasing martingale convergence. Therefore by the standard stopping theorem and the increasing martingale convergence theorem

$$M_T^- = \lim_{n\to\infty} M_{T_n}^- = \lim_{n\to\infty} M_{T_n}^+ = \lim_{n\to\infty} \mathbb{E}\left(M_\infty \mid \boldsymbol{F}_{T_n}^+\right)$$
$$= \mathbb{E}\left(M_\infty \mid \sigma\left(\bigcup_{n=1}^{\infty} \boldsymbol{F}_{T_n}^+\right)\right) \text{ a.s.}.$$

It remains to prove

$$\boldsymbol{F}_T^- = \sigma\left(\bigcup_{n=1}^{\infty} \boldsymbol{F}_{T_n}^+\right).$$

First we show $\boldsymbol{F}_{T_n}^+ \subset \boldsymbol{F}_T^-$. If $A \in \boldsymbol{F}_{T_n}^+$ then

$$A \cap \{T_n < r\} \in \boldsymbol{F}_r \text{ for every } r \geq 0$$

and hence because of $T_n < T$ we have

$$A = \bigcup_{r\in\mathbb{Q}_+} \left(A \cap \{T_n < r\}\right) \cap \{r < T\}\big) \in \boldsymbol{F}_T^-.$$

Conversely, every set $F \cap \{T > t\}$ where $F \in \boldsymbol{F}_t$ may be written as

$$\bigcup_{n=1}^{\infty} F \cap \{T_n > t\}$$

where each $F \cap \{T_n > t\}$ is in $\boldsymbol{F}_{T_n}$ and in particular in $\boldsymbol{F}_{T_n}^+$. This completes the proof. ∎

This implies the following continuity result on predictable martingales.

6.4.11 Theorem. Every right-continuous predictable local martingale is a.s. continuous.

Proof. Suppose that M is a right-continuous predictable local martingale. Using localization we may assume that M is a uniformly integrable martingale closed by some random variable M_∞. Given $\varepsilon > 0$ it is sufficient to prove that M has a.s. no jumps greater than ε. Let T be the time of the first jump greater than ε, i.e.

$$T = \inf\left\{t \geq 0 : |M_t - M_t^-| > \varepsilon\right\}.$$

If $t_n \downarrow t$ and $t_n > t$ then $|M_{t_n} - M_{t_n}^-| \longrightarrow 0$ by the right-continuity of M.

Therefore the infimum in the definition of T is attained, i.e.

$$|M_T - M_T^-| > \varepsilon \text{ on } \{T < \infty\}.$$

If we can show that $M_T = M_T^-$ a.s. then $T = \infty$ a.s. and we are done. We have $|M_t - M_t^-| \le \varepsilon$ on $[0,T[$. Since M and M^- both are predictable we get

$$[T] = [0,T] \cap \{|M - M^-| > \varepsilon\} \in \boldsymbol{P}.$$

Thus T is predictable. Therefore, the last theorem implies

$$M_T^- = \mathbb{E}\left(M_\infty \mid \boldsymbol{F}_T^-\right). \tag{6.4.3}$$

The classical stopping theorem 3.2.7 implies

$$M_T = \mathbb{E}\left(M_\infty \mid \boldsymbol{F}_T^+\right). \tag{6.4.4}$$

But M is predictable and hence by remark 6.4.9(b) the variable M_T is $\boldsymbol{F}_T^-$-measurable. Hence in (6.4.4) we can replace $\boldsymbol{F}_T^+$ by $\boldsymbol{F}_T^-$. Together with (6.4.3) this shows $M_T = M_T^-$ a.s. thus completing the proof. ■

We shall use this result mainly in connection with theorem 5.3.2:

6.4.12 Corollary. A right-continuous predictable local martingale of locally finite variation is a.s. constant.

6.5 Predictable Projection and Desintegration

The main result of this section is theorem 6.5.5. It relates a finite measure μ on the σ-field $\boldsymbol{P}$ to an increasing process A in the form

$$\mu([0,T]) = \mathbb{E}(A_T).$$

The monotone convergence theorem shows that the condition on μ given in the following definition is necessary for such a relation. Theorem 6.5.5 will show that this condition is also sufficient.

6.5.1 Definition. A finite measure μ on $\boldsymbol{P}$ is called **admissible** if for every localizing sequence (T_k),

$$\mu(]T_k,\infty]) \xrightarrow[k\to\infty]{} 0\ .$$

Admissibility of μ is a condition on the structure of μ-nullsets. For every stopping time S the measure μ_S defined by

$$\mu_S(P) = \mathbb{P}\left(\left\{\omega : (S(\omega),\omega) \in P\right\}\right)$$

is admissible. These measures are used in the following criterion for properties which hold μ-a.s. for all admissible measures μ simultaneously.

6.5.2 Proposition. Let X be a predictable process. Then $X_S = 0$ a.s. for every finite predictable stopping time S if and only if $X = 0$ μ-a.s. for every admissible measure μ.

Proof. By way of contradiction, let μ be an admissible measure and assume $\mu(X \neq 0) > 0$. According to the regularity lemma 6.4.6 there is a continuous adapted process Y such that $\{Y = 0\} \subset (]0,\infty[\times \Omega) \cap \{X \neq 0\}$ and $\mu(Y = 0) > 0$. Let T be the entrance time of Y into 0. Then T is predictable. Let (T_k) be a sequence of stopping times such that $T_k \uparrow T$ and $T_k < T$ (cf. proposition 2.3.5). Then

$$\left\{Y = 0\right\} \subset \bigcap_{k \geq 1}]T_k,\infty[$$

If $\mathbb{P}(T < \infty) = 0$ then (T_k) is a localizing sequence and hence $\mu(Y = 0) = 0$ since μ is admissible. Thus $\mathbb{P}(T < \infty) > 0$. Choose $t > 0$ such that $\mathbb{P}(T \leq t) > 0$. Then $S = t \wedge T$ is a finite predictable stopping time such that $Y_S = 0$ on the set $\{T = S\}$ and hence in contrast to our assumption

$$\mathbb{P}(X_S \neq 0) \geq \mathbb{P}(Y_S = 0) \geq \mathbb{P}(T = S) > 0.$$

Conversely, if $X = 0$ μ-a.s. for every admissible measure then this is true in particular for the measures μ_S. This shows that $X_S = 0$ $\mathbb{P}$-a.s. for every positive stopping time and thus completes the proof of the proposition. ■

The predictable projection given by theorem 6.5.4 below is something like a conditional expectation operator on the level of processes rather than of random variables. We just use it as a tool in the construction of the predictable desintegration in theorem 6.5.5.

6.5.3 Definition. Let X be a bounded $\mathscr{B}(\mathbb{R}_+)\otimes\boldsymbol{F}$- or $\mathscr{B}(]0,\infty[)\otimes\boldsymbol{F}$-measurable process. A predictable process X^P is called a **predictable projection** of X if for every finite predictable stopping time T,

$$X_T^P = \mathbb{E}\left(X_T \;\middle|\; \boldsymbol{F}_T^-\right). \tag{6.5.1}$$

The uniqueness statement in part (a) of the following theorem can be strenghtened to uniqueness up to a.s. equality if one uses the techniques mentioned after proposition 2.3.4.

6.5.4 Theorem. (a) Every bounded $\mathscr{B}(\mathbb{R}_+)\otimes\boldsymbol{F}$-measurable process X has a predictable projection X^P. It is unique up to μ-nullsets for every admissible measure μ.

(b) Let $X(t,\omega) = f(t)g(\omega)$ where the bounded functions f and g are Borel on $\mathbb{R}_+$ and $\boldsymbol{F}$-measurable, respectively. Then

$$X^P(t,\omega) = f(t)\, M_t^-(\omega) \tag{6.5.2}$$

where $(M_t^-)_{t>0}$ is the left limit process associated with the martingale $M_t = \mathbb{E}(g \mid \boldsymbol{F}_t)$.

(c) For all bounded $\mathscr{B}(\mathbb{R}_+)\otimes\boldsymbol{F}$-measurable processes X and all predictable processes Y one has $(XY)^P = X^P Y$.

Proof. We first prove the explicit parts (b) and (c). The process X^P defined in (b) is predictable since M^- is predictable by lemma 6.1.3 and the factor f even is measurable with respect to the σ-field generated by the rectangles $]s,t] \times \Omega$. Let T be a predictable stopping time. Then the predictable stopping theorem and remark 6.4.9(b) and (c) imply

$$X_T^P = f(T)M_T^- = f(T)\,\mathbb{E}(g \mid \boldsymbol{F}_T^-) = \mathbb{E}(f(T)g \mid \boldsymbol{F}_T^-) = \mathbb{E}(X_T \mid \boldsymbol{F}_T^-)$$

and thus X^P is a predictable projection of X.

(c) For a predictable process Y and a finite stopping time T the random variable Y_T is $\boldsymbol{F}_T^-$-measurable according to remark 6.4.9(b). Therefore

$$(X^P Y)_T = X_T^P\, Y_T = \mathbb{E}(X_T \mid \boldsymbol{F}_T^-))Y_T = \mathbb{E}((XY)_T \mid \boldsymbol{F}_T^-)$$

which proves (c).

(a) We start with the uniqueness. If Y and Z are predictable projections of X then $Y_T = Z_T$ a.s. for every finite predictable stopping time and hence lemma 6.5.2 shows that $\mu(Y \neq Z) = 0$ for every admissible measure μ. The monotone class theorem 6.1.4 allows to extend the existence statement from the processes X considered in part (b) to general bounded product-measurable processes X.■

The following is the central result of this section. It is fundamental for a deeper understanding of semimartingales. It may be viewed as a "predictable desintegration theorem". It is also called the theorem of "dual predictable projection". An essential point in the statement is the predictability of the process A. The theorem is used in the Doob-Meyer decomposition of sub-martingales (and supermartingales). A simpler proof of a version of these two results which uses the concept of 'natural' processes rather than predictability has been given by K.M. Rao. It is reproduced e.g. in Karatzas and Shreve (1988), section 1.4.

6.5.5 Theorem. For every admissible measure μ on ***P*** there is a predictable right-continuous a.s. nondecreasing locally bounded process A starting at 0 such that

$$\int X \, d\mu = \mathbb{E}\Big(\int_0^\infty X \, dA\Big) \tag{6.5.3}$$

for every bounded predictable process X. The process A is unique up to a.s. equality.

We divide the proof into several separate assertions.

6.5.6 Lemma. The process A in the theorem is unique up to a.s. equality.

Proof. Suppose that A and A' are two processes of the indicated type. Then $\mathbb{E}(A_\infty) < \infty$ and $\mathbb{E}(A'_\infty) < \infty$. Thus $A - A'$ is uniformly integrable and

$$\mathbb{E}((A - A')_T) = \mu(]0,T]) - \mu(]0,T]) = 0$$

for every stopping time T. Then by proposition 3.2.9 the process A - A' is a martingale and a fortiori a local martingale. It is right-continuous, predictable and of finite variation. Thus A - A' = 0 by corollary 6.4.12. ■

6.5.7 Lemma. For every admissible measure μ on $\boldsymbol{P}$ there is a unique measure μ^P on $\mathscr{B}(\mathbb{R}_+)\otimes\boldsymbol{F}$ such that for every bounded $\mathscr{B}(\mathbb{R}_+)\otimes\boldsymbol{F}$-measurable process X

$$\int X \, d\mu^P = \int X^P \, d\mu . \tag{6.5.4}$$

If $X_S = 0$ a.s. for every finite predictable stopping times (a fortiori if $X = 0$ a.s.) then $\mu^P(X \neq 0) = 0$.

Proof. The number $\int X^P d\mu$ is well defined since the μ-equivalence class of X^P is uniquely determined by X (cf. theorem 6.5.4(a)). The map $X \longmapsto \int X^P d\mu$ is additive and σ-continuous as a function of X. Therefore it defines a measure μ^P on $\mathscr{B}(\mathbb{R}_+)\otimes\boldsymbol{F}$ satisfying (6.5.4). Suppose $X_S = 0$ a.s. for all finite predictable S (This holds in particular if $X = 0$ a.s.). Then $|X|^P = 0$ μ-a.s. by definition and uniqueness of $|X|^P$. Then (6.5.4) applied to $|X|$ implies $X = 0$ μ^P-a.s. which completes the proof. ■

Sometimes a measure $\mathscr{B}(\mathbb{R}_+)\otimes\boldsymbol{F}$ is called **predictable measure** if it is of the form μ^P where μ is admissible on $\boldsymbol{P}$.

6.5.8 Lemma. Let ν be a finite measure on $\mathscr{B}(\mathbb{R}_+)\otimes\boldsymbol{F}$ such that $\nu(\mathbb{R}_+ \times N) = 0$ for every $\mathbb{P}$-nullset N. There is a $\mathscr{B}(\mathbb{R}_+)\otimes\boldsymbol{F}$-measurable right-continuous non-decreasing process $A : \mathbb{R}_+ \times \Omega \longrightarrow \mathbb{R}_+$ such that $A_0 = 0$ and

$$\int_{\mathbb{R}_+\times\Omega} X \, d\nu = \mathbb{E}\Big(\int_0^\infty X \, dA\Big) \tag{6.5.5}$$

for all bounded $\mathscr{B}(\mathbb{R}_+)\otimes\boldsymbol{F}$-measurable processes X.

Proof. This is a reformulation of a standard fact about desintegration or existence of conditional distributions. Here is the main argument: For every $s \in \mathbb{Q}_+$ the measure $\mu([0,s] \times \cdot)$ on $\boldsymbol{F}$ is absolutely continuous with respect to $\mathbb{P}$. Let A'_s be a version of its Radon-Nikodym derivative. Then $A_s \le A_{s'}$ a.s. for $s \le s'$ and

$$A(t,\omega) = \inf_{s\in\mathbb{Q},s>t} A'_s(\omega)$$

is right-continuous and satisfies

$$\mu\big([0,t] \times F\big) = \mathbb{E}\big(\mathbf{1}_F \, A(t,\cdot)\big)$$

for all $F \in \boldsymbol{F}$ and $t \ge 0$. This implies (6.5.5). ■

The main part of the proof consists of the following proposition.

6.5.9 Proposition. Let μ be a finite measure on $\boldsymbol{P}$ and let A be a right-continuous nondecreasing process such that

$$\int X^P \, d\mu = \mathbb{E}\Big(\int_0^\infty X \, dA\Big)$$

for every bounded $\mathscr{B}(\mathbb{R}_+)\otimes \boldsymbol{F}$-measurable process X. Then A is $({}^\circ\boldsymbol{F}_t)$-predictable.

Proof. (1) First we claim that for every finite predictable stopping time S the random variable A_S is ${}^\circ\boldsymbol{F}_S^-$-measurable. First of all A_S is $\boldsymbol{F}$-measurable since A is right-continuous (apply e.g. proposition 2.3.8(b) to the constant filtration $\boldsymbol{G}_t \equiv \boldsymbol{F}$). Therefore it suffices to show that $A_S = \mathbb{E}(A_S \mid \boldsymbol{F}_S^-)$ a.s. or equivalently that

$$\mathbb{E}(A_S Z) = \mathbb{E}\big(\mathbb{E}(A_S \mid \boldsymbol{F}_S^-)Z\big)$$

for every bounded random variable Z. Given Z consider

$$Y = Z - \mathbb{E}(Z \mid \boldsymbol{F}_S^-)$$

and the martingale

$$M_t = \mathbb{E}(Y \mid \boldsymbol{F}_t).$$

Then M^- is the predictable projection of the process $\tilde{Y} : (t,\omega) \longmapsto Y(\omega)$ by theorem 6.5.4(b). In particular, for the predictable stopping times $t \wedge S$ we have

$$M_{t\wedge S}^- = \mathbb{E}(Y \mid \boldsymbol{F}_{t\wedge S}^-) = 0$$

since $\mathbb{E}(Y \mid \boldsymbol{F}_S^-) = 0$ and $\boldsymbol{F}_{t\wedge S}^- \subset \boldsymbol{F}_S^-$. The stochastic interval $]0,S]$ is predictable and therefore according to theorem 6.5.4(c)

$$\big(\tilde{Y}\,\mathbf{1}_{]0,S]}\big)^P = \tilde{Y}^P\,\mathbf{1}_{]0,S]} = M^-\,\mathbf{1}_{]0,S]} = 0.$$

According to the hypothesis this implies

$$\mathbb{E}\big(A_S\, Y\big) = \mathbb{E}\Big(Y \int_0^S dA\Big) = \mathbb{E}\Big(\int_0^\infty \tilde{Y}\,\mathbf{1}_{[0,S]}\, dA\Big) = \int \big(\tilde{Y}\,\mathbf{1}_{[0,S]}\big)^P d\mu = 0$$

and thus by definition of Y

$$\mathbb{E}\big(A_S\, Z\big) = \mathbb{E}\big(A_S\, \mathbb{E}(Z \mid \boldsymbol{F}_S^-)\big) = \mathbb{E}\big(\mathbb{E}(A_S \mid \boldsymbol{F}_S^-)Z\big).$$

Hence A_S is ${}^\circ\boldsymbol{F}_S^-$-measurable. In particular, A_t is ${}^\circ\boldsymbol{F}_t^-$-measurable for every t and the process A is $({}^\circ\boldsymbol{F}_t)$-adapted.

(2) In order to show that A is ($\boldsymbol{F}_t$)-predictable we have to deal with the jumps of A. For this we first prove the following two facts about the graph measures μ_T defined after definition 6.5.1. The hypothesis implies that μ is admissible. Consider the associated measure μ^P introduced in lemma 6.5.7.

(i) If $T : \Omega \longrightarrow [0,\infty]$ is $\boldsymbol{F}$ measurable such that $[T] \subset \{A^- \neq A\}$ then $\mu^P([T] \cap B) = 0$ implies $\mu_T(B) = 0$ for every $B \in \mathscr{B}(\mathbb{R}_+) \otimes \boldsymbol{F}$.

(ii) If $\mu_S^P(B) = 0$ for every predictable S then $\mu^P(B) = 0$.

For the proof of (i) note that by (6.5.5)

$$\mu^P([T]\cap B) = \int (1_{[T]\cap B})^P \, d\mu = \mathbb{E}\Big(\int_0^\infty 1_{[T]} 1_B \, dA\Big) = \mathbb{E}\Big((1_B)_T \, 1_{\{T<\infty\}} \big(A_T - A_T^-\big)\Big) .$$

Since by assumption on T the random variable $A_T - A_T^-$ is strictly positive on $\{T < \infty\}$ this expression can vanish only if $(1_B)_T \, 1_{\{T<\infty\}} = 0$ $\mathbb{P}$-a.s., i.e. if $\mu_T(B) = 0$.

(ii) is a direct consequence of the nullset statement in lemma 6.5.7 applied to $X = 1_B$.

(3) The problem about the jumps of A is that we do not know yet that they happen at predictable stopping times. So we cover as much as possible of the jump set $\{A^- \neq A\}$ by graphs of predictable stopping times: Among the sets of the form

$$\bigcup_{n\in\mathbb{N}} [S_n] \cap \{A^- \neq A\}$$

where the S_n are predictable stopping times there is a set of maximal μ^P-measure. (This is the μ^P-essential supremum of the sets $[S] \cap \{A^- \neq A\}$.) Since

$$A_{S_n} = \mathbb{E}\big(A_{S_n} \mid \boldsymbol{F}_{S_n}^-\big) = A_{S_n}^P \text{ a.s.} \tag{6.5.6}$$

for every n and since μ is admissible this set is μ^P-equivalent to the $({}^\circ\boldsymbol{F}_t)$-*predictable* set

$$\bigcup_{n\in\mathbb{N}} [S_n] \cap \{A^- \neq A^P\}$$

which we denote by G.

(4) Let $T : \Omega \longrightarrow [0,\infty]$ be $\boldsymbol{F}$-measurable such that $[T] \subset \{A^- \neq A\}$. We claim that $\mu_T(G^c) = 0$. Let S be any predictable stopping time. By construction of G

we have

$$\mu^P\left([S] \cap \{A^- \neq A\} \cap G^c\right) = 0$$

and hence $\mu^P([S] \cap [T] \cap G^c) = 0$. Therefore by (i) in part (2) we conclude $\mu_T([S] \cap G^c) = 0$ or equivalently $\mu_S([T] \cap G^c) = 0$. Since this holds for every predictable S it follows by (ii) in part (2) that $\mu^P([T] \cap G^c) = 0$ and hence again by (i) that $\mu_T(G^c) = 0$.

(5) For $m \in \mathbb{N}$ let $(T_k^m)_{k\geq 1}$ be the sequence of $\frac{1}{m}$-jumping times of A: Set $T_0^m \equiv 0$ and

$$T_{k+1}^m = \inf\left\{t > T_k^m : (A - A^-)_t > \frac{1}{m}\right\}.$$

Then

$$\left\{A^- \neq A\right\} = \bigcup_{k, m \in \mathbb{N}} [T_k^m]$$

since every jump of A happens at one of the times T_k^m. By step (4) their graphs are a.s. contained in G. Thus A is a.s. equal to A^- outside G, i.e.

$$A = \mathbf{1}_G\, A + \mathbf{1}_{G^c}\, A^- \text{ a.s..}$$

Because of (6.5.6) we have $\mathbf{1}_G\, A = \mathbf{1}_G\, A^P$ a.s. and

$$A = \mathbf{1}_G\, A^P + \mathbf{1}_{G^c}\, A^- \text{ a.s..}$$

The process on the right-hand side is $({}^\circ\boldsymbol{F}_t)$-predictable and hence so is A. The proof of the proposition is now complete. ■

The final lemma shows that the theorem does not rely on the "usual conditions".

6.5.10 Lemma. For a right-continuous and a.s. nondecreasing process A the following are equivalent:

(a) A is $({}^\circ\boldsymbol{F}_t^+)$-predictable,

(b) A is a.s. equal to a right-continuous a.s. nondecreasing $(\boldsymbol{F}_t)$-predictable locally bounded process $\breve{A}$.

Proof. The implication (b) → (a) is easy since for every $\mathbb{P}$-nullset $N \in \boldsymbol{F}$ the process $\mathbf{1}_{\mathbb{R}_+\times N^c}$ is $({}^\circ\boldsymbol{F}_t^+)$-predictable.

For the converse implication (a) → (b) we may assume without loss of generality that all paths of A are nondecreasing and that $A_0 = 0$. For $r \in \mathbb{Q}$

and $r > 0$ let $T_r = \inf\{A \geq r\}$. Then $T_r > 0$ and the set $]0,T_r[$ equals $\{A < r\}$ and hence is $({}^\circ\boldsymbol{F}_t^+)$-predictable. Thus T_r is a $({}^\circ\boldsymbol{F}_t^+)$-predictable stopping time. Hence by theorem 6.4.5 there is a $(\boldsymbol{F}_t)$-predictable stopping time S'_r such that $S'_r = T_r$ a.s.. Now define

$$S_r = \sup_{\substack{t \in \mathbb{Q} \\ 0<t\leq r}} S'_{r'}, \quad A^* = \sup_{r\in\mathbb{Q}, r>0} r\, \mathbf{1}_{[S_r,\infty[} .$$

Then A^* has the following properties:

(i) A^* is $(\boldsymbol{F}_t)$-predictable. In fact, for every r we have

$$[S_r,\infty[\;=\; \bigcap_{\substack{0<t\leq r \\ t\in\mathbb{Q}}} [\,S'_{r'},\infty[\;\in \boldsymbol{P};$$

Being a countable supremum of predictable processes A^* itself is predictable.

(ii) A^* is nondecreasing. This is clear since each $r\,\mathbf{1}_{[S_r,\infty[}$ is nondecreasing.

(iii) A^* is right-continuous. For this suppose $t_n \downarrow t$. Then $A_t^* \leq \lim A_{t_n}^*$ by monotonicity. For the converse inequality let $r \in \mathbb{Q}_+$ be less that $\lim A_{t_n}^*$. Then $r < A_{t_n}^*$ and hence $S_r \leq t_n$ for all n because the family (S_r) is nondecreasing. Therefore, $S_r \leq t$ and $A_t^* \geq r$. Thus $A_t^* \geq \lim A_{t_n}^*$.

(iv) $A^* = A$ a.s.. By construction, $S'_r = T_r$ a.s.. The definition of T_r implies

$$T_r = \sup_{0<t\leq r,\, t\in\mathbb{Q}} T_t$$

and thus $S_r = T_r$ a.s. for all $r \in \mathbb{Q}$. Hence

$$A^* = \sup_{r\in\mathbb{Q}, r>0} r\,\mathbf{1}_{[T_r,\infty[} = A \text{ a.s.} .$$

It remains to modify A^* such that it becomes finite and locally bounded. The stopping times $(S_k)_{k\geq 1}$ converge a.s. to infinity since $A_t < \infty$ for all t. But $A_{S_k}^*$ may be unbounded. For any integer k the predictable stopping time S_k can be approximated by a sequence $(S_{k,l})_{l\geq 1}$ of a.s. smaller stopping times according to theorem 6.4.5. We diagonalize and let $R_m = \inf_{k\geq m} S_{k,l_k}$ with a sufficiently fast sequence (l_k) such that (R_m) is a localizing sequence. Since $R_m < S_m$ a.s. we have $A_{R_m}^* \leq m$ a.s.. Therefore

$$\tilde{A} = \sum_{m=1}^{n} \left(A^* \wedge m\right) \mathbf{1}_{]R_{m-1},R_m]}$$

defines a predictable process which is a.s. equal to A^*. In particular, $\tilde{A}$ is a.s. nondecreasing. It is finite and right-continuous since A^* is right-continuous and $\tilde{A}$ vanishes on $[\sup_m R_m, \infty[$. Finally, $\tilde{A}$ is locally bounded since $\tilde{A} \leq m$ on $[0,R_m]$. Thus $\tilde{A}$ has all properties required in (b). ■

Proof of theorem 6.5.5. Let the admissible measure μ be given. Consider the measure μ^P of lemma 6.5.7. Clearly $\mu^P(\mathbb{R}_+ \times N) = 0$ for every $\mathbb{P}$-nullset. Hence lemma 6.5.8 gives a process A which fulfills the assumptions of lemma 6.5.9. This process is $({}^\circ\boldsymbol{F}_t)$-predictable according to this lemma and it can be modified so as to be $(\boldsymbol{F}_t)$-predictable and locally bounded by lemma 6.5.10. This is the desired process A. ■

6.6 Decomposition of Submartingales

Let $(\boldsymbol{F}_n)_{n\geq 0}$ be a filtration with index set $\mathbb{N}_0$. A process $(X_n)_{n\geq 0}$ may be called "$(\boldsymbol{F}_n)$-predictable" if X_n is $\boldsymbol{F}_{n-1}$-measurable for every n. Every $(\boldsymbol{F}_n)$-submartingale (X_n) can be written in the form

$$X_n = M_n + A_n$$

where $M_0 = X_0$, $A_0 \equiv 0$ and

$$A_n = A_{n-1} + \mathbb{E}\left(X_n \mid \boldsymbol{F}_{n-1}\right) - X_{n-1}$$
$$M_n = X_n - A_n .$$

The process (A_n) is $(\boldsymbol{F}_n)$-predictable and by the very definition of a submartingale it is also nondecreasing. Moreover,

$$\mathbb{E}\left(M_n - M_{n-1} \mid \boldsymbol{F}_{n-1}\right) = \mathbb{E}\left(X_n \mid \boldsymbol{F}_{n-1}\right) - X_{n-1} - \left(A_n - A_{n-1}\right) = 0,$$

i.e. (M_n) is a $(\boldsymbol{F}_n)$-martingale. Thus X is the sum of a martingale and an increasing predictable process. This is the Doob decomposition of a submartingale in discrete time.

The analogous decomposition in continuous time due to P.A. Meyer is more difficult. It uses the desintegration theorem 6.5.5. Note that we do not assume the "usual conditions". The proof contains an argument (proposition 6.6.2) which seems to be crucial for the characterization of semimartingales in chapter 10.

6.6.1 Theorem. (Doob-Meyer decomposition) Let X be a right-continuous adapted process. Then the following are equivalent:

(a) X is a uniformly integrable submartingale.

(b) X can be decomposed as

$$X = M + A$$

where M is a uniformly integrable martingale and A is a right-continuous predictable a.s. nondecreasing locally bounded $\mathcal{L}^1$-bounded process.

The decomposition in (b) is unique up to a.s. equality.

Proof. (b) $\rightarrow$ (a). An integrable a.s. increasing and adapted process A is a submartingale since for $s < t$,

$$\mathbb{E}\left(A_t \mid \boldsymbol{F}_s\right) \geq \mathbb{E}\left(A_s \mid \boldsymbol{F}_s\right) = A_s.$$

Moreover, it is uniformly integrable if it is $\mathcal{L}^1$-bounded. In fact, $A_t \leq A_\infty$ for all t where $A_\infty = \lim_{t\uparrow\infty} A_t$ is integrable by monotone convergence. From these remarks the implication (b) $\rightarrow$ (a) follows immediately.

(a) $\rightarrow$ (b). If we can find an admissible measure μ on $\boldsymbol{P}$ such that

$$\mathbb{E}\left(X_T - X_0\right) = \mu\left(]0,T]\right) \tag{6.6.1}$$

for all finite stopping times T then we are done. In fact, let A be the increasing process of theorem 6.5.5 and $M = X - A$. Then M is uniformly integrable since $\mathbb{E}(A_\infty) < \infty$ and

$$\mathbb{E}\left(M_T\right) = \mathbb{E}\left(X_T - A_T\right) = \mathbb{E}\left(X_T - X_0\right) - \mathbb{E}\left(A_T\right) + \mathbb{E}\left(X_0\right) = \mathbb{E}\left(X_0\right) = \mathbb{E}\left(M_0\right)$$

for every finite stopping time T. Therefore M is a martingale.

If $(X_t)_{t\geq 0}$ is a uniformly integrable submartingale then it is closed by theorem 3.2.4, i.e. there is a random variable X_∞ such that $(X_t)_{0\leq t\leq\infty}$ still is a (uniformly integrable) submartingale.

We define the set function m for stochastic intervals $]S,T]$ with $S \leq T$ by

$$m\left(]S,T]\right) = \mathbb{E}\left(X_T - X_S\right).$$

The submartingale property implies that m is nonnegative. Moreover, the uniform integrability implies that X_{T_k} converges in $\mathcal{L}^1$ to X_T and X_∞ for $T_k \downarrow T$ and $T_k \uparrow \infty$, respectively. Hence $m(]T,T_k]) \downarrow 0$ and $m(]T_k,\infty]) \downarrow 0$ in these cases. Therefore the following proposition implies the existence of the measure μ. ■

6.6.2 Proposition. Let m be a nonnegative finitely additive set function on the system of all stochastic intervals $]S,T]$. Suppose that m has the following continuity properties:

$$(6.6.2) \qquad m\big(]T, T_n]\big) \xrightarrow[n\to\infty]{} 0 \qquad \text{for } T_n \downarrow T,$$

$$(6.6.3) \qquad m\big(]T_k, \infty]\big) \xrightarrow[k\to\infty]{} 0 \text{ for every localizing sequence } (T_k).$$

Then m can be extended to an admissible measure μ on $\boldsymbol{P}$.

Proof. First consider the field $\boldsymbol{G}$ generated by the sets $]S,T]$. We show that every element G of $\boldsymbol{G}$ is a finite union of disjoint stochastic intervals:

$$(6.6.4) \qquad G = \bigcup_{i=1}^{r}]R_i, S_i] .$$

The indicator function 1_G on $\mathbb{R}^+ \times \Omega$ is an elementary process since the set operations in $\boldsymbol{G}$ correspond to function algebra operations on the indicators. The corresponding jumping times defined in lemma 4.4.3 yield a representation as in (6.6.4). The formula

$$m(G) = \sum_{i=1}^{r} m\big(]R_i, S_i]\big)$$

extends m to a content on $\boldsymbol{G}$.

We have to show that m is σ-continuous. Let (G_n) be a sequence in $\boldsymbol{G}$ such that $G_n \downarrow \emptyset$. We claim $m(G_n) \to 0$. Each G_n has a representation

$$(6.6.5) \qquad G_n = \bigcup_{i=1}^{r_n}]R_i^n, S_i^n] .$$

Given $\varepsilon > 0$ choose positive numbers δ_i^n, $1 \le i \le r_n$, $n \in \mathbb{N}$, and consider the sets

$$(6.6.6) \qquad E_k = \bigcap_{n=1}^{k} \Big(\bigcup_{i=1}^{r_n}]\, R_i^n + \delta_i^n, S_i^n] \Big).$$

Using (6.6.2) the numbers δ_i^n can be chosen small enough such that

$$(6.6.7) \qquad m(E_k) \ge m(G_k) - \varepsilon \sum_{n=1}^{k} 2^{-n}.$$

Let $F_k(\omega)$ be the closure of the set $\{t : (t,\omega) \in E_k\}$. Then from (6.6.6) one gets

$$F_k(\omega) \subset \bigcup_{i=1}^{r_n}]\, R_i^n(\omega), S_i^n(\omega)\,]$$

and thus $F_k(\omega) \downarrow \emptyset$ because of (6.6.5) and $G_k \downarrow \emptyset$. Therefore the compact set $F_k(\omega) \cap [0,t]$ is empty for sufficiently large k. This implies that

$$T_k(\omega) = \inf F_k(\omega)$$

defines a localizing sequence. (Note that T_k is a stopping time being the

first jumping time of the elementary process 1_{E_k}.) Moreover, $E_k \subset]T_k,\infty[$ and (6.6.3) implies $m(E_k) \to 0$. Hence $m(G_k) \le 2\varepsilon$ for sufficiently large k because of (6.6.7).
That the resulting measure is admissible follows directly from (6.6.3). ■

6.6.3 Corollary For $M \in \boldsymbol{M}$ there is a predictable a.s. increasing process $\langle M\rangle \in \boldsymbol{A}$ such that $M^2 - \langle M\rangle$ is a local martingale. This process is unique up to a.s. equality.

Proof. For $M \in \boldsymbol{M}$ there is a localizing sequence $(T_k)_{k\ge1}$ such that $T = 0$ and $M_{.\wedge T_k}$ is a $\boldsymbol{L}^2$-bounded martingale. Then $M^2_{.\wedge T_k}$ is a submartingale which is uniformly integrable by proposition 5.6.3. Let A^k be the process which corresponds to this submartingale by theorem 6.6.1. Then

$$M^2_{.\wedge T_k} - A^k$$

is a uniformly integrable martingale. Letting

$$\langle M\rangle_t = \sum_k A^k_{t\wedge T_{k+1}} - A^k_{t\wedge T_k}$$

we get a process with the desired properties. ■

6.6.4 Remark. This process $\langle M\rangle$ is called the **compensator** or **predictable increasing process** of M. By the uniqueness part of theorem 5.6.2(b) we have $\langle M\rangle = [M]$ for $M \in \boldsymbol{M}_c$.

Examples. (a) For Brownian motion $\langle B\rangle_t = [B]_t = t$.
(b) For the Poisson martingale $M = (N_t - t)_{t\ge0}$ we proved in 5.6.7 that $[M]_t = N_t$ and that $(M_t^2 - t)_{t\ge0}$ is a martingale. Thus $\langle M\rangle_t = t$ since $t \longmapsto t$ is predictable.

6.6.5 Proposition. For every predictable $H \ge 0$ and every $M \in \boldsymbol{M}$ one has

$$\mathbb{E}\Big(\int_0^\infty H\, d[M]\Big) = \mathbb{E}\Big(\int_0^\infty H\, d\langle M\rangle\Big). \tag{6.6.8}$$

Proof. Let (T_k) be a localizing sequence such that $(M^2 - [M])_{t\wedge T_k}$ and $(M^2 - \langle M\rangle)_{.\wedge T_k}$ both are uniformly integrable martingales for each k. Then $([M] - \langle M\rangle)_{.\wedge T_k}$ is a martingale and hence (6.6.8) holds for $H = 1_{[0,T]}$, $T \le T_k$. The rest is a straightforward measure theoretic extension argument. ■

CHAPTER 7

SEMIMARTINGALES AND STOCHASTIC DIFFERENTIALS

In order to develop an efficient calculus for stochastic integration we need a common framework for pathwise Stieltjes integration and integration with respect to local $\boldsymbol{L}^2$-martingales. This is most naturally given by the concept of a semimartingale. In the next chapter and in chapter 10 we shall see that semimartingales are very flexible also in other respects.

The first section contains the definitions of a semimartingale and of the corresponding integral. The integral calculus leads to an associated differential calculus. A key role in this calculus is played by an extension of the quadratic variation of an $\boldsymbol{L}^2$-martingale: the mutual variation of semimartingales. This is developed to some extent in section 7.3. The resulting rules are formalized in section 7.4 and extended to higher dimensions in 7.5. There are several alternatives for the description of noncontinuous semimartingales the equivalence of which is nontrivial. Section 7.2 is devoted to these questions. It shows the power of the methods in chapter 6 but it will be used only in chapter 10.

7.1 Integration with Respect to Semimartingales

Semimartingales are composed of local $\boldsymbol{L}^2$-martingales and processes of finite variation:

7.1.1 Definition. A process X is a **semimartingale** if it admits a decomposition $X = M + A$ where M is a (right-continuous) local $\boldsymbol{L}^2$-martingale and A is right-continuous, adapted and of locally finite variation, i.e. $M \in \boldsymbol{M}$ and $A \in \boldsymbol{A}$. The space of all semimartingales will be denoted by $\boldsymbol{S}$. In analogy to $\boldsymbol{M}_c$, the symbols $\boldsymbol{S}_c$ and $\boldsymbol{A}_c$ denote the spaces of all a.s. continuous semimartingales and of all a.s. continuous elements of $\boldsymbol{A}$, respectively.

In general, the decomposition of a semimartingale is not unique. In fact, the Poisson martingale $(N_t - t)_{t \ge 0}$ is both in $\boldsymbol{M}$ and in $\boldsymbol{A}$ (cf. example 5.6.7). Restriction to $\boldsymbol{A}_c$ (or $\boldsymbol{M}_c$), however, enforces uniqueness.

7.1.2 Proposition. There is (up to a.s. equality) at most one decomposition $X = M + A$ of a semimartingale X into $M \in \boldsymbol{M}$ and $A \in \boldsymbol{A}_c$.

Proof. Assume that $X = M' + A'$ is another decomposition where $M' \in \boldsymbol{M}$ and $A' \in \boldsymbol{A}_c$. Then the process $A - A' = M' - M$ is of locally finite variation, a.s. continuous and a local martingale. By theorem 5.3.2 it vanishes a.s. which proves the result. ■

An integral with respect to a semimartingale $X = M + A$ with $M \in \boldsymbol{M}$ and $A \in \boldsymbol{A}$ of course will be given by the formula

$$\int_0^t H\,dX = \int_0^t H\,dM + \int_0^t H\,dA$$

for those H for which the right hand side already has been defined.

In the integral calculus several integrators will arise at the same time - for instance if one uses the linearity of the integral as a function of the integrating process. Hence we shall restrict ourselves most of the time to integrands H for which the integral $\int_0^t H\,dX$ is defined for *all* semimartingales X simultaneously. Given the previous chapters the largest natural class of this kind is the class of those processes H which are locally bounded and predictable or at least a.s. equal to such a process.

7.1.3 Definition. Let $\boldsymbol{B}$ denote the class of those processes which are a.s. equal to a locally bounded predictable process.

If you did not go through sections 6.1 and 6.2
you may in the sequel take instead of $\boldsymbol{B}$ *the class* $\boldsymbol{H}$

of adapted and (separable and a.s.) regular left-continuous processes. For $H \in \boldsymbol{H}$ the integrals $\int_0^{\cdot} H\,dM$ and $\int_0^{\cdot} H\,dA$ have been defined in 5.4.5 and 5.3.6, respectively. Sometimes we can (and shall) give simpler arguments for integrands in $\boldsymbol{H}$ because then the approximation arguments do not depend on the particular integrator X.

7.1.4 Proposition. (a) For every $H \in \boldsymbol{H}$ the version H^- is predictable and locally bounded, i.e. $\boldsymbol{H} \subset \boldsymbol{B}$.
(b) Every semimartingale X is a.s. regular and hence $X^- \in \boldsymbol{H}$.

Proof. (a) For $H \in \boldsymbol{H}$ the process H^- is locally bounded and predictable by proposition 4.1.3 and lemma 6.1.3 respectively. Since $H = H^-$ a.s. this implies $H \in \boldsymbol{B}$.
(b) The a.s. regularity of the $\boldsymbol{M}$-part was mentioned after definition 4.2.1. For the $\boldsymbol{A}$-part note that a function of finite variation is regular. The rest follows from remark 4.4.6. ∎

For a locally bounded predictable process H the integral $\int_0^{\cdot} H\,dA$ was defined in 6.1.5 and the integral $\int_0^{\cdot} H\,dM$ in 6.2.3 (cf. 6.2.11(a)).

7.1.5 Definition. For every H in $\boldsymbol{B}$ and every semimartingale $X = M + A$ with $M \in \boldsymbol{M}$ and $A \in \boldsymbol{A}$ the integral process

$$\int_0^{\cdot} H\,dX = \left(\int_0^t H\,dX\right)_{t \geq 0}$$

is defined by

$$\int_0^t H\,dX = \int_0^t \tilde{H}\,dM + \int_0^t \tilde{H}\,dA. \tag{7.1.1}$$

where $\tilde{H}$ is any predictable locally bounded version of H.
If S and T are finite stopping times and I is the semi-open stochastic interval $]S,T]$ then we write

$$\int_I H\,dX \text{ or } \int_S^T H\,dX \text{ instead of } \int_0^T H\,dX - \int_0^S H\,dX.$$

Remarks. (a) In corollary 7.1.11 below we shall prove that this definition does not depend on the particular decomposition $X = M+A$ of the semimartingale X.
(b) This definition is consistent with the one for $H \in \boldsymbol{H}$. For the $\boldsymbol{A}$-part this is obvious from the pathwise interpretation. For the $\boldsymbol{M}$-part the consistency remark following definition 6.2.3 applies.
(c) As mentioned above this definition can be extended to a larger class of integrands once the semimartingale $X = M + A$ is fixed. Whenever

$$H \in \boldsymbol{L}^{2,loc}(\mu_M) \text{ and } \mathbb{P}\left(\int_0^t |H_s|\,|dA_s| < \infty\right) = 1$$

the formula (7.1.1) makes sense (cf. proposition 6.1.5 and definition 6.2.3).

The following is obvious since $\int_0^{\cdot} H\, dM \in \boldsymbol{M}$ and $\int_0^{\cdot} H\, dA \in \boldsymbol{A}$ (cf. 5.3.6 and 5.4.5 for $H \in \boldsymbol{H}$ and 6.1.5 and 6.2.2 for $H \in \boldsymbol{B}$).

7.1.6 Theorem. For every semimartingale X and every $H \in \boldsymbol{B}$ the integral process $\int_0^{\cdot} H\, dX$ is a semimartingale.

Now let us turn to the continuity of semimartingale integrals. We consider first integrands in $\boldsymbol{H}$. Recall the following continuity of the integral which was proved in proposition 5.5.3 for the martingale part and which holds also for the Stieltjes-integral because of the inequality

$$\sup_{t \le T} \left| \int_0^t H\, dA \right| \le \int_0^T |H|\, |dA| \le \left(\sup_{t \le T} |H_t| \right) \int_0^T |dA|.$$

7.1.7 Proposition. Let $(H^n)_{n \ge 1}$ be a sequence in $\boldsymbol{H}$ and let $X = M + A$ be a semimartingale. If $H^n \longrightarrow H$ l.u.p. then $\int_0^{\cdot} H^n dX \longrightarrow \int_0^{\cdot} H\, dX$ l.u.p..

Comment. This apparently weak continuity (which easily extends to integrands in $\boldsymbol{B}$) is much at the heart of semimartingale integration. There is a surprising result of C. Dellacherie and K. Bichteler to the effect that semimartingales are the *only* processes for which the elementary integral on the space of elementary processes has this continuity property. We shall prove this theorem in chapter 10. It may be considered as an analogue to proposition 5.1.5: If m is a function whose integral on the left-continuous step functions is continuous for the sup-norm, then m automatically is of bounded variation.

For the next sections, the approximation of stochastic integrals by the integrals of elementary processes is of particular importance. The main underlying idea in the proof is that the fluctuation of a semimartingale $X = M + A$ is measured in terms of the increasing process $[M]_t + \int_0^t |dA|$.

7.1.8 Theorem. Let $H \in \boldsymbol{B}$ and a finite number $X^1, \dots, X^N$ of semimartingales be given. Then there is a sequence $(H^n)_{n \ge 1}$ of elementary processes such that for every $i = 1, \dots, N$ the stochastic integrals $\int_0^{\cdot} H^n dX^i$ converge l.u.p. to $\int_0^{\cdot} H\, dX^i$. Moreover, the sequence $(H^n)_{n \ge 1}$ can be chosen such that for every process $K \in \boldsymbol{B}$ the processes $\int_0^{\cdot} H^n K\, dX^i$ converge l.u.p. to $\int_0^{\cdot} HK\, dX^i$.

The strengthened form will be used only in section 7.4.

Proof. In the case $H \in \boldsymbol{H}$ one simply choses a sequence $(H^n)_{n\geq 1}$ which converges l.u.p. to H. This is possible by proposition 4.4.4. Then also for every $K \in \boldsymbol{B}$ the sequence $(H^n K)$ converges l.u.p. to HK and the assertion follows from the preceding proposition. The reader who wants to restrict his attention to integrands in $\boldsymbol{H}$ may pass from here directly to corollary 7.1.9.

For the proof of the theorem in the case $H \in \boldsymbol{B}$ let $X^i = M^i + A^i$ be any fixed decomposition of the semimartingale X^i and let H be predictable and locally bounded. First choosing appropriate localizing sequences and then taking the minimum we find a localizing sequence $(T_k)_{k\geq 1}$ such that for every $k \geq 1$

(i) $\mathbb{E}([M^i]_{T_k}) < \infty$ for every i,

(ii) $\int_0^t |dA^i| \leq k$ for $t < T_k$ and every i,

(iii) $\sup\limits_{0<t\leq T_k} |H_t| \leq k$

(concerning (i) cf. lemma 6.2.6). For $k \in \mathbb{N}$ consider the locally integrable increasing process

$$\tilde{A}_t^k = \sum_{i=1}^N [M^i]_t + \Big(\int_0^t |dA^i|\Big) \wedge k.$$

Then (i) - (iii) imply

$$\mathbb{E}\Big(\int_0^{T_k} H^2 d\tilde{A}^k\Big) < \infty.$$

According to lemma 6.2.8 there is an elementary process H^k such that

$$\mathbb{E}\Big(\int_0^{T_k} (H - H^k)^2 \, d\tilde{A}^k\Big) < \frac{1}{k}. \tag{7.1.2}$$

Since $T_k \leq T_n$ for $n \geq k$ this implies for each i

$$\mathbb{E}\Big(\int_0^{T_k} (H - H^n)^2 \, d[M^i]\Big) \xrightarrow[n\to\infty]{} 0 \text{ for every } k \geq 1 \tag{7.1.3}$$

and then the "Doob" estimate (6.2.3) yields

$$\sup_{0\leq t\leq T_k} \Big|\int_0^t H \, dM^i - \int_0^t H^n \, dM^i\Big| \xrightarrow[n\to\infty]{} 0 \text{ in } \boldsymbol{L}^2 \text{ for every } k \geq 1. \tag{7.1.4}$$

For $t < T_k$ and every measurable integrand G the Cauchy-Schwarz estimate

$$\Big|\int_0^t G \, dA^i\Big| \leq \int_0^{T_k} |G| \, d\tilde{A}^k \leq \Big(\int_0^{T_k} d\tilde{A}^k\Big)^{1/2} \cdot \Big(\int_0^{T_k} G^2 \, d\tilde{A}^k\Big)^{1/2}$$

holds. Hence (7.1.2) implies also

$$\sup_{0\leq t\leq T_k} \Big|\int_0^t H \, dA^i - \int_0^t H^n \, dA^i\Big| \xrightarrow[n\to\infty]{} 0 \text{ in probability for every } k. \tag{7.1.5}$$

The two limit statements (7.1.4) and (7.1.5) yield the assertion for $K \equiv 1$. For general $K \in \boldsymbol{B}$ let $(R_k)_{k\geq 1}$ be a localizing sequence such that K is uniformly bounded on $]0,R_k]$. Then (7.1.2) implies

$$\mathbb{E}\Big(\int_0^{R_k\wedge T_k}(HK - H^nK)^2\, d\tilde{A}^k\Big) \xrightarrow[n\to\infty]{} 0 \text{ for every } k \geq 1 \tag{7.1.2'}$$

and hence the integrals of H^nK converge l.u.p. to the integrals of HK by the same arguments as before. ■

Let us note a number of immediate consequences. The first one shows that the definition of the martingale integral is consistent with the pathwise definition.

7.1.9 Corollary. Let $M \in \boldsymbol{M} \cap \boldsymbol{A}$ and let H be in $\boldsymbol{B}$. Then the stochastic integral process $\left(\int_0^t H\, dM\right)_{t\geq 0}$ can also be computed as a pathwise Stieltjes integral.

Proof. In the theorem take $X^1 = M$ considered as an element of $\boldsymbol{M}$ and $X^2 = M$ considered as an element of $\boldsymbol{A}$. The theorem allows to extend the equality of the corresponding integrals from elementary integrands to general $H \in \boldsymbol{B}$. ■

Here is linearity in the integrating processes:

7.1.10 Corollary. The integral is linear in the integrator: for every $H \in \boldsymbol{B}$, all $X,Y \in \boldsymbol{S}$ and $a,b \in \mathbb{R}$,

$$\int_0^{\cdot} H\, d\big(aX + bY\big) = a\int_0^{\cdot} H\, dX + b\int_0^{\cdot} H\, dY \text{ a.s.}\,. \tag{7.1.6}$$

Proof. For $H \in \boldsymbol{B}$ choose a sequence (H^n) of elementary processes such that the integrals of H^n with respect to the three semimartingales X, Y and $aX + bY$ converge l.u.p. to the corresponding integral processes of H. Since the identity (7.1.6) holds for each H^n it carries over to H. ■

By now we can show the independence of $\int_0^{\cdot} H\, dX$ of the semimartingale decomposition $X = M + A$.

7.1.11 Corollary. Let $X = M + A$ and $X = M' + A'$ be two decompositions of the semimartingale X. Then for every H in $\boldsymbol{B}$ one has a.s.

$$\int_0^{\cdot} H\, dM + \int_0^{\cdot} H\, dA = \int_0^{\cdot} H\, dM' + \int_0^{\cdot} H\, dA'.$$

Proof. By linearity 7.1.10, the assertion is equivalent to the identity

$$\int_0^t H\, d\left(M - M'\right) = \int_0^t H\, d\left(A' - A\right).$$

which follows from 7.1.9 since $M - M' = A' - A$. ■

Further, we show that the integral inherits continuity from the integrand.

7.1.12 Corollary. If $H \in \boldsymbol{B}$ and the semimartingale X is a.s. continuous then $\int_0^{\cdot} H\, dX$ is a.s. continuous.

Proof. If H is elementary then this is part of proposition 5.2.2. Moreover, a.s.-continuity is preserved under l.u.p. convergence according to theorem 4.3.3(c). ■

A final straightforward application of the approximation by elementary integrands is

7.1.13 Corollary. For $X \in \boldsymbol{S}$, $H \in \boldsymbol{B}$, stopping times S,T with $S \le T$ and a $\boldsymbol{F}_S^+$-measurable random variable G one has

$$\int_S^T GH\, dX = G \int_S^T H\, dX \text{ a.s.}$$

Proof. Due to theorem 7.1.8 one may assume that H is an elementary process, i.e. that H is of the form

$$H = \sum_{i=1}^{n} H_i \mathbf{1}_{]T_i, T_{i+1}]}.$$

In addition we may suppose that $S = T_j$ and $T = T_{k+1}$ for some $j \le k$. Then the asserted equality holds since both sides are equal to

$$\sum_{i=j}^{k} GH_i(X_{T_{i+1}} - X_{T_i}). \blacksquare$$

7.2 The Structure of Noncontinuous Semimartingales

This section is concerned with some questions arising from the nonuniqueness of the semimartingale decomposition. The results will be used only in section 10.1.

A closer look at noncontinuous semimartingales makes essential use of predictability. A typical question is the following one. For the definition of stochastic integrals in the previous section it was convenient to require in a decomposition $X = M + A$ of a semimartingale the process M to be a local $\mathcal{L}^2$-martingale rather than a general local martingale. But up to now it is not clear that a martingale is a semimartingale. In theorem 7.2.1 below we prove this using the machinery developed in sections 6.4 to 6.6. We also characterize those semimartingales for which A can be chosen to be predictable.

7.2.1 Theorem. Every local martingale M is a semimartingale in our sense; more precisely, there is a locally bounded local martingale M' such that M - M' is of locally integrable variation.

For the proof we need the following lemma.

7.2.2 Lemma. Suppose that (T_k) is a localizing sequence. Let X be a right-continuous adapted process such that each stopped process $X_{\cdot\wedge T_k}$ can be written as a sum of a locally bounded local martingale and a process in $\boldsymbol{A}$ of locally integrable variation. Then X itself has such a representation.

Proof. For each k one has a representation as indicated:

$$X_{\cdot\wedge T_k} = M^k + A^k .$$

For convenience of notation we set $T_O \equiv 0$. Since $T_k \uparrow \infty$ a.s. we get a.s. a representation of X by

$$\begin{aligned} X_t &= X_O + \sum_{k=1}^{\infty} X_{t\wedge T_k} - X_{t\wedge T_{k-1}} \\ &= X_O + \sum_{k=1}^{\infty} M^k_{t\wedge T_k} - M^k_{t\wedge T_{k-1}} + \sum_{k=1}^{\infty} A^k_{t\wedge T_k} - A^k_{t\wedge T_{k-1}} . \end{aligned}$$

In each series for every (t,ω) there is at most one nonvanishing term. The first series in the last line defines a locally bounded local martingale since

this is true for its partial sums. Similarly, the second series is in $\boldsymbol{A}$ and has locally integrable variation.
If the above representation holds everywhere the lemma is proved. Otherwise the difference of the two sides is right-continuous, adapted and vanishes a.s.. Thus it can be added to the $\boldsymbol{A}$-part of the representation. ■

Proof of theorem 7.2.1. Let M be a local martingale. Lemma 7.2.2 shows that for the proof it suffices to consider a uniformly integrable martingale. Let $T = \inf\{t : |M_t| > c\}$. Using the lemma once more it suffices to prove the theorem for the uniformly integrable martingale $M_{\cdot \wedge T}$. Thus one may assume that M is of the form

$$M_t = \mathbb{E}(M_T \mid \boldsymbol{F}_t)$$

and that M is bounded on $[0,T[$ by the constant c. Because of the decomposition

$$M_t = \mathbb{E}\left(M_T \vee 0 \mid \boldsymbol{F}_t\right) + \mathbb{E}\left(M_T \wedge 0 \mid \boldsymbol{F}_t\right)$$

we may also suppose $M \geq 0$. Then

$$M = M \wedge c + \left(M_T - M_T \wedge c\right) 1_{[T,\infty[}.$$

The second term on the right is a process in $\boldsymbol{A}$ and it is locally integrable since M_T is integrable.

It remains to show that $M \wedge c$ has a representation as indicated. Since $M \geq 0$ it is uniformly bounded. Since $x \longmapsto x \wedge c$ is concave $M \wedge c$ is a supermartingale by lemma 3.0.2. Therefore by theorem 6.6.1 it has a representation

$$M \wedge c = N - A$$

with a martingale N and a right-continuous a.s. increasing predictable locally bounded process A starting at 0. Let (S_k) be a localizing sequence such that $0 \leq A \leq k$ on $[0,S_k]$. Then

$$0 \leq M \wedge c + A = N \leq c + k$$

on $[0,S_k]$. In particular, N is a locally bounded martingale. Thus $M \wedge c$ and hence also M have the desired representations. ■

As a consequence, one can replace in the definition of a semimartingale the local $\boldsymbol{L}^2$-martingales by local martingales or by locally bounded local martingales.

7.2.3 Corollary. The following are equivalent:

(i) $X = M + A$ where $M \in \boldsymbol{M}$ and $A \in \boldsymbol{A}$, i.e. X is a semimartingale.

(ii) $X = M + A$ where M is a local martingale and $A \in \boldsymbol{A}$.

(iii) $X = M + A$ where M is a locally bounded local martingale and $A \in \boldsymbol{A}$.

The previous section was based on the fact that a stochastic integral with respect to a local $\boldsymbol{L}^2$-martingale defines again a local $\boldsymbol{L}^2$-martingale. An analogous result holds for local martingales.

7.2.4 Proposition. Let M be a local martingale. Then for every $H \in \boldsymbol{B}$ the integral process $\int_0^{\cdot} H\, dM$ is a local martingale.

Proof. First assume that $H = \sum_{i=1}^{n} H_i \mathbf{1}_{]T_i, T_{i+1}]}$ is an elementary process. If M is a martingale then according to theorem 5.4.1 the process $\int_0^{\cdot} H\, dM$ is a martingale, too. The explicit representation (5.2.3) of the elementary integral shows that uniform integrability carries over from M to $\int_0^{\cdot} H\, dM$. Thus proposition 5.2.3(b) shows that $\int_0^{\cdot} H\, dM$ is a local martingale if M is one. For a general $H \in \boldsymbol{B}$ one may assume by localization that H is bounded and that M is a closed martingale. If M is locally bounded then M and hence $\int_0^{\cdot} H\, dM$ are local $\boldsymbol{L}^2$-martingales. Therefore using the decomposition of theorem 7.2.1 it suffices to consider the case where M has integrable total variation. By the approximation theorem 7.1.8 there is a sequence (H^n) of elementary processes such that $\int_0^{\cdot} (H^n - H)\, dM \longrightarrow 0$ l.u.p.. Since H is bounded we may assume that the sequence (H^n) also is uniformly bounded. Then there is a constant c such that for each $t^* \geq 0$

$$\sup_{t \leq t^*} \left| \int_0^t H\, dM - \int_0^t H^n\, dM \right| \leq c \int_0^t |dM|.$$

Since the random variable on the right has finite expectation and the left hand side converges to 0 in probability this convergence is even in $\boldsymbol{L}^1$. Therefore the martingale property of the elementary integrals $\int_0^{\cdot} H^n\, dM$ carries over to $\int_0^{\cdot} H\, dM$. ■

Finally, we study the question whether the $\boldsymbol{A}$-part of a semimartingale can be chosen to be predictable. The main point is the following observation.

7.2.5 Proposition. A process A in $\mathbf{A}$ is of locally integrable total variation if and only if there is a predictable process A' in $\mathbf{A}$ starting at 0 such that A - A' is a local martingale. The process A' is unique up to a.s. equality. If A is a.s. increasing then A' is a.s. increasing. If $\mathbb{E}(\int_0^\infty |dA|) < \infty$ then A - A' is a uniformly integrable martingale.

Proof. A process $A \in \mathbf{A}$ is the difference of the two processes

$$A^+ = \tfrac{1}{2}\big(v(A) + A\big)$$

$$A^- = \tfrac{1}{2}\big(v(A) - A\big)$$

where $v(A)_t = \int_0^t |dA|$ if the latter is finite and $v(A)_t = 0$ on the remaining nullset. They are adapted (see 5.3.1f) right-continuous and a.s. nondecreasing (see proposition 5.1.8 and the preceding remarks). Suppose now that $\int_0^\cdot |dA|$ is locally integrable. Since $0 \le A^+ + A^- = \int_0^\cdot |dA|$ a.s. both A^+ and A^- are locally integrable.

So for the existence of A' one may suppose that A is increasing. Then the associated measure μ on $\boldsymbol{P}$ defined by

$$\int X\, d\mu = \mathbb{E}\Big(\int_0^\infty X\, dA\Big)$$

is admissible since A is locally integrable. Hence it satisfies the hypothesis of the predictable desintegration theorem 6.5.4, i.e. there is a locally bounded predictable process $A' \in \mathbf{A}$ such that

$$\int X\, d\mu = \mathbb{E}\Big(\int_0^\infty X\, dA'\Big)$$

for all predictable processes $X \ge 0$.

Let us show that A - A' is a local martingale. Let (T_k) be a localizing sequence such that $\mathbb{E}(A_{T_k}) < \infty$ and $A'_{\cdot \wedge T_k}$ is bounded for each k. Then $(A - A')_{\cdot \wedge T_k}$ is a uniformly integrable process such that

$$\mathbb{E}\Big((A - A')_{T \wedge T_k}\Big) = \mu\big(]0, T \wedge T_k]\big) - \mu\big(]0, T \wedge T_k]\big) = 0$$

for every bounded stopping time T. Hence $(A - A')_{\cdot \wedge T_k}$ is a martingale according to the criterion 3.2.9 and A - A' is a local martingale. Corollary 6.4.12 implies the uniqueness.

For the converse, we first show that every predictable $A \in \boldsymbol{A}$ is of locally integrable variation. The left-continuous versions A^- of A and $\left(\int_0^\cdot |dA|\right)^-$ of $\int_0^\cdot |dA|$ are predictable and thus the identity

$$(7.2.1) \qquad \int_0^\cdot |dA| = \left(\int_0^\cdot |dA|\right)^- + |A - A^-|$$

shows that the process $\int_0^\cdot |dA|$ is predictable. Therefore, by lemma 6.5.10 it is "a.s. locally bounded", i.e. it is a.s. equal to a locally bounded process. In particular, it is locally integrable.

The integrability statement at the end of the assertion is straightforward in the increasing case and the general case is easily reduced to the latter.

Finally, suppose that A and A' are in $\boldsymbol{A}$, that $M = A - A'$ is a local martingale starting at 0 and that A' is of locally integrable variation. It remains to show that A also is of locally integrable variation. For this let T be any stopping time. Then

$$\int_0^T |dA| \le \int_0^T |dA'| + |M_T|$$

and the right hand side has finite expectation for all members of a suitable localizing sequence. ■

The announced result reads:

7.2.6 Theorem. The following are equivalent:

(i) X = M + A where M is a local martingale and A is in $\boldsymbol{A}$ and of locally integrable variation.

(ii) X = M + A where M is a locally bounded local martingale and A is in $\boldsymbol{A}$ and of locally integrable variation.

(iii) X = M + A where M is a local martingale and $A \in \boldsymbol{A}$ is predictable.

The representation in (iii) is unique.

Proof. The equivalence of (i) and (ii) follows from theorem 7.2.1 and the last paragraph in the previous proof. The equivalence of (i) and (iii) from the preceding proposition. The uniqueness of the decomposition in (iii) follows again from corollary 6.4.12. ■

Note that in the decomposition (iii) the martingale part M generally is not in $\boldsymbol{M}$ i.e. it is not locally in $\boldsymbol{L}^2$. In fact, if M is any right-continuous local martingale not contained in $\boldsymbol{M}$ - for instance the process constructed in example 4.2.6(a) - then $X = M + 0$ is its unique decomposition with a predictable $\boldsymbol{A}$-part.

The uniqueness statement shows e.g. that the Poisson process N admits only one predictable process $A \in \boldsymbol{A}$ such that $N - A$ becomes a local martingale, namely $A_t \equiv t$.

7.2.7 Definition. A process X is called a **special semimartingale** if it satisfies one of the conditions of the preceding theorem.

The special semimartingales appear in the later text only in the proof of theorem 10.1.8 in conjunction with

7.2.8 Corollary. A locally bounded semimartingale is a special semimartingale.

Proof. Let X be a locally bounded semimartingale. It has a representation $X = M + A$ with a locally bounded local martingale M and $A \in \boldsymbol{A}$. Then the process $A = X - M$ is also locally bounded. It remains to prove that its variation process is locally bounded and a fortiori locally integrable. Since A^- and $\left(\int_0^\cdot |dA|\right)^-$ are locally bounded this follows from the representation (7.2.1) in the proof of proposition 7.2.5. ■

7.3 Mutual Variation of Semimartingales

The increasing process of a local $\boldsymbol{L}^2$-martingale M gauges the quadratic variation of M. This quadratic variation can be extended to a bilinear operation on the space of semimartingales: the mutual variation. Note that for $X \in \boldsymbol{S}$ the process X^- is in $\boldsymbol{H}$ and hence it is an admissible integrand (cf. proposition 7.1.4(b).)

7.3.1 Definition. The **mutual variation process** of two semimartingales X and Y is defined by

$$(7.3.1) \qquad [X,Y]_t = X_tY_t - X_0Y_0 - \int_0^t X^-\,dY - \int_0^t Y^-\,dX.$$

For $X = Y$ this reads

$$[X,X]_t = X_t^2 - X_0^2 - 2\int_0^t X^- \, dX$$

and hence in generalization of definition 5.6.1 the process $[X,X]$ is called the **quadratic variation** process of X and is denoted by $[X]$. In extension to the convention following definition 5.6.1 we also write $\langle X\rangle$ instead of $[X]$ if $X \in \mathcal{S}_c$. Then by corollary 7.1.12 the process $\langle X\rangle$ is in $\mathcal{S}_c$. Note that the symbol $\langle M\rangle$ was introduced in corollary 6.6.3 and remark 6.6.4 for local $\mathcal{L}^2$-martingales. Since $\langle M\rangle = [M]$ for $M \in \mathcal{M}_c$ the notation is consistent.

The expression "mutual variation" will find an explanation in theorem 7.3.7 below. The formula (7.3.1) in this definition can be considered as the *formula of partial integration* for stochastic integrals. The term $[X,Y]$ indicates how much the classical partial integration formula has to be modified.

Since the stochastic integral is linear both in the integrand and in the integrator (cf. corollary 7.1.10) we have

7.3.2 Lemma. Mutual variation is **symmetric** and **bilinear**, i.e. for all semimartingales X,Y,Z and all a,b in $\mathbb{R}$ one has a.s.

$$\begin{aligned} [X,Y] &= [Y,X], \\ [aX+bY,Z] &= a[X,Z] + b[Y,Z], \\ [X,aY+bZ] &= a[X,Y] + b[X,Z]. \end{aligned}$$

The formulation of many results in the sequel is simplified considerably if we introduce stochastic differentials. This is nothing mystic but mirrors the fact that very often the increments of a process are more important than the process itself. In the next section we shall study stochastic differentials more systematically.

7.3.3 Definition. The **differential** dX of a stochastic process X is the random function on the set of semi-open stochastic intervals $]S,T]$ between finite stopping times $S \le T$ given by

$$dX(]S,T])(\omega) = X_T(\omega) - X_S(\omega). \tag{7.3.2}$$

First let us extend the "Riemann approximation" of proposition 5.5.6 to semimartingales in a form which uses these differentials. Let I be a finite stochastic interval $]S,T]$ and let ρ denote a partition of I by a finite sequence of stopping times

$$S = T_0 \le T_1 \le \ldots \le T_k = T.$$

Write I_i instead of $]T_i,T_{i+1}]$. For an integrand H and a semimartingale X consider the sum

$$\sum_i H^+_{T_i}\, dX(I_i)$$

which can be rewritten as $\int_S^T H^\rho \, dX$ where $H^\rho = \sum_i H^+_{T_i} 1_{]T_i,T_{i+1}]}$. A similar relation holds if H^+ is replaced by H. In analogy to proposition 5.5.6 the Riemann type approximation can now be formulated as follows. However here we use finite partitions of $]S,T]$ rather than infinite partitions of $\mathbb{R}_+ \times \Omega$.

7.3.4 Theorem. If $H \in \boldsymbol{H}$ and $X \in \boldsymbol{S}$ then for every stochastic interval $I =]S,T]$,

$$(7.3.3)\quad \int_I H\, dX = \lim_{|\rho|\to 0} \sum_{i=1}^{k_n} H^+_{T_i}\, dX(I_i) = \lim_{|\rho|\to 0} \sum_{i=1}^{k_n} H_{T_i}\, dX(I_i) \text{ in probability.}$$

More precisely, this means: If $(\rho_n)_{n\ge1}$ is a sequence of partitions of I given by finite sequences $(T_i^n)_{1\le i\le k_n}$, $n \ge 1$, such that $|\rho_n| \longrightarrow 0$ a.s. then one has

$$(7.3.4)\quad \int_I H\, dX = \lim_{n\to\infty} \sum_{i=1}^{k_n} H^+_{T_i^n}\, dX(I_i) = \lim_{n\to\infty} \sum_{i=1}^{k_n} H_{T_i^n}\, dX(I_i) \text{ in probability.}$$

Proof. Let the sequence of partitions $(\rho_n)_{n\ge1}$ such that $|\rho_n| \longrightarrow 0$ be given. In a straightforward manner these partitions may be extended to infinite partitions of $\mathbb{R}_+$ for which still $|\rho_n| \longrightarrow 0$. Then

$$(7.3.5)\qquad \int_0^t H\, dX = \lim_{n\to\infty} \sum_{i=1}^{\infty} H^+_{t\wedge T_i^n}\left(X_{t\wedge T_{i+1}^n} - X_{t\wedge T_i^n}\right) \text{ l.u.p..}$$

We may also replace H^+ by H in the sum. In fact, both approximations follow from proposition 5.1.9 if $X \in \boldsymbol{A}$ and from proposition 5.5.6 if $X \in \boldsymbol{M}$. Thus (7.3.5) holds also for $X \in \boldsymbol{S}$.

The definition of l.u.p. convergence (cf. 4.3.1(a)) implies that if the right-hand side is evaluated at the finite stopping time T it converges to $\int_0^T H\, dX$ in probability. This yields the assertion for the stochastic interval $]0,T]$. Doing the same for S and taking differences one arrives at (7.3.4) in the general form. Note that the right-hand side in (7.3.4) is an additive function of I as can be seen from (7.3.2). ∎

7.3.5 Remark. The formulation of the theorem does not fully reflect the fact that the convergence in (7.3.5) is locally *uniform* in probability. In terms of stochastic differentials this means the following stronger version of (7.3.4): Let T be a finite stopping time. Let further (ρ_n) be a sequence of partitions of the interval $]0,T]$ such that $|\rho_n| \longrightarrow 0$. Then for every $\varepsilon > 0$ one has

$$(7.3.6)\quad \mathbb{P}\Big(\sup_{s\le t\le T}\Big|\int_s^t H\,dX - \sum_i H^+_{T_i^n}\,dX\big(]s,t]\cap]T_i^n,T_{i+1}^n]\big)\Big| > \varepsilon\Big) \underset{n\to\infty}{\longrightarrow} 0\,.$$

The limit statements in the sequel have corresponding strengthened versions but we omit them.

The following elementary "Leibniz rule" is the key to most properties of mutual variation.

7.3.6 Lemma. Let X and Y be stochastic processes and $I =]S,T]$. Then

$$dX(I)\cdot dY(I) = d(XY)(I) - X_S\cdot dY(I) - Y_S\cdot dX(I).$$

Proof. The formula is verified by the elementary calculation

$$\begin{aligned}(X_T - X_S)(Y_T - Y_S) &= X_TY_T - X_SY_S + 2X_SY_S - X_SY_T - Y_SX_T\\ &= (XY)_T - (XY)_S - X_S(Y_T - Y_S) - Y_S(X_T - X_S).\ \blacksquare\end{aligned}$$

The announced representation of mutual variation follows now easily.

7.3.7 Theorem. Let X and Y be semimartingales. Then

$$d[X,Y](I) = \lim_{|\rho|\to 0}\sum_i dX(I_i)dY(I_i)\quad \text{in probability}.$$

Proof. The process X^- is in $\boldsymbol{H}$ and it satisfies $(X^-)^+ = X$ a.s. by remark 4.4.6 since X is right-continuous. The same applies to Y^-. The elementary rule 7.3.6 and theorem 7.3.4 imply

$$\begin{aligned}\lim_{|\rho|\to 0}\sum_i dX(I_i)dY(I_i) &= \lim_{|\rho|\to 0}\Big(\sum_i d(XY)(I_i) - \sum_i X_{T_i}dY(I_i) - \sum_i Y_{T_i}dX(I_i)\Big)\\ &= dXY(I) - \int_I X^-\,dY - \int_I Y^-\,dX = d[X,Y](I)\end{aligned}$$

in probability which proves the assertion. ■

It was already mentioned in section 5.6 that the increasing process of a local martingale is a measure of the quadratic variation. In fact, we have more generally:

7.3.8 Corollary. If X is a semimartingale then

$$d[X](I) = \lim_{|\rho|\to 0} \sum_i dX(I_i)^2 \quad \text{in probability.}$$

In particular, $d[X](I) \geq 0$ a.s. for every stochastic interval and the process $[X]$ has a.s. nondecreasing paths and $[X] \in \boldsymbol{A}$.

Let us collect some useful properties of quadratic variation.

7.3.9 Proposition. If X and Y are semimartingales then

$$d[X,Y] = \frac{1}{4}\left(d[X+Y] - d[X-Y]\right).$$

In particular, the process $[X,Y]$ is in $\boldsymbol{A}$. If X and Y are in $\boldsymbol{S}_c$ then $[X,Y]$ is in $\boldsymbol{A}_c$.

In proposition 7.3.18 below we shall see that only one factor needs to be continuous for $[X,Y] \in \boldsymbol{A}_c$.

Proof. From bilinearity (lemma 7.3.2) follows

$$[X+Y] - [X-Y] = [X] + 2[X,Y] + [Y] - \left([X] - 2[X,Y] + [Y]\right) = 4[X,Y].$$

which implies the formula for $d[X,Y]$. Also it shows that $[X,Y]$ is a difference of nondecreasing processes and therefore locally of finite variation.
The continuity assertion follows from corollary 7.1.12. ■

Remark. Since $XY = [X,Y] + \int_0 X^- dY + \int_0 Y^- dX$ and since stochastic integrals define semimartingales it follows that the (pointwise) product of two semimartingales is a semimartingale. A much more general statement is part of Itô's formula.

If in theorem 7.3.7 one applies the Cauchy-Schwarz inequality on the right-hand side the limit gives together with 7.3.8:

7.3.10 Proposition. If X and Y are semimartingales then

$$\left|d[X,Y]\right| \le \left(d[X]\right)^{1/2}\left(d[Y]\right)^{1/2}$$

and

$$\left|[X,Y]\right| \le [X]^{1/2}\,[Y]^{1/2}.$$

Mutual variation is consistent with stopping.

7.3.11 Proposition. Suppose that $X, Y \in \boldsymbol{S}$ and T is a stopping time. Then

$$[X, Y_{\cdot\wedge T}] = [X,Y]_{\cdot\wedge T} \text{ a.s..}$$

Proof. Since both processes are right-continuous we have only to verify the a.s. equality $[X,Y_{\cdot\wedge T}]_t = [X,Y]_{t\wedge T}$ for a fixed time t. We apply theorem 7.3.7 to

(a) the interval $]0,t]$ and the processes X and $Y_{\cdot\wedge T}$,

(b) the interval $]0,t\wedge T]$ and the processes X and Y.

We may assume that the partitions of $]0,t]$ are extensions of the partitions of $]0,t\wedge T]$. Then

$$dY_{\cdot\wedge T}(I_i) = \begin{cases} 0 & \text{if } I_i \cap\,]0,t\wedge T] = \emptyset \\ dY(I_i) & \text{if } I_i \subset\,]0,t\wedge T] \end{cases}.$$

Therefore

$$\begin{aligned}[X,Y_{\cdot\wedge T}]_t &= \lim_{|\rho|\to 0} \sum_i dX(I_i)dY_{\cdot\wedge T}(I_i) \\ &= \lim_{|\rho|\to 0} \sum_i \left\{dX(I_i)dY(I_i) : I_i \subset\,]0,t\wedge T]\right\} = [X,Y]_{t\wedge T}. \blacksquare\end{aligned}$$

Now let us turn to some special properties of mutual variation under continuity assumptions. The main point is the following simple fact.

7.3.12 Proposition. If $A \in \boldsymbol{A}$ and $X \in \boldsymbol{S}_c$ then for every finite stochastic interval $I =]S,T]$ one has almost surely

$$\lim_{|\rho|\to 0} \sum_i dA(I_i)\, dX(I_i) = 0.$$

Proof. We have a.s.

$$\lim_{|\rho|\to 0} \sum_i |dA(I_i)\, dX(I_i)| \le \lim_{|\rho|\to 0} \left(\sup_i |dX(I_i)|\right) \sum_i |dA(I_i)| = 0$$

because of uniform continuity of almost all paths of X and finite variation of the paths of A on I. ■

7.3.13 Corollary. If $A \in \boldsymbol{A}$ and $X \in \boldsymbol{S}$ and one of the two processes is a.s. continuous then $[A,X] \equiv 0$.

Proof. If X is a.s. continuous this follows from the proposition and theorem 7.3.7. In particular $[A] = [A,A] = 0$ for $A \in \boldsymbol{A}_c$. According to proposition 7.3.10 we thus have for $A \in \boldsymbol{A}_c$ and any $X \in \boldsymbol{S}$

$$|[A,X]| \leq [X]^{1/2}[A]^{1/2} = [X]^{1/2} \cdot 0 = 0. \blacksquare$$

Comment. Recalling the definition of $[X,Y]$ and its interpretation as the formula of partial integration thereafter, this fact may be rephrased as follows: For the integrals $\int A^- \, dX$ and $\int X \, dA$ (or $\int A \, dX$ and $\int X^- \, dA$) you can use ordinary partial integration if either X or A is a.s. continuous. The definition of the stochastic integral $\int A^- \, dX$ therefore can be reduced via partial integration to the pathwise Stieltjes integral $\int_0^t X \, dA$. This is the "stochastic Stieltjes integral" mentioned in section 1.2.(B).

Next we see that quadratic variation of a continuous semimartingale depends only on the martingale part. More generally

7.3.14 Corollary. Suppose that $X = Y + A$ is a semimartingale where $A \in \boldsymbol{A}_c$. Then $[X] = [Y]$.

Proof. We have by linearity and the preceding corollary

$$[X] = [X,X] = [Y + A, Y + A] = [Y,Y] + 2[Y,A] + [A,A] = [Y,Y] = [Y]. \blacksquare$$

In extension of theorem 5.6.2 we have

7.3.15 Theorem. (a) Suppose that $M, N \in \boldsymbol{M}$. Then $MN - [M,N] \in \boldsymbol{M}$ and if $M, N \in \boldsymbol{M}_c$ then $MN - [M,N] \in \boldsymbol{M}_c$.
(b) If either M or N is in $\boldsymbol{M}_c$ then $[M,N]$ is (up to a.s. equality) the only process in $\boldsymbol{A}_c$ such that $MN - [M,N] \in \boldsymbol{M}$.

Proof. (a) From the definition,

$$\left(MN - [M,N]\right)_t - M_0N_0 = \int_0^t N^- \, dM + \int_0^t M^- \, dN$$

which is in $\mathcal{M}$ or $\mathcal{M}_c$, being the sum of two stochastic integrals whose integrators are in $\mathcal{M}$ or $\mathcal{M}_c$, respectively.
(b) If M and N are in $\mathcal{M}_c$ then [M,N] is a.s. continuous by the above formula. In the general case the a.s. continuity of [M,N] will be proved in 7.3.18(b). Now the uniqueness statement follows like the uniqueness of the increasing process (cf. 5.6.2) from theorem 5.3.2. ■

Similar to correlation, mutual variation of independent local martingales should vanish. We show this under a continuity assumption.

7.3.16 Theorem. Let M and N be independent local martingales (w.r.t. the same filtration) and starting at zero. Then MN is a local martingale. If either M or N is a.s. continuous then [M,N] = 0.

Proof. Assume first that M and N are martingales. Then M and N are not only martingales w.r.t. to the original filtration but also w.r.t. their natural filtrations

$$\mathcal{G}_t^M = \sigma(M_s : s \le t) \text{ and } \mathcal{G}_t^N = \sigma(N_s : s \le t),$$

respectively. We claim that MN is a martingale w.r.t. $\mathcal{G}_t = \sigma(\mathcal{G}_t^M \cup \mathcal{G}_t^N)$. For $s < t$ choose a bounded $\mathcal{G}_s^M$-measurable function f and a bounded $\mathcal{G}_s^N$-measurable function g. Then by independence

$$\begin{aligned} \mathbb{E}\left(fgM_tN_t\right) &= \mathbb{E}\left((fM_t)(gN_t)\right) = \mathbb{E}\left(fM_t\right)\mathbb{E}\left(gN_t\right) \\ &= \mathbb{E}\left(fM_s\right)\mathbb{E}\left(gN_s\right) = \mathbb{E}\left(fgM_sN_s\right) \end{aligned}$$

which proves the martingale property of MN.
Let now M and N be local martingales. We use the natural localizing stopping times

$$S_n = \inf\left\{t \ge 0 : |M_t| > n\right\} \text{ and } T_n = \inf\left\{t \ge 0 : |N_t| > n\right\}.$$

These are stopping times w.r.t. the natural filtrations of M and N. Therefore the two martingales $M_{\cdot \wedge S_m}$ and $N_{\cdot \wedge T_n}$ are independent. Hence the first part of this proof shows that $MN_{\cdot \wedge S_n \wedge T_n}$ is a martingale and MN is a local martingale. The process $[M,N]$ is also adapted to the filtration $(\boldsymbol{G}_t)_{t\geq 0}$, hence it vanishes by the uniqueness assertion in the last theorem. ■

7.3.17 Example. In particular, suppose that $B^1, \ldots, B^d$ are d independent $(\boldsymbol{F}_t)$-Brownian motions for the same filtration $(\boldsymbol{F}_t)$. Then $\langle B^i,B^i\rangle_t = t$ and $\langle B^i,B^j\rangle_t = 0$ if $i \neq j$. (This special case can be proved directly.)

The last result of this section is needed only in the optional section 8.4. It contains a modification of the uniqueness statement in theorem 7.3.15 which carries over to the noncontinuous case. Define for an a.s. regular process X the associated **jump process** ΔX by

$$\left(\Delta X\right)_t = X_t^+ - X_t^-.$$

Thus an a.s. regular process X is a.s. continuous if and only if $\Delta X = 0$ a.s..

7.3.18 Proposition. (a) For every $H \in \boldsymbol{H}$ and $X \in \boldsymbol{S}$ one has

$$\Delta\left(\int_0^{\cdot} H\,dX\right) = H\,\Delta X \quad \text{a.s.} \tag{7.3.7}$$

(b) Let X and Y be in $\boldsymbol{S}$. Then

$$\Delta[X,Y] = \Delta X\cdot\Delta Y \text{ a.s..} \tag{7.3.8}$$

If either X or Y is a.s. continuous then so is $[X,Y]$.

(c) Let M and N be in $\boldsymbol{M}$. If $A \in \boldsymbol{A}$ is such that $\Delta A = \Delta M\cdot\Delta N$ a.s. and $MN - A \in \boldsymbol{M}$ then $A = [M,N]$ a.s. .

Proof. (a) The formula (7.3.7) is obvious for the elementary integral. For general $H \in \boldsymbol{H}$ we may find a sequence of elementary processes H^n, $n \geq 1$, such that

$$\sup_{s\leq t}\left|H_s^n - H_s\right| \xrightarrow[n\to\infty]{} 0 \text{ a.s.}, \tag{7.3.9}$$

$$\sup_{s\leq t}\left|\int_0^s H^n\,dX - \int_0^s H\,dX\right| \xrightarrow[n\to\infty]{} 0 \text{ a.s.} \tag{7.3.10}$$

The convergence (7.3.9) implies $H^n\,\Delta X \longrightarrow H\,\Delta X$ a.s. and (7.3.10) implies

$$\Delta\left(\int_0^{\cdot} H^n\,dX\right) \longrightarrow \Delta\left(\int_0^{\cdot} H\,dX\right) \text{ a.s..}$$

This completes the proof of (a).

(b) Using (a) and the definition of mutual variation one gets

$$\Delta[X,Y] = \Delta(XY) - X^- \Delta Y - Y^- \Delta X .$$

The right-hand side of this equation equals $\Delta X \, \Delta Y$ by a computation as in the proof of the elementary product rule:

$$X^+ Y^+ - X^- Y^- - X^- \left(Y^+ - Y^-\right) - Y^- \left(X^+ - X^-\right) = \left(X^+ - X^-\right)\left(Y^+ - Y^-\right).$$

(c) Now let M and N be in $\boldsymbol{M}$ and assume that the process $A \in \boldsymbol{A}$ satisfies $\Delta A = \Delta M \cdot \Delta N$ a.s.. If $MN - A$ is in $\boldsymbol{M}$ then $[M,N] - A$ is also in $\boldsymbol{M}$ and since the Δ-operation is linear one has by (7.3.7)

$$\Delta\left([M,N] - A\right) = \Delta M \cdot \Delta N - \Delta M \cdot \Delta N = 0 \text{ a.s.},$$

i.e. $[M,N] - A$ is a.s. continuous. Hence theorem 5.3.2 proves $[M,N] = A$. ■

7.4 The Space of Stochastic Differentials

In this section, we collect a couple of useful rules concerning the stochastic differentials introduced in definition 7.3.3. The symbolical calculus is remarkably simple. The present exposition follows essentially the lines of K. Itô (1975). We associate to every space of integrating processes the corresponding space of differentials and introduce algebraic operations on these spaces.

7.4.1 Definition. The space of all stochastic differentials dX where $X \in \boldsymbol{S}$ is denoted by $d\boldsymbol{S}$. The subspaces $d\boldsymbol{S}_c$, $d\boldsymbol{M}$, $d\boldsymbol{M}_c$, $d\boldsymbol{A}$ and $d\boldsymbol{A}_c$ are defined in the same way. We introduce three operations on $d\boldsymbol{S}$. The symbol I denotes a finite stochastic interval of the form $]S,T]$.

(A) **Addition.** If dX and dY are in $d\boldsymbol{S}$ set

$$(dX + dY)(I) = dX(I) + dY(I).$$

(M) **Multiplication by "scalars".** If $H \in \boldsymbol{B}$ and $X \in \boldsymbol{S}$ set

$$H \bullet dX(I) = \int_I H \, dX.$$

(P) **("Algebra") product.** If $dX, dY \in d\boldsymbol{S}$ set

$$dX \bullet dY(I) = d(XY)(I) - X^- \bullet dY(I) - Y^- \bullet dX(I).$$

By slight abuse of notation we have chosen the same symbol "$\bullet$" both for the operation (M) on $\boldsymbol{B} \times d\boldsymbol{S}$ and (P) on $d\boldsymbol{S} \times d\boldsymbol{S}$. Hopefully, no ambiguity will arise in the sequel. The algebra product (P) is closely connected to mutual variation since

$$dX \bullet dY(]S,T]) = [X,Y]_T - [X,Y]_S.$$

And here are the algebraic rules at a single glance.

7.4.2 Theorem. The operations (A), (M) and (P) obey the following rules:
(a) The spaces $d\boldsymbol{S}$, $d\boldsymbol{M}$ and $d\boldsymbol{A}$ are groups w.r.t. the addition (A).

(b) If $H \in \boldsymbol{B}$ and $X \in \boldsymbol{S}, \boldsymbol{S}_c, \boldsymbol{M}, \boldsymbol{M}_c, \boldsymbol{A}$ or $\boldsymbol{A}_c$ then
$H \bullet dX \in d\boldsymbol{S}, d\boldsymbol{S}_c, d\boldsymbol{M}, d\boldsymbol{M}_c, d\boldsymbol{A}$ or $d\boldsymbol{A}_c$, respectively.

(c) $d\boldsymbol{S} \bullet d\boldsymbol{S} \subset d\boldsymbol{A}$, $d\boldsymbol{S}_c \bullet d\boldsymbol{S} \subset d\boldsymbol{A}_c$, $d\boldsymbol{A}_c \bullet d\boldsymbol{S} = \{0\}$, $d\boldsymbol{S}_c \bullet d\boldsymbol{S} \bullet d\boldsymbol{S} = \{0\}$.

(d) If $G, H \in \boldsymbol{B}$ and $X, Y, Z \in \boldsymbol{S}$ then

(1)	$H \bullet (dX+dY)$	=	$H \bullet dX + H \bullet dY$,
(2)	$(G+H) \bullet dX$	=	$G \bullet dX + H \bullet dX$,
(3)	$G \bullet (H \bullet dX)$	=	$(G \cdot H) \bullet dX$,
(4)	$dX \bullet dY$	=	$dY \bullet dX$,
(5)	$H \bullet (dX \bullet dY)$	=	$(H \bullet dX) \bullet dY$,
(6)	$dX \bullet (dY + dZ)$	=	$dX \bullet dY + dX \bullet dZ$
(7)	$(dX \bullet dY) \bullet dZ$	=	$dX \bullet (dY \bullet dZ)$.

Remark. Except for part (c) the theorem may be rephrased as follows:
The space $d\boldsymbol{S}$ together with the operations (A), (M) and (P) is a commutative algebra over the ring $\boldsymbol{B}$; the spaces $d\boldsymbol{S}_c$, $d\boldsymbol{A}$ and $d\boldsymbol{A}_c$ are subalgebras of $d\boldsymbol{S}$ and $d\boldsymbol{M}$ and $d\boldsymbol{M}_c$ are submoduli of $d\boldsymbol{S}$.

Proof. In view of the preparations in the preceding sections the proof essentially boils down to the verification of (3) and (5) in (d). Nevertheless, let us be thorough and check off the items on the list
The statement (a) is in fact clear; (b) is a reformulation of theorem 7.1.6, the remark preceding that theorem and corollary 7.1.12. In part (c) the inclusion $d\boldsymbol{S} \bullet d\boldsymbol{S} \subset d\boldsymbol{A}$ follows from proposition 7.3.9. If one factor is a.s. continuous then so is the product in view of proposition 7.3.18(b). If one factor is in $\boldsymbol{A}_c$ then the product vanishes (cf. 7.3.13).

We turn now to the rules collected under (d). Rule (1) rephrases the linearity in the integrating processes proved in corollary 7.1.10. Similarly, rule (2) reformulates the additivity of both the martingale integral and the Lebesgue-Stieltjes integral (cf. sections 5.1, 5.4 and 6.2).

(3) The identity $G\bullet(H\bullet dX) = (GH)\bullet dX$ is a substantial assertion. In more explicite terms it states : For $H \in \boldsymbol{B}$ and $X \in \boldsymbol{S}$ define the semimartingale Y by

$$Y_t = \int_0^t H\, dX.$$

Then for every $G \in \boldsymbol{B}$ the following identity holds

$$\int_I G\, dY = \int_I (GH)\, dX. \tag{7.4.1}$$

If $X \in \boldsymbol{A}$ then the formula $Y_t = \int_0^t H\, dX$ expresses that for almost all $\omega \in \Omega$ the function $H_.(\omega)$ is the Radon-Nikodym derivative of the signed measure on $\mathbb{R}_+$ associated with $Y_.(\omega)$. Hence in this case the assertion can be viewed as part of the Radon-Nikodym theorem.

Assume now $X \in \boldsymbol{S}$. Suppose first that G and H are elementary processes. Then there is a partition $0 = T_0 \leq \ldots \leq T_n = \infty$ of $[0,\infty]$ such that for each ω the paths $G(\cdot,\omega)$ and $H(\cdot,\omega)$ are constant on each interval $]T_i, T_{i+1}]$. Let the interval $I =]S,T]$ be contained in $]T_i,T_{i+1}] \times \Omega$ for some i. Then

$$\int_I G\, dY = G_S(Y_T - Y_S) = G_S \int_I H\, dX = G_S\big(H_S(X_T - X_S)\big)$$
$$= (GH)_S(X_T - X_S) = \int_I GH\, dX.$$

Hence the desired identity (7.4.1) holds in this case. Because the elementary integral is additive the identity extends to general intervals I.

Suppose now that G is an elementary process and $H \in \boldsymbol{B}$. By theorem 7.1.8 there is a sequence (H^n) of elementary processes such that the integrals with respect to dX of H^n and of GH^n converge l.u.p. to the integrals of H and GH, respectively. If we denote $\int_0^t H^n\, dX$ by Y_t^n then the corresponding processes Y^n converge l.u.p. to Y and since G is elementary it follows that

$$\int_0^\cdot G\, dY^n \longrightarrow \int_0^\cdot G\, dY \text{ l.u.p..}$$

On the other hand, according to the first step

$$\int_0^\cdot G\, dY^n = \int_0^\cdot GH^n\, dX \longrightarrow \int_0^\cdot GH\, dX \text{ l.u.p..}$$

This proves (7.4.1) for elementary G and $H \in \boldsymbol{B}$. Fix now $H \in \boldsymbol{B}$. The extension of (7.4.1) from elementary G to $G \in \boldsymbol{B}$ is then proved similarly with the help of theorem 7.1.8.

(4) The commutativity follows from the definition.

(5) The rule $H\bullet(dX\bullet dY) = (H\bullet dX)\bullet dY$ can be verified as follows. Set $Z_t = \int_0^t H\, dX$. If H is an elementary process then as in the proof of (3) it is sufficient to consider intervals $I =]S,T]$ on which the paths of H are constant. Then

$$dZ\bullet dY(I) = \lim_{|\rho|\to 0} \sum_i dZ(I_i)dY(I_i)$$

$$= \lim_{|\rho|\to 0} H_S \sum_i dX(I_i)dY(I_i) = \int_I H\, d[X,Y].$$

For general $H \in \boldsymbol{B}$ choose a sequence (H^n) of elementary processes such that the integrals with respect to X and $[X,Y]$ of H^nK converge l.u.p. to the corresponding integral processes of HK for every $K \in \boldsymbol{B}$.

The process Y^- is in $\boldsymbol{H}$ and hence in $\boldsymbol{B}$. Thus according to (3)

$$Y^-\bullet dZ = (Y^-H)\bullet dX \quad \text{and} \quad Y^-\bullet dZ^n = (Y^-H^n)\bullet dX$$

where Z^n denotes the process $\int_0^\cdot H^n\, dX$. Then

$$\lim_n \int_0^\cdot Y^-\, dZ^n = \lim_n \int_0^\cdot H^n Y^-\, dX = \int_0^\cdot HY^-\, dX = \int_0^\cdot Y^-\, dZ \text{ l.u.p.}$$

and since $(Z^n)^-$ converges l.u.p. to Z^- also

$$\int_0^\cdot (Z^n)^-\, dY \longrightarrow \int_0^t Z^-\, dY \text{ l.u.p..}$$

Therefore, by partial integration

$$\int_0^\cdot H\, d[X,Y] = \lim \int_0^\cdot H^n\, d[X,Y] = \lim\, [Z^n,Y]_t$$

$$= \lim \Big(d(Z^nY) - Y^-\bullet dZ^n - (Z^n)^-\bullet dY\Big)(]0,t])$$

$$= \Big(d(ZY) - Y^-\bullet dZ - Z^-\bullet dY\Big)(]0,t])$$

$$= dY\bullet dZ\,(]0,t]).$$

This completes the proof of (5).

(6) The distributive law is again just a reformulation of an already known fact, namely, that mutual variation is bilinear.

(7) The associative law is trivial in the continuous case since the product of three factors in $d\boldsymbol{S}_c$ vanishes. In the general case the reader easily verifies that both $(dX\bullet dY)\bullet dZ$ and $dX\bullet(dY\bullet dZ)$ are equal to

$$d(XYZ)-X^-\bullet d(YZ)-Y^-\bullet d(XZ)-Z^-\bullet d(XY)+(YZ)^-\bullet dX+(XZ)^-\bullet dY+(XY)^-\bullet dZ.$$

This completes the proof of theorem 7.4.2. ■

As an exercise we can now easily compute the quadratic variation of integral processes (cf. propositon 5.6.9).

7.4.3 Corollary. The quadratic variation of a stochastic integral is given by the formula

$$[\int_0^{\cdot} H\, dX]_t = \int_0^t H^2\, d[X].$$

and more generally

$$[\int_0^{\cdot} G\, dX, \int_0^{\cdot} H\, dY]_t = \int_0^t GH\, d[X,Y].$$

Proof. This is verified by the symbolic calculation

$$\left(G\bullet dX\right)\bullet\left(H\bullet dY\right) = G\bullet\left(dX\bullet(H\bullet dY)\right) = G\bullet\left((H\bullet dY)\bullet dX\right)$$

$$= G\bullet\left(H\bullet\left(dY\bullet dX\right)\right) = G\bullet\left(H\bullet(dX\bullet dY)\right) = GH\bullet(dX\bullet dY). \blacksquare$$

The following "Riemann approximations" will be convenient for instance in proving the Itô-formula:

Proposition 7.4.4. If $H \in \boldsymbol{H}$ and $X,Y \in \boldsymbol{S}$, then

(a) $$H\bullet dX(I) = \lim_{|\rho|\to 0} \sum_i H^+_{T_i}\, dX(I_i) \quad \text{in probability,}$$

(b) $$H\bullet dX\bullet dY(I) = \lim_{|\rho|\to 0} \sum_i H^+_{T_i}\, dX(I_i)\, dY(I_i) \quad \text{in probability.}$$

The limits in (a) and (b) are taken along any sequence of partitions (ρ_n) of I with $|\rho_n| \longrightarrow 0$.

Proof. The approximation (a) has been proved in 7.3.4. For the proof of (b) let H be in $\boldsymbol{H}$. Then as in the proof of the approximation of mutual variation in theorem 7.3.7 the elementary rule 7.3.6 and theorem 7.3.4 imply together with the rule (3)

$$\lim_{|\rho|\to 0} \sum_i H^+_{T_i}\, dX(I_i) dY(I_i)$$

$$= \lim_{|\rho|\to 0} \left(\sum_i H^+_{T_i} d(XY)(I_i) - \sum_i (H^+X)_{T_i} dY(I_i) - \sum_i (H^+Y)_{T_i} dX(I_i)\right)$$

$$= \int_I H\, d(XY) - \int_I HX^-\, dY - \int_I HY^-\, dX$$

$$= H\bullet\left(d(XY) - X^-\bullet dY - Y^-\bullet dX\right)(I) = H\bullet(dX\bullet dY)\,(I)$$

in probability which proves the assertion. ■

7.5 Multidimensional Semimartingale Operations

Up to now, we have restricted ourselves to stochastic processes with values in the real line. Needless to say, a multidimensional theory is needed for most applications. We have already met multidimensional Brownian motion in this text. The extension of previous concepts and results to finite dimensional vector spaces is more or less performed by straightforward application of rules from linear algebra - no additional stochastic arguments are needed.

We introduce here some notation which does not refer explicitely to the coordinates. The formulas of stochastic analysis have a tendency to look unnecessarily heavy when written out explicitely. Browse through this section quickly and return to it only if necessary.

It is convenient to consider matrix-valued processes, vectors being considered as $1 \times n$ or $n \times 1$-matrices.

7.5.1 Definition. Let p, q and r be positive integers. Denote by $\boldsymbol{H}^{p,q}$, $\boldsymbol{B}^{p,q}$, $\boldsymbol{M}^{p,q}$, $\boldsymbol{A}^{p,q}$ and $\boldsymbol{S}^{p,q}$ the spaces of $p \times q$-matrixvalued processes whose components are in $\boldsymbol{H}$, $\boldsymbol{B}$, $\boldsymbol{M}$, $\boldsymbol{A}$ or $\boldsymbol{S}$, respectively. Furthermore, let $d\boldsymbol{H}^{p,q}, \ldots, d\boldsymbol{S}^{p,q}$ be the corresponding spaces of differentials. The multidimensional operations are defined component by component:

(A) If dX and dY are in $d\boldsymbol{S}^{p,q}$ define $dX+dY \in d\boldsymbol{S}^{p,q}$ by

$$(dX + dY)_{ij} = dX_{ij} + dY_{ij}\,, \; 1 \le i \le p,\; 1 \le j \le q.$$

We now have to distinguish between left and right multiplication:

(LM) If $H \in \boldsymbol{B}^{p,q}$ and $dX \in d\boldsymbol{S}^{q,r}$ define $H \bullet dX \in d\boldsymbol{S}^{p,r}$ by

$$\left(H \bullet dX\right)_{ij} = \sum_{k=1}^{q} H_{ik} \bullet dX_{kj}, \quad 1 \le i \le p, \quad 1 \le j \le r.$$

(RM) If $dY \in d\boldsymbol{S}^{p,q}$ and $K \in \boldsymbol{B}^{q,r}$ define $(dY) \bullet K \in d\boldsymbol{S}^{p,r}$ by

$$\left((dY) \bullet K\right)_{ij} = \sum_{k=1}^{q} K_{kj} \bullet dY_{ik}, \; 1 \le i \le p, \quad 1 \le j \le r.$$

For every stochastic interval I define

$$\int_I H\bullet dX = H\bullet dX(I) \text{ and } \int_I (dY)\bullet K = (dY)\bullet K(I).$$

(P) If $X \in \boldsymbol{S}^{p,q}$ and $Y \in \boldsymbol{S}^{q,r}$ then

$$dX\bullet dY = d(XY) - X\bullet dY - (dX)\bullet Y .$$

The operation (P) may also be written component by component; e.g. we compute

$$\left(dX\bullet dY\right)_{ij} = \sum_{k=1}^{q} dX_{ik}\bullet dY_{kj} .$$

7.5.2 Proposition. All rules of theorem 7.4.2 about stochastic differentials remain valid as long as they make sense in terms of the appropriate choice of the dimension. For example for $H \in \boldsymbol{B}^{m,p}$, $X \in \boldsymbol{S}^{p,q}$ and $Y \in \boldsymbol{S}^{q,r}$ one has

$$\left(H\bullet dX\right)\bullet dY = H\bullet\left(dX\bullet dY\right) \in d\boldsymbol{A}^{m,r} .$$

The rest of this section is somewhat pedantic and also not essential for the later text. It just indicates that the preceding concepts of vectorvalued stochastic calculus do not depend in any way on the particular choice of linear coordinates. Of course, this is what you would expect (and need e.g. if you start to consider semimartingales with values in any kind of differentiable manifolds like spheres, tori etc., etc.). But some things in stochastic analysis are different from what we are used to, so we better are careful.

Let us first note a formula responsible for change of linear coordinates.

7.5.3 Corollary. Suppose that $H \in \boldsymbol{B}^{p,q}$ and $X \in \boldsymbol{S}^{p,r}$. Let A, B and C be invertible matrices of rank p, q and r, respectively. Then

$$H\bullet dX = A\left(A^{-1}HB\right) \bullet d\left(B^{-1}XC\right)C^{-1} .$$

Proof. Straightforward use of the associativity laws provided by theorem 7.4.2 and the preceding proposition. ■

This allows to pass from Euclidean p-space to arbitrary real vector spaces of finite dimension. Let us agree that $\mathbb{R}^p$ consists of column vectors.

7.5.4 Definition. Let **E** be a finite-dimensional vector space, of dimension p. Consider a map $X : \mathbb{R}_+ \times \Omega \longrightarrow \mathbf{E}$. Fix a basis $\{e_1,\ldots,e_p\}$ of **E**. Let the **coordinate map** $X' : \mathbb{R}_+ \times \Omega \longrightarrow \mathbb{R}^p$ with respect to this base be the map whose components X_i are given by

$$X(t,\omega) = \sum_{i=1}^{p} X_i(t,\omega)e_i.$$

Then X is said to be an **E-valued semimartingale**, **continuous local martingale**, ... if the components X_i are semimartingales (continuous local martingales,..), i.e. if $X' \in \boldsymbol{S}^{p,1}$ $(\boldsymbol{M}_c^{p,1},\ldots)$.

Now observe

7.5.5 Proposition. The definition of an **E**-valued semimartingale (local martingale etc.) is independent of the choice of the base in **E**.

Proof. The coordinate map X' with respect to one base is related to the coordinate map X'^* with respect to another base by the formula

$$X'^* = SX'$$

where $\boldsymbol{S}$ is an invertible matrix. Since the spaces of semimartingales, local martingales etc. are linear spaces the result follows. ■

The operations of stochastic calculus are defined by means of the coordinate processes. Given two linear spaces **E** and **F** with bases $\{e_1,\ldots,e_p\}$ and $\{f_1,\ldots,f_q\}$, respectively, one gets a canonical base $\{A_{ij}\}$ for the space **L(E,F)** of linear maps from **E** to **F** by $A_{ij}(e_i) = f_j$.

7.5.6 Definition. Let **E**, **F** and **G** be finite dimensional vector spaces. Suppose that H has values in **L(F,G)** and X has values in **F** or **L(E,F)**. Consider fixed bases in **E**, **F** and **G** and the corresponding bases of **L(E,F)** and **L(F,G)** and set

$$\left(H \bullet dX\right)' = H' \bullet dX'.$$

Then $H \bullet dX$ is the **G**-valued (or **L**(**E**,**G**)-valued) differential with coordinate differential $(H \bullet dX)'$. The operations $(dY) \bullet H$, $dX \bullet dY$, $X \bullet dY$ and $(dX) \bullet Y$ are defined in the same way.

Again, we observe

7.5.7 Proposition. The definitions above are independent of the choice of the bases.

Proof. For $H \bullet dX$ the assertion follows from corollary 7.5.3 and the rest from similar formulas. ■

CHAPTER 8

ITO-CALCULUS

Itô's formula is the most important tool in the theory of stochastic integration. It plays the counterpart of the fundamental theorem of classical calculus or rather its application to change of variables. It differs notably from the latter due to presence of quadratic variation.

Let us consider the differential $df(X)$ for a continuously differentiable function f and a continuous semimartingale X. If $X = A \in \boldsymbol{A}$ has finite variation then proposition 5.1.6 states that

$$f(A_t) = f(A_0) + \int_0^t f'(A_s)\, dA_s,$$

from which we conclude

$$df(A) = f'(A)\bullet dA .$$

This is the customary rule for change of variables. If $X = M$ is a continuous local martingale we are not yet in the position to carry out the corresponding calculations. Let us content ourselves with the special case $f(x) = x^2$. We know that

$$\langle M\rangle_t = M_t^2 - M_0^2 - 2\int_0^t M\, dM.$$

This equation may be rewritten as

$$df(M) = f'(M)\bullet dM + ½f''(M)\bullet d\langle M\rangle$$

(we have used the fact that $f'(x) = 2x$ and $f''(x) = 2$ in a somehow strange way). We will see immediately that this formula stays correct in general. One can in fact get rid of the additional term $½f''(M)\bullet d\langle M\rangle$ if one modifies the stochastic integral - the "scalar multiplication" - appropriately . This leads to the Stratonovich-Fisk integral which will be introduced in section 8.5. But one has to pay a heavy price for this formal advantage: the integral will no longer be a (local) martingale.

8.1 Itô's Formula: The Continuous Case

The Itô formula (8.1.1) will be proved for smooth real functions of multidimensional a.s. continuous semimartingales. We write $\partial_i f$ for the partial derivative w.r.t. the i-th coordinate.

8.1.1 Theorem. Suppose that $X^1, \dots, X^d$ are a.s. continuous semimartingales. Assume further that U is an open subset of $\mathbb{R}^d$ and that $(X_t^1, \dots, X_t^d) \in U$ a.s. for every $t \geq 0$. If $f : U \longrightarrow \mathbb{R}$ is a twice continuously differentiable function then $f(X^1, \dots, X^d)$ is a semimartingale with differential

$$df(X^1, \dots, X^d)$$

$$(8.1.1) \qquad = \sum_{j=1}^{d} \partial_j f(X^1, \dots, X^d) \bullet dX^j + ½ \sum_{j,k=1}^{d} \partial_j \partial_k f(X^1, \dots, X^d) \bullet \langle X^j, X^k \rangle .$$

In the proof we use the following version of the multidimensional Taylor formula. Since it plays such an important role for Itô's formula we reproduce the proof.

8.1.2 Lemma. Suppose that U is an open subset of $\mathbb{R}^d$ and that f is a real twice continuously differentiable function on U. For every pair (x,v) such that $x \in U$ and $x + v \in U$ let r(x,v) be the error of the second order Taylor approximation for $f(x + v) - f(x)$, i.e.

$$(8.1.2) \qquad r(x,v) = f(x+v) - f(x) - \sum_{j=1}^{d} \partial_j f(x) v_j - ½ \sum_{j,k=1}^{d} \partial_j \partial_k f(x) v_j v_k .$$

Then r(x,v) can be estimated by

$$r(x,v) \leq \eta(x,v) \, |v|^2$$

where $|\cdot|$ denotes Euclidean norm and η is a function such that

$$\sup \left\{ |\eta(x,v)| : x \in C, |v| \leq \varepsilon \right\} \xrightarrow[\varepsilon \to 0]{} 0$$

for every compact subset C of U .

Proof of the lemma. Let $C \subset U$ be compact. Then there is some $\varepsilon > 0$ such that the compact set

$$C_\varepsilon = \left\{ y \in \mathbb{R}^d : \operatorname{dist}(y,C) \leq \varepsilon \right\}$$

is contained in U. Assume now $x \in C$ and $|v| < \varepsilon$. Consider the function

$$h : [0,1] \longrightarrow \mathbb{R},\ t \longmapsto f(x + tv).$$

Then

$$h'(t) = \sum_{j=1}^{d} \partial_j f(x + tv)v_j,\ h''(t) = \sum_{j,k=1}^{d} \partial_j \partial_k f(x + tv)v_j v_k .$$

The one-dimensional Taylor formula yields for some $s = s(x,v) \in [0,1]$ that

$$\begin{aligned} f(x + v) - f(x) &= h(1) - h(0) = h'(0) + \tfrac{1}{2}h''(s) \\ &= \sum_{j=1}^{d} \partial_j f(x)v_j + \tfrac{1}{2} \sum_{j,k=1}^{d} \partial_j \partial_k f(x + sv)v_j v_k \end{aligned}$$

and hence

$$r(x,v) = \tfrac{1}{2} \sum_{j,k=1}^{d} \big(\partial_j \partial_k f(x + sv) - \partial_j \partial_k f(x)\big)v_j v_k .$$

Since

$$\sum_{j,k=1}^{d} |v_j v_k| = \Big(\sum_{j=1}^{d} |v_j|\Big)^2 \le d\,|v|^2$$

and the derivatives $\partial_j \partial_k f$ are uniformly continuous on the set C_ε we get the desired estimate with

$$\eta(x,v) = \frac{d}{2} \max_{\substack{j,k \\ s \in [0,1]}} \big|\partial_j \partial_k f(x + sv) - \partial_j \partial_k f(x)\big| .$$

This completes the proof. ■

Proof of theorem 8.1.1. Set $X = (X^1, \ldots, X^d)$. Choose $I =]S,T]$ and consider a partition of I into intervals $I_i =]T_i, T_{i+1}]$, where $S = T_0 < \cdots < T_{q+1} = T$. We use the above Taylor formula interpreting the data there as follows

$$x = \big(X^1_{T_i}, \ldots, X^d_{T_i}\big) = X_{T_i},$$

$$v = \big(dX^1(I_i), \ldots, dX^d(I_i)\big) = dX(I_i)$$

$$x + v = \big(X^1_{T_{i+1}}, \ldots, X^d_{T_{i+1}}\big) = X_{T_{i+1}} .$$

Interchanging the order of summation we get

$$df(X)(I) = \sum_{i=0}^{q} \left(f(X_{T_{i+1}}) - f(X_{T_i}) \right)$$

$$(8.1.3) \qquad = \sum_{j=1}^{d} \Big[\sum_{i=0}^{q} \partial_j f(X_{T_i}) \cdot dX^j(I_i) \Big]$$

$$+ \tfrac{1}{2} \sum_{j,k=1}^{d} \Big[\sum_{i=0}^{q} \partial_j \partial_k f(X_{T_i}) dX^j(I_i) \cdot dX^k(I_i) \Big] + \sum_{i=0}^{q} r\left(X_{T_i}, dX(I_i) \right).$$

If ρ runs through a sequence (ρ_n) of partitions of I with $|\rho_n| \longrightarrow 0$ we apply the approximation formulas of proposition 7.4.4 to the a.s. left-continuous processes $H = \partial_j f(X^-)$ and $H' = \partial_j \partial_k f(X^-)$. Then $H^+ = \partial_j f(X)$ and $H'^+ = \partial_j \partial_k f(X)$ and hence we have in probability

$$(8.1.4) \qquad \lim_{|\rho| \to 0} \sum_{i=0}^{q} r\left(X_{T_i}, dX(I_i) \right)$$

$$= df(X)(I) - \sum_{j=1}^{d} \partial_j f(X^-) \bullet dX^j(I) - \tfrac{1}{2} \sum_{j,k=1}^{d} \partial_j \partial_k f(X^-) \bullet d\langle X^j, X^k \rangle (I).$$

We are done if the error sum on the left converges to zero. Now

$$(8.1.5) \qquad \sum_{i=1}^{q} \left| r(X_{T_i}, dX(I_i)) \right| \le \sup_i \eta\left(X_{T_i}, dX(I_i) \right) \Big[\sum_{j=1}^{d} \sum_{i=1}^{q} dX^j(I_i)^2 \Big].$$

For almost every $\omega \in \Omega$ the continuous function $X(\cdot,\omega)$ maps the compact interval I onto a compact subset of U and is uniformly continuous on I. For these ω the convergence to zero of the first factor follows now from lemma 8.1.2. The square brackets have a limit in probability by corollary 7.3.8. Thus we see that the right-hand side in (8.1.5) and hence also the error sum converge to zero which concludes the proof. ■

Let us see how Itô's formula looks like for Brownian motion.

8.1.3 Example. Suppose that $(B^1, \dots, B^d)$ are d independent $(\boldsymbol{F}_t)$-Brownian motions and that $f : \mathbb{R}^d \longrightarrow \mathbb{R}$ has continuous partial derivatives of second order. Let Δ denote the Laplacian $\sum_{i=1}^{d} \partial_i^2$ and ∇ the gradient. In example 7.3.17

it was shown that $d\langle B^i, B^j\rangle = \delta_{ij}dt$. The Itô formula reads

$$df(B) = \sum_{i=1}^{d} \partial_i f(B) \bullet dB^i + \tfrac{1}{2} \sum_{i=1}^{d} \partial_i^2 f(B) \cdot dt = \nabla f(B) \bullet dB^{\top} + \tfrac{1}{2}\Delta f(B)dt .$$

It is not purely by chance that the formula reminds us of the heat equation. We shall pursue this aspect in sections 9.3 and 12.1.

8.2 An Application: The Exponential of a Continuous Semimartingale

The differential equation $dz/dx = z$ under the initial condition $z(0) = 1$ is solved by $z = e^x$. Similarly, given a semimartingale X, we may consider the stochastic differential equation

$$dZ = Z \bullet dX, \; Z_0 \equiv 1, \tag{8.2.1}$$

i.e. we may look for a stochastic process Z satisfying (8.2.1). By the way, each process Z for which (8.2.1) makes sense is a semimartingale.

8.2.1 Theorem. Suppose that X is an a.s. continuous semimartingale starting at 0. The process Z given by

$$Z_t = \exp\left(X_t - \tfrac{1}{2}\langle X\rangle_t\right)$$

solves (8.2.1) and it is the only solution up to a.s. equality. If X is in $\boldsymbol{M}_c$ then so is Z.

The solution Z is sometimes called the (**Itô**) **exponential of** X. Fig.8.2.1 below shows a couple of paths of the exponential of Brownian motion.

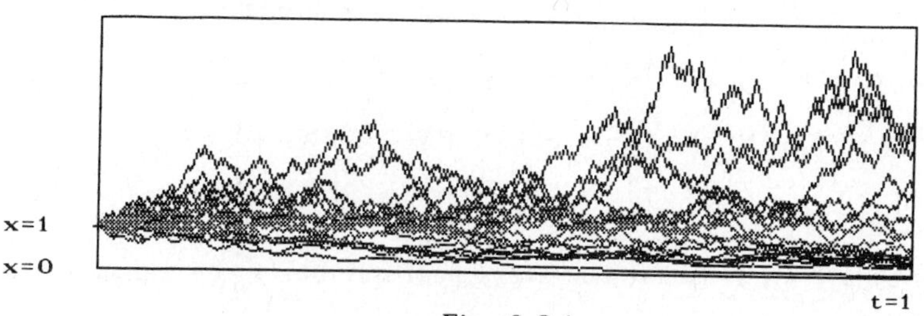

Fig. 8.2.1

Proof of theorem 8.2.1. Rewrite (8.2.1) as the integral equation

$$Z_t = 1 + \int_0^t Z_s \, dX_s .$$

Being a stochastic integral w.r.t. X, every solution Z of (8.2.1) is a local martingale if X is a local martingale (cf. section 5.4.). Consider now $Z_t = \exp(X_t - \frac{1}{2}\langle X\rangle_t)$. Itô's formula applied to $f(x,t) = \exp(x - t)$ gives

$$dZ = \exp(X - \tfrac{1}{2}\langle X\rangle)\bullet(dX - d\tfrac{1}{2}\langle X\rangle) + \tfrac{1}{2}\exp(X - \tfrac{1}{2}\langle X\rangle)\bullet d\langle X\rangle = Z\bullet dX$$

since the remaining term vanishes by 7.3.13. Clearly $Z_0 \equiv 1$. Hence Z solves (8.2.1).

To prove uniqueness, let Z' be another solution of (8.2.1). Since Z never vanishes we may choose $f(z) = 1/z$ in Itô's formula to get

$$d\frac{1}{Z} = -\frac{1}{Z^2}\bullet dZ + \frac{1}{Z^3}\bullet d\langle Z\rangle = -\frac{1}{Z^2}\bullet\left(Z\bullet dX\right) + \frac{1}{Z^3}\bullet\left(Z^2\bullet d\langle X\rangle\right)$$
$$= \frac{1}{Z}\bullet\left(-dX + d\langle X\rangle\right).$$

You see it is really simple to work with Itô's formula and the rules from sections 7.3 and 7.4. The very definition of the inner product gives now

$$d\frac{Z'}{Z} = Z'\bullet d\frac{1}{Z} + \frac{1}{Z}\bullet dZ' + dZ'\bullet d\frac{1}{Z}$$
$$= \frac{Z'}{Z}\bullet\left(-dX + d\langle X\rangle + dX\bullet(-dX + d\langle X\rangle)\right) = 0.$$

Therefore, Z'/Z is constantly equal to 1 and consequently Z' = Z completing the proof. ■

The Itô exponential of X is frequently denoted by $\mathcal{E}(X)$. The name exponential suggests the rule $\mathcal{E}(X)\mathcal{E}(Y) = \mathcal{E}(X + Y)$ which plainly does not hold. In fact,

$$\mathcal{E}(X)\mathcal{E}(Y) = \mathcal{E}(X + Y + \langle X,Y\rangle)$$

since using

$$\langle X + Y + \langle X,Y\rangle\rangle = \langle X\rangle + \langle Y\rangle + 2\langle X,Y\rangle$$

one gets

$$\exp\left(X - \tfrac{1}{2}\langle X\rangle\right)\exp\left(Y - \tfrac{1}{2}\langle Y\rangle\right) = \exp\left(X + Y - \tfrac{1}{2}\left(\langle X\rangle + \langle Y\rangle\right)\right)$$
$$= \exp\left(X + Y + \langle X,Y\rangle - \tfrac{1}{2}\langle X + Y + \langle X,Y\rangle\rangle\right).$$

Replacing X by aX we see that for every real number a the unique solution of the differential equation

$$dZ = aZ\bullet dX, \quad Z_0 \equiv 1,$$

is $\exp(aX - \frac{1}{2} a^2\langle X\rangle)$. Thus each element of $\boldsymbol{M}_c$ generates a whole family of exponential local martingales.Though looking at first glance rather abstract, exponential martingales prove to be a useful technical tool. For instance, they help to compute the Fourier transforms $a \longmapsto \mathbb{E}(e^{iaX_t})$ and thus to identify the distribution of semimartingales. For this we want to allow the number a to be complex. So let us have a look at the complex case. Needless to say, complex-valued processes inherit the names martingale, ... from their real and imaginary parts.

8.2.2 Proposition. For every $M \in \boldsymbol{M}_c$ and each complex number a the complex exponential $\exp(aM - \frac{1}{2}a^2\langle M\rangle)$ is a local martingale.

A criterion which implies that an exponential is even a martingale will be provided in proposition 9.2.4.

The systematic (and straightforward) "complexification" of the last proof needs the complex Itô-formula which we discuss in a larger context in section 8.6. Here we use another argument which gives a first indication of the close connection between stochastic calculus and parabolic differential operators. In the next formulas, it will be more conspicuous to use the notation f_t and f_x for partial derivatives instead of ∂_i. Differentials are to be read component by component.

8.2.3 Lemma. If $X \in \boldsymbol{S}_c$ and $f(t,x)$ is a twice continuously differentiable complex-valued function on $[0,\infty[\times \mathbb{R}$ then

$$df(\langle X\rangle,X) = \left[f_t(\langle X\rangle,X) + \tfrac{1}{2} f_{xx}(\langle X\rangle,X)\right] \bullet d\langle X\rangle + f_x(\langle X\rangle,X) \bullet dX.$$

Proof. Recall from theorem 7.4.2 that $d\boldsymbol{S}_c \bullet d\boldsymbol{A}_c = \{0\}$. The Itô formula - applied first to the real and imaginary parts separately - then yields

$$\begin{aligned} df(\langle X\rangle,X) &= f_t(\langle X\rangle,X) \bullet d\langle X\rangle + f_x(\langle X\rangle,X) \bullet dX + \tfrac{1}{2} f_{xx}(\langle X\rangle,X) \bullet d\langle X\rangle \\ &= \left[f_t(\langle X\rangle,X) + \tfrac{1}{2} f_{xx}(\langle X\rangle,X)\right] \bullet d\langle X\rangle + f_x(\langle X\rangle,X) \bullet dX. \blacksquare \end{aligned}$$

8.2.4 Example. For a $(\boldsymbol{F}_t)$-Brownian motion, the formula of the lemma reads

$$df(t,B) = \left(f_t(t,B) + \tfrac{1}{2} f_{xx}(t,B)\right)dt + f_x(t,B) \bullet dB.$$

Let us now add the missing proof.

Proof of proposition 8.2.2. The function $f(t,x) = \exp(ax - \frac{1}{2} a^2 t)$ satisfies the partial differential equation $f_t + \frac{1}{2} f_{xx} = 0$. By the last lemma we have

$$f(\langle M\rangle, M) = \int_0^{\cdot} f_x(\langle M\rangle, M)\, dM$$

which is a complex a.s. continuous local martingale. ∎

8.3 Another Aspect of Itô's Formula: Local Time

It is useful and instructive to try Itô's formula for functions which are not differentiable. Of particular interest are the functions $f(x) = |x - \alpha|$; they lead to the construction of the local time process at α. This process measures how much time the path of a semimartingale spends in small intervals around the level α.

The limit procedures below produce nonregular integrands and hence we have to refer to the extended integral from section 6.2.

For notational simplicity we consider only the case $\alpha = 0$. So let $f(x) = |x|$. In order to bring the Itô formula into the game we approximate f around zero by twice continuously differentiable functions f_n: Let $\varepsilon_n > 0$, $n \in \mathbb{N}$, tend to 0 as $n \to \infty$ and let f_n be a $\mathscr{C}^2$-function on $\mathbb{R}$ with the following properties:

(i) $f_n(x) = |x|$ for $|x| \geq \varepsilon_n$,

(ii) $f_n(x) = f_n(-x)$ for every x,

(iii) $\ddot{f}_n(x) \geq 0$ for every x,

i.e. each f_n extends the function $x \longmapsto |x|$ inside $[-\varepsilon_n, \varepsilon_n]$ in a convex symmetric way. Then

(a) $f_n(x) \longrightarrow |x|$,

(b) $\dot{f}_n(x) \longrightarrow \operatorname{sgn}(x)$,

(c) $|\dot{f}_n| \leq 1$,

(d) $\ddot{f}_n(x) = 0$ for $|x| \geq \varepsilon_n$,

(e) $\int \ddot{f}_n(x)\, dx = 2$

(we define $\mathrm{sgn}(x)$ as -1, 0 or 1 if x is smaller, equal or greater than zero, respectively). The statements (a) and (b) follow from (i) for $x \neq 0$; (b) holds also for $x = 0$ since by symmetry $f_n'(0) = 0$. By (iii) the derivative f_n' is non-decreasing and since it coincides with $\mathrm{sgn}(x)$ outside $[-\varepsilon_n, \varepsilon_n]$ this implies (c). Therefore, the functions f_n are uniformly continuous and hence the convergence in (a) holds also for $x = 0$. Finally, (d) is clear from (i) and (e) follows from $f_n'(\varepsilon_n) - f_n'(-\varepsilon_n) = \mathrm{sgn}(\varepsilon_n) - \mathrm{sgn}(-\varepsilon_n) = 2$.

We write down Itô's formula for a semimartingale $X \in \boldsymbol{S}_c$ and these functions f_n:

(8.3.1) $$f_n(X_t) - f_n(X_0) = \int_0^t f_n'(X_s)\, dX_s + \tfrac{1}{2}\int_0^t f_n''(X_s)\, d\langle X_s\rangle.$$

The main point is the limit behaviour of the integrals on the right-hand side. The convergence of the first integral is covered by the following **dominated convergence theorem** for stochastic integrals.

8.3.1 Proposition. Let $(H^n)_{n\geq 1}$ and H be in $\boldsymbol{B}$ such that $H^n(t,\omega) \longrightarrow H(t,\omega)$ for all t and ω. Suppose that $|H^n| \leq G$ for a locally bounded predictable process G. Then for every $X \in \boldsymbol{S}$

(8.3.2) $$\int_0^\cdot H^n\, dX \longrightarrow \int_0^\cdot H\, dX \text{ l.u.p.}\,.$$

Proof. Choose a semimartingale decomposition $X = M + A$. Let T be a stopping time such that G is uniformly bounded on $]0,T]$ and

$$\mathbb{E}\left(\langle M\rangle_T + \int_0^T |dA|\right) < \infty.$$

The integrability condition implies $\mu_M([0,T]) < \infty$ (cf. lemma 6.2.6). Therefore by ordinary dominated convergence $\mathbf{1}_{[0,T]}(H^n - H) \longrightarrow 0$ in $\boldsymbol{L}^2(\mu_M)$ and the continuity of the stochastic integral (cf. 6.2.4) yields

$$\mathbb{E}\left(\sup_{t\leq T}\left|\int_0^T H^n\, dM - \int_0^T H\, dM\right|^2\right) \longrightarrow 0.$$

Similarly, for the Stieltjes part by dominated convergence

$$\mathbb{E}\left(\sup_{t\leq T}\left|\int_0^T H^n\, dA - \int_0^T H\, dA\right|\right) \longrightarrow 0.$$

These two limit statements hold also in probability for all members of a localizing sequence $(T_k)_{k\geq 1}$ and this gives the convergence in (8.3.2). ■

Returning to the discussion of (8.3.1) the proposition shows that

$$\int_0^t f_n'(X_s)\, dX_s \longrightarrow \int_0^t \operatorname{sgn}(X_s)\, dX_s \text{ l.u.p.}\,.$$

The left-hand side in (8.3.1) converges to $|X_t| - |X_0|$ and therefore the second integral in (8.3.1) converges l.u.p. to a process L given by

$$L_t = \lim_{n\to\infty} \tfrac{1}{2}\int_0^t f_n''(X)\, d\langle X\rangle = |X_t| - |X_0| - \int_0^t \operatorname{sgn}(X)\, dX. \tag{8.3.3}$$

Because of the second representation, the process L does not depend on the choice of the $\mathscr{C}^2$-functions f_n satisfying (i) through (iii). The process L is called **local time at zero**. Observe that as we increase t the value of L_t will not change as long as X_t stays strictly positive or strictly negative. Hence a change in L_t will indicate that X enters zero. This can be made more explicit:

8.3.2 Theorem. For every a.s. continuous semimartingale X let L be the local time process given by (8.3.3). Then

$$L_t = \lim_{\varepsilon\to 0} \frac{1}{2\varepsilon}\int_0^t \mathbf{1}_{\{|X|\le\varepsilon\}} d\langle X\rangle \tag{8.3.4}$$

in probability. In particular, for a $(\boldsymbol{F}_t)$-Brownian motion B,

$$L_t = \lim_{\varepsilon\to 0} \frac{1}{2\varepsilon}\, \lambda\left\{s \le t : |B_s| \le \varepsilon\right\}. \tag{8.3.5}$$

Proof. Let (ε_n) be any null-sequence and choose $\varepsilon_n' < \varepsilon_n$ such that $\varepsilon_n' \varepsilon_n^{-1} \xrightarrow[n\to\infty]{} 1$. Then there is a symmetric continuous function φ_n for which

$$\int_{-\varepsilon_n}^{\varepsilon_n} \varphi_n(x)dx = 2\ ,\ \varepsilon_n^{-1}\mathbf{1}_{[-\varepsilon_n',\varepsilon_n']} \le \varphi_n \le \varepsilon_n'^{-1}\mathbf{1}_{[-\varepsilon_n,\varepsilon_n]}\,.$$

Let $f_n \in \mathscr{C}^2$ satisfy $f_n'' = \varphi_n$, (i) and (ii). Then

$$\frac{1}{2\varepsilon_n}\int_0^t \mathbf{1}_{\{X_s \in [-\varepsilon_n',\varepsilon_n']\}} d\langle X\rangle_s \le \tfrac{1}{2}\int_0^t f_n''(X_s) d\langle X\rangle_s \le \frac{1}{2\varepsilon_n'}\int_0^t \mathbf{1}_{\{X_s \in [-\varepsilon_n,\varepsilon_n]\}} d\langle X\rangle_s$$

and hence (8.3.3) implies (8.3.4). Since for Brownian $d\langle X\rangle_s = ds$ the relation (8.3.5) is indeed a special case of (8.3.4). ■

We conclude in particular that continuous semimartingales spend only very little time at level 0 (or at any other level for that matter).

8.3.3 Corollary. For every $X \in \boldsymbol{S}_c$,

$$\mathbb{P}\left(\int_0^t 1_{\{X=0\}}\, d\langle X\rangle = 0\right) = 1 \tag{8.3.6}$$

and for an $(\boldsymbol{F}_t)$-Brownian motion B,

$$\mathbb{P}\left(\lambda\left\{ s : X_s = 0 \right\} = 0\right) = 1.$$

Proof. We have by (8.3.4) for small $\varepsilon > 0$,

$$\int_0^t 1_{\{X=0\}}\, d\langle X\rangle \le \int_0^t 1_{\{|X| \le \varepsilon\}}\, d\langle X\rangle \approx 2\varepsilon L_t .$$

Since L_t is a.s. finite by its definition (8.3.3) this 'estimate' implies (8.3.6). ■

From the representation (8.3.3) it follows that $|X| \in \boldsymbol{S}_c$ for every $X \in \boldsymbol{S}_c$:

8.3.4 Corollary. For a semimartingale $X \in \boldsymbol{S}_c$ the following holds:

(a) The process $|X|$ is a semimartingale.

(b) For every $H \in \boldsymbol{B}$ the integral w.r.t. $|X|$ may be written as

$$\int_0^{\cdot} H\, d|X| = \int_0^{\cdot} H \operatorname{sgn}(X)\, dX + \int_0^{\cdot} H\, dL . \tag{8.3.7}$$

In particular, for every locally bounded Borel function $g : \mathbb{R} \longrightarrow \mathbb{R}$ one has

$$\int_0^{\cdot} g(X)\, d|X| = \int_0^{\cdot} g(X) \operatorname{sgn}(X)\, dX + g(0)\cdot L . \tag{8.3.8}$$

(c) The process L is adapted to the nullset augmentation of the natural filtration of $|X|$.

Proof. (a) The equation (8.3.3) implies

$$|X| = |X_0| + L + \int_0^{\cdot} \operatorname{sgn}(X)\, dX .$$

The process L is adapted and a.s. increasing, hence in $\boldsymbol{A}_c$. Since the integral process $\int_0^{\cdot} \operatorname{sgn}(X)\, dX$ is a semimartingale so is $|X|$.

(b) The identity (8.3.7) follows from theorem 7.4.2. For the proof of (8.3.8) let first g be a bounded continuous function. Recall: If a sequence (F_n) of continuous increasing functions converges to F on a dense set then the respective integrals of bounded continuous h converge, i.e.

$$\int_0^t h\, dF_n \longrightarrow \int_0^t h\, dF .$$

By the theorem and a straightforward diagonal procedure one can choose a nullsequence (ε_n) such that a.s. for all rational $s \leq t$,

$$L_s = \lim_{n\to\infty} \frac{1}{2\varepsilon_n} \int_0^s \mathbf{1}_{\{|X|\leq\varepsilon_n\}} \, d\langle X\rangle .$$

Outside the corresponding $\mathbb{P}$-nullset one has

$$\int_0^t g(X_s)\, dL_s = \lim_{n\to\infty} \frac{1}{2\varepsilon_n} \int_0^t g(X_s)\, \mathbf{1}_{\{|X|\leq\varepsilon_n\}} \, d\langle X\rangle$$

by the above mentioned convergence result. For every n this integral is an average of the values of g in the interval $[-\varepsilon_n, \varepsilon_n]$. Since g is continuous these values are almost equal to $g(0)$. Therefore the limit may be rewritten as

$$g(0) \lim_{n\to\infty} \frac{1}{2\varepsilon_n} \int_0^t \mathbf{1}_{\{|X|\leq\varepsilon_n\}} \, d\langle X\rangle = g(0)\, L_t .$$

This together with (8.3.7) implies (8.3.8) for $g \in \mathscr{C}_b(\mathbb{R})$. It is now easy to extend (8.3.8) to bounded Borel functions g by a monotone class argument using the dominated convergence proposition 8.3.1. The boundedness of g is removed by localization.

(c) Let $(\boldsymbol{G}_t)$ denote the natural filtration of $|X|$. For $g = \mathbf{1}_{\{0\}}$ one has $g(x)\,\mathrm{sgn}(x) = 0$ and for this special choice of g the equation (8.3.8) reads

$$L_t = \int_0^t \mathbf{1}_{\{|X|=0\}} \, d|X| . \tag{8.3.9}$$

Hence L is a stochastic integral process for a $(\boldsymbol{G}_t)$-adapted integrand w.r.t. to the $(\boldsymbol{G}_t)$-semimartingale $|X|$. This implies the assertion. ■

8.4 Itô's Formula With Jumps

A $\mathscr{C}^2$-function of a noncontinuous semimartingale still is a semimartingale. Only the transformation formula looks more complicated than in the continuous case. But it is quite natural. In the proof of theorem 8.1.1 continuity of the semimartingale X was used in order to show that the total error of the Taylor approximation converges to zero as the partition becomes finer and finer. For semimartingales with jumps one just sums up these errors at the jumps and adds the result to the formula.

Continuing with the notation from lemma 8.1.2 and proposition 7.3.18, let $r(x,v)$ be the error of the second order Taylor approximation for $f(x+v) - f(x)$ and $(\Delta X)_t = X_t^+ - X_t^-$ the jump process of X.

8.4.1 Theorem. Let X be a $\mathbb{R}^d$-valued semimartingale. Assume that $U \subset \mathbb{R}^d$ is open and

$$\mathbb{P}(X_t \in U \text{ and } X_t^- \in U \text{ forall } t) = 1 .$$

Let f be a twice continuously differentiable function on U. Then there is a process $R \in \boldsymbol{A}$ such that

$$R_t = \sum_{0 \le s \le t} r\left(X_s^-, \Delta X_s\right) \quad \text{a.s.} \tag{8.4.1}$$

(in particular, this series is a.s. absolutely convergent) and the process $\left(f(X_t)\right)_{t \ge 0}$ is a semimartingale with differential

$$df(X) = \sum_{j=1}^{d} \partial_j f(X^-) \bullet dX^j + ½ \sum_{j,k=1}^{d} \partial_j \partial_k f(X^j) \bullet d[X^j, X^k] + dR . \tag{8.4.2}$$

Remark. Of course one can write down the process $r(X^-, \Delta X)$ explicitly. Since X is right-continuous, we have $X_t^+ = X_t$ and hence

$$\begin{aligned} r(X_s^-, \Delta X_s) &= f(X_s) - f(X_s^-) \\ &- \sum_{j=1}^{d} \partial_j f(X_s^-) \Delta X_s^j - ½ \sum_{j,k=1}^{d} \partial_j \partial_k f(X^-) \, \Delta X_s^j \Delta X_s^k . \end{aligned} \tag{8.4.3}$$

Thus the error r vanishes except for $\Delta X \neq 0$. In particular, for every regular path of X one has $r(X_s^-, \Delta X_s) = 0$ except for countably many s.

Another way to look at these formulas is to extract the jumps from the semimartingales X and f(X). Then Itô's formula becomes almost like in the continuous case. However this modification requires the jump process ΔX of X to be in $\boldsymbol{A}$.

8.4.2. Corollary. Assume in addition to the assumptions of theorem 8.4.1 that

$$\sum_{s \le t} |\Delta X_s| < \infty \quad \text{a.s. for each } t . \tag{8.4.4}$$

Then

$$\sum_{s \le t} |\Delta f(X)_s| < \infty \text{ a.s. for each } t \tag{8.4.5}$$

and for the continuous semimartingales

$$X_t^c = X_t - \sum_{s\le t} \Delta X_s \quad \text{and} \quad f(X_t)^c = f(X_t) - \sum_{s\le t} \Delta f(X)_s$$

one has the modified Itô formula

$$(8.4.6)\quad df(X)^c = \sum_{j=1}^{d} \partial_j f(X^-) \bullet d\langle X^{c,j}\rangle + \tfrac{1}{2} \sum_{j,k=1}^{d} \partial_j\partial_k f(X^-) \bullet d\langle X^{c,j}, X^{c,k}\rangle .$$

Proof of theorem 8.4.1. We jump back into the proof of theorem 8.1.1. Recall the formula

$$(8.1.4)\quad \begin{aligned} &\lim_{|\rho|\to 0} \sum_{i=0}^{q} r\big(X_{T_i}, dX(I_i)\big) \\ &= df(X)(I) - \sum_{j=1}^{d} \partial_j f(X^-) \bullet dX_j(I) - \tfrac{1}{2} \sum_{j,k=1}^{d} \partial_j\partial_k f(X^-) \bullet d[X^j, X^k](I) . \end{aligned}$$

This was established without using the continuity of the semimartingales. Comparing this formula with (8.4.2) we see that we need to transform the limit in probability on the left of (8.1.4) into the series in (8.4.1). For this we consider the stochastic interval [0,t] and choose a partition sequence such that

$$(8.4.7)\quad \sum_i \left|dX(I_i^n)\right|^2 \xrightarrow[n\to\infty]{} \sum_{j=1}^{d} [X^j]_t \text{ a.s.} .$$

In particular, for almost every ω there is a finite $K(\omega)$ such that

$$(8.4.8)\quad \sup_n \sum_i \left|X_{T^n_{i+1}} - X_{T^n_i}\right|^2 (\omega) \le K(\omega) < \infty .$$

Since X is a semimartingale almost all paths of X are regular and right-continuous, and moreover, they satisfy a.s. $X_s \in U$ and $X_s^- \in U$ for all s. We fix ω such that this and (8.4.8) hold. For such ω all limit points of the set $\{X_s(\omega) : s \in [0,t]\}$ are contained in U and this set is relatively compact in U. We can apply lemma 8.1.2 to see that for every $\delta > 0$ there is some $\varepsilon(\omega) > 0$ such that

$$(8.4.9)\quad \begin{cases} \left|r\big(X_s(\omega), X_{s'}(\omega) - X_s(\omega)\big)\right| \le \frac{\delta}{K(\omega)} \cdot \left|X_{s'}(\omega) - X_s(\omega)\right|^2 \\ \text{whenever} \\ s,s' \le t \text{ and } \left|X_{s'}(\omega) - X_s(\omega)\right| \le \varepsilon(\omega) . \end{cases}$$

Now we analyse the error sum on the left of (8.1.4). Let $Q(\omega)$ be the set

$$\{s \in [0,t] \mid |\Delta X_s(\omega)| \geq \varepsilon(\omega)\}.$$

Since the path of ω is regular the set $Q(\omega)$ is finite. The summation can be divided into two parts:

$$(8.4.10) \qquad \sum_i r\left(X_{T_i^n}(\omega), X_{T_{i+1}^n}(\omega) - X_{T_i^n}(\omega)\right) = \sum\nolimits_1 + \sum\nolimits_2 .$$

The first one (denoted by $\sum_1$) is extended over those i for which the interval $]T_i^n(\omega), T_{i+1}^n(\omega)]$ contains one and only one point in $Q(\omega)$. For sufficiently large n all remaining indices i satisfy

$$\left|X_{T_{i+1}^n}(\omega) - X_{T_i^n}(\omega)\right| \leq \varepsilon(\omega)$$

and hence by (8.4.8) and (8.4.9)

$$(8.4.11) \qquad \sum\nolimits_2 \left|r\left(X_{T_i^n} dX(I_i^n)\right)(\omega)\right| \leq \frac{\delta}{K} \sum_i \left|dX(I_i^n)\right|^2 \leq \delta .$$

For large n each term of the first sum corresponds to a unique point s in $Q(\omega)$ and $X_{T_i^n}(\omega)$ converges to $X_s^-(\omega)$ and $X_{T_{i+1}^n}(\omega) - X_{T_i^n}(\omega)$ to $\Delta X_s(\omega)$. Thus

$$\sum\nolimits_1 \longrightarrow \sum_{s \in Q(\omega)} r(X_s^-, \Delta X_s).$$

Moreover

$$\sum_{s \leq t} \left|r(X_s^-, \Delta X_s)\right| \leq \sum_{s \in Q} \left|r(X_s^-, \Delta X_s)\right| + \frac{\delta}{K} \sum_{0 < |\Delta X^j|^2 < \varepsilon} |\Delta X_s|^2 < \infty ,$$

the second sum being smaller than δ according to (8.4.8). This shows that the series is absolutely summable and that the sum in (8.4.10) actually converges a.s. to the series. More precisely, one has for an ω with the above properties

$$(8.4.12) \qquad \sum r\left(X_{\tau \wedge T_i^n}, X_{\tau \wedge T_{i+1}^n} - X_{\tau \wedge T_i^n}\right)(\omega) \xrightarrow[n \to \infty]{} \sum_{s \leq t} \left(r(X_s^-, \Delta X_s)\right)(\omega)$$

uniformly in $\tau \leq t$. The series defines a process of locally finite variation because of the absolute summability. It has an adapted right-continuous version R because it is the l.u.p. limit of a sequence of adapted right-continuous processes. Thus a version $R \in \boldsymbol{A}$ exists as required. ■

Proof of Corollary 8.4.2. Since the initial segments $\{X_s(\omega) : 0 \leq s \leq t\}$ of the paths are relatively compact in U the function f is Lipschitz continuous on these sets. So (8.4.4) implies (8.4.5). In proving the modified Itô formula (8.4.6) we restrict ourselves to the one-dimensional case. The formulas (7.3.7) and (7.3.8) concerning the jumps of stochastic integrals and of mutual variation imply

(8.4.13) $$\Delta\Big(\int_0^\cdot f'(X^-)\,dX\Big) = f'(X^-)\,\Delta X \text{ a.s.}$$

and

(8.4.14) $$\Delta\Big(\tfrac{1}{2}\int_0^\cdot f''(X^-)\,d[X]\Big) = \tfrac{1}{2} f''(X^-)\,\Delta[X] = \tfrac{1}{2} f''(X^-)\,(\Delta X)^2 \text{ a.s. .}$$

From (8.4.13) we conclude

$$\Big(\int_0^\cdot f'(X^-)\,dX\Big)^c = \int_0^\cdot f'(X^-)\,dX - \sum_{s\le\cdot} f'(X_s^-)\,\Delta X_s = \int_0^\cdot f'(X^-)\,dX^c$$

and hence

(8.4.15) $$\int_0^t f'(X^-)\,dX = \int_0^t f'(X^-)\,dX^c + \sum_{s\le t} f'(X_s^-)\,\Delta X_s\,.$$

Similarly, using (8.4.15) we get a.s.

(8.4.16) $$\tfrac{1}{2}\int_0^t f''(X^-)\,d[X] = \tfrac{1}{2}\int_0^t f''(X^-)\,d\langle X^c\rangle + \tfrac{1}{2}\sum_{s\le t} f''(X_s^-)(\Delta X_s)^2\,.$$

The two sums in (8.4.15) and (8.4.16) cancel with parts of the error terms in (8.4.2) and (8.4.3) in such a way that

$$\begin{aligned} f(X_t) - f(X_0) &= \int_0^t f'(X^-)\,dX + \tfrac{1}{2}\int_0^t f''(X^-)\,d[X] + \sum_{s\le t} r(X_s^-, \Delta X_s) \\ &= \int_0^t f'(X^-)\,dX^c + \tfrac{1}{2}\int_0^t f''(X^-)\,d\langle X^c\rangle + \sum_{s\le t}\Delta f(X_s). \end{aligned}$$

This is equivalent to the assertion. ■

Remark. The requirement in the corollary that the jumps of the semimartingale X are a.s. absolutely summable over finite intervals is not satisfied by every $\mathcal{L}^2$-martingale. As an example consider a sequence $(N^i)_{i\ge1}$ of i.i.d. Poisson processes and let M be the process

$$M_t = \sum_{i=1}^\infty M_t^i \qquad \text{where} \qquad M_t^i = \frac{1}{i}(N_t^i - t)\,.$$

The terms in the series are orthogonal elements of $\mathcal{L}^2$ and $\sum_{i=1}^\infty \|M_t^i\|_2^2 < \infty$. This implies that M is a $\mathcal{L}^2$-martingale. However, the jumps of the M^i are all directed upwards and they coincide in magnitude with the jumps of the shrunk Poisson processes N^i/i . Therefore and since a series of independent nonnegative random variables with divergent series of expectations diverges a.s. one has a.s.

$$\sum_{s\le t} \Delta M_s = \sum_i \sum_{s\le t} \Delta M^i_s = \sum_i \frac{N^i_t}{i} = \infty .$$

Hence (8.4.4) does not hold for M.

8.5 The Stratonovich-Fisk Integral

In Itô's concept, the integrals w.r.t. to continuous local martingales are again continuous local martingales. This fact is of great advantage both during the previous constructions and in applications to stochastic differential equations and diffusion processes. This pleasant property was - roughly spoken - enforced by evaluating the integrand at the left endpoints of the intervals which were chosen to define the integral (cf. the introduction of chapter 5). On the other hand, we had to put up with a modification of the chain rule.

Averaging the values at both endpoints (i.e. using a "trapezoid rule") instead of evaluating left endpoints remedies this disadvantage - and eliminates all the above mentioned advantages. The result is the Stratonovich-Fisk integral.

We will not repeat the whole procedure of defining elementary integrals and then approximate. Instead, we will formally define the integral as a new operation on the space $d\boldsymbol{S}_c$; This will at the same time provide the rule how to switch from one integral to the other. The intuitive averaging definition will appear as propositon 8.5.2(a).

8.5.1 Definition. Let X and Y be a.s. continuous semimartingales. The **Stratonovich-Fisk differential** and **integral** are given by

$$Y \circ dX = Y \bullet dX + \tfrac{1}{2}\, dY \bullet dX ,$$

$$\int_0^t Y \circ dX = \int_0^t Y\, dX + \tfrac{1}{2}\langle Y,X\rangle_t .$$

Note that not only the integrator X but also the integrand Y has to be a semi-martingale. For a.s. continuous integrands with finite variation the Itô and the Stratonovich-Fisk integrals coincide since the mutual variation vanishes. If a continuous deterministic integrand is of finite variation then the integral of section 1.2(C) was introduced by a $\boldsymbol{L}^2$-approximation; hence it can be considered as a Itô-integral and therefore also as a Stratonovich integral. We prove now the above mentioned approximation by Riemann-Stieltjes sums. We adopt the notation from section 7.3.

8.5.2 Proposition. Let X and Y be in $\boldsymbol{S}_c$. Then
(a) The Stratonovich-Fisk integral has the representation

$$\int_0^t Y \circ dX = \lim_{|\rho| \to \infty} \sum_i \frac{Y_{T_i} + Y_{T_{i+1}}}{2} \left(X_{T_{i+1}} - X_{T_i} \right)$$

in probability.
(b) If either X or Y is in $\boldsymbol{A}_c$ then

$$\int_0^t Y \circ dX = \int_0^t Y \, dX .$$

Proof. (a) Suppose that $I_i =]T_i, T_{i+1}]$ is the i-th interval of the partition ρ of the stochastic interval I. Write

$$\frac{Y_{T_i} + Y_{T_{i+1}}}{2} \, dX(I_i) = Y_{T_i} \cdot dX(I_i) + \tfrac{1}{2} \, dY(I_i) \cdot dX(I_i) .$$

If $|\rho| \longrightarrow 0$, then the sum of the terms on the right side converges in probability to $Y \bullet dX(I)$ and $\tfrac{1}{2} \, dY \bullet dX(I)$ by theorem 7.3.4 and 7.3.7.
(b) is clear since $\langle Y, X \rangle = 0$ if either X or Y is in $\boldsymbol{A}_c$ (theorem 7.4.2(c)). ■

We simply incorporate the new integral into the previous calculus and consider it as a new operation on the space $d\boldsymbol{S}_c$.

8.5.3 Definition. Define an operation $\circ : \boldsymbol{S}_c \times d\boldsymbol{S}_c \longrightarrow d\boldsymbol{S}_c$ by
(SM) **Symmetric multiplication.** If $X, Y \in \boldsymbol{S}_c$ set

$$Y \circ dX(I) = Y \bullet dX(I) + \tfrac{1}{2} \, dX \bullet dY(I) .$$

The associated simple rules for calculus hold in addition to those from theorem 7.5.2:

8.5.4 Theorem. If X,Y and Z are in $\boldsymbol{S}_c$

1^0 $$X \circ (dY + dZ) = X\circ dY + X\circ dZ,$$

2^0 $$(X+Y)\circ dZ = X\circ dZ + Y\circ dZ,$$

3^0 $$X\circ(dY \bullet dZ) = (X\circ dY)\bullet dZ.$$

4^0 $$X\circ(dY\bullet dZ) = X \bullet (dY\bullet dZ) \text{ and}$$
$$X\circ(Y\circ dZ) = (XY)\circ dZ.$$

Proof. These rules are easily verified using those stated in theorem 7.4.2. ■

As already mentioned, the Stratonovich integral is introduced in particular because the chain rule takes the customary form.

8.5.5 Theorem. Suppose that $X^1, \dots, X^d$ are a.s. continuous semimartingales. Assume further that U is an open subset of $\mathbb{R}^d$ and that $(X_t^1, \dots, X_t) \in U$ for every t with probability one. Let $f : U \longrightarrow \mathbb{R}$ have continuous partial derivatives of third order. Then

$$df(X^1, \dots, X^d) = \sum_{i=1}^{d} \partial_i f(X^1, \dots, X^d) \circ dX^i .$$

Proof. Set $X = (X^1, \dots, X^d)$. Apply Itô's formula to the partial derivatives of first order:

$$d\partial_i f(X) = \sum_{j=1}^{d} \partial_j\partial_i f(X) \bullet dX^j + \tfrac{1}{2} \sum_{j,k=1}^{d} \partial_j \partial_k \partial_i f(X) \bullet dX^k \bullet dX^j.$$

Symmetric multiplication by dX_i gives

$$\sum_{j=1}^{d} \partial_i f(X) \circ dX^i = \sum_{j=1}^{d} \partial_i f(X) \bullet dX^i + \tfrac{1}{2} \sum_{j=1}^{d} d\partial_i f(X) \bullet dX^i$$
$$= \sum_{j=1}^{d} \partial_i f(X) \bullet dX^i + \tfrac{1}{2} \sum_{j=1}^{d} \partial_j\partial_i f(X) \bullet dX^j \bullet dX^i = df(X).$$

In fact, the terms $dX^k \bullet dX^j \bullet dX^i$ vanish by the continuity assumption. Thus the proof is complete. ■

Remark. One could have taken the identity (a) from proposition 8.5.2 as a definition of the Stratonovich-Fisk operation. Then the assertion of the theorem of course also makes sense and can be shown to remain valid if f has continuous partial derivatives of second order.

As an exercise, we review the differential equation (8.2.1).

8.5.6 Example. Let $X \in \mathcal{S}_c$ start at zero. Then the "Stratonovich" differential equation

$$(8.5.1) \qquad dZ = Z \circ dX, \; Z_0 \equiv 1 ,$$

has the solution $Z = \exp(X)$. This is the only one up to a.s. equality. One way is to rewrite the Stratonovich as an Itô equation: Assume that (8.5.1) holds. Then using 4^0 in theorem 8.5.4 we arrive at

$$dZ = Z \circ dX = Z \bullet dX + \tfrac{1}{2} dZ \bullet dX = Z \bullet dX + \tfrac{1}{2}(Z \circ dX) \bullet dX$$

$$= Z \bullet dX + \tfrac{1}{2} Z \bullet d\langle X \rangle + \tfrac{1}{2}(\tfrac{1}{2} dZ \bullet d\langle X \rangle) = Z \bullet d(X + \tfrac{1}{2}\langle X \rangle).$$

Hence Z is a solution of the Itô-equation (8.2.1) with X replaced by $X + \tfrac{1}{2}\langle X \rangle$. Theorem 8.2.1 gives - up to a.s. equality - the unique solution

$$Z = \exp\left(X + \tfrac{1}{2}\langle X \rangle - \tfrac{1}{2}\langle X + \tfrac{1}{2}\langle X \rangle \rangle\right) = \exp(X).$$

That this is a solution of (8.5.1) can be seen by reading this calculation backwards or simply by appealing to theorem 8.5.5.
Do it now the other way round: Prove the uniqueness by means of the Stratonovich calculus and then derive the solution of (8.2.1)!

Another motivation for Stratonovich calculus is given by the approximation of the solutions of stochastic differential equations. We sketch this problem in the introduction to chapter 11.

8.6 The Vector-Valued Itô and Stratonovich Formulas

We give two versions of the Itô (resp. Stratonovich formula) for df(X) where f is a vectorvalued map. The first one is merely a reformulation of the scalar version. For us it is an intermediate step in the proof of the second version. This second form allows to use functional calculus of matrices in computations with semimartingales and it also contains the Itô calculus for holomorphic maps to be applied in section 9.5. Again this is a section which you may skip in a first reading and come back to it when needed.

Let U be an open subset of $\mathbb{R}^p$ and let $f = (f^1, \ldots, f^q) : U \longrightarrow \mathbb{R}^q$ be twice continuously differentiable. Then for each $x \in U$ the symbol $Df(x)$ denotes the total derivative in $L(\mathbb{R}^p, \mathbb{R}^q)$, i.e. the linear map corresponding to the Jacobian $(\partial_j f_i)_{i,j}$. The second total derivative $DDf(x)$ may be described in different ways; we consider it as a bilinear map from $\mathbb{R}^p \times \mathbb{R}^p$ to $\mathbb{R}^q$ such that

$$Df(x+u)(v) = Df(x)(v) + DDf(x)(u,v) + o(|u|) \text{ as } u \longrightarrow 0,$$

or equivalently,

$$\left[DDf(x)(u,v)\right]_j = \sum_{i,k=1}^{p} \partial_k \partial_i f_j(x) u_i v_j .$$

For simplicity, we restrict ourselves to a.s. continuous semimartingales. We use the notation of section 7.5.

8.6.1 Theorem. Suppose that U is an open subset of $\mathbb{R}^p$, that X is an a.s. continuous p-dimensional semimartingale with $\mathbb{P}(X_t \in U) = 1$ for all t and that $f : U \longrightarrow \mathbb{R}^q$ is twice continuously differentiable. Then $f(X)$ is a semimartingale and

$$df(X) = Df(X) \bullet dX + \tfrac{1}{2} DDf(X) \bullet (dX, dX) .$$

If f is three times continuously differentiable then

$$df(X) = Df(X) \circ dX$$

where the symbol $\circ$ is to be read component by component.

Proof. Assume first that $q = 1$. Then

$$Df(X) \bullet dX = \sum_{i=1}^{p} \partial_i f(X) \bullet dX^i ,$$

and

$$DDf(X) \bullet (dX, dX) = \sum_{i,j=1}^{p} \partial_{ij} f(X) \bullet dX^i \bullet dX^j .$$

Hence for this case we have simply rewritten the Itô formula from theorem 8.1.1 in another form. If $q > 1$ then the formula is established by putting together the components. The Stratonovich formula can be derived along the same lines. ■

Another - more interesting - version is the following one.

8.6.2 Theorem. Let **E** be the commutative algebra of real $p \times p$-matrices. Suppose that U is an open subset of the cartesian product space $\mathbf{E}^n$ and that $X^1, \dots, X^n$ are continuous **E**-valued semimartingales such that $P\big((X^1, \dots, X^n)_t \in U\big) = 1$ for every $t \geq 0$. Consider a map $f : U \longrightarrow \mathbf{E}$ which is locally equal to a power series, i.e. f has locally the form

$$f(A_1,\dots,A_n) = \sum_{\substack{i_1+\dots+i_n=0 \\ i_k \geq 0}}^{\infty} a_{i_1,\dots,i_n}(A_1 - A_1^*)^{i_1} \dots (A_n - A_n^*)^{i_n}$$

of an absolutely convergent series of matrix products. Then

$$df(X) = \sum_{i=1}^{n} D_i^* f(X) \bullet dX^i + \tfrac{1}{2} \sum_{i,j=1}^{n} D_i^* D_j^* f(X) \bullet dX^i \bullet dX^j = \sum_{i=1}^{n} D_i^* f(X) \circ dX^i$$

where $D_i^* f : U \longrightarrow \mathbf{E}$ and $D_i^* D_j^* \; f : U \longrightarrow \mathbf{E}$ are derived from f by formal partial differentiation of the (local) power series.

Proof. Let $D_i f : U \longrightarrow \mathbf{L}(\mathbf{E},\mathbf{E})$ and $D_i D_j : U \longrightarrow \mathbf{Bil}(\mathbf{E} \times \mathbf{E}, \mathbf{E})$ be the partial "total" derivatives w.r.t. the components of $\mathbf{E}^n$ of first and second order, respectively. If f is twice continuously differentiable one verifies like in the proof of theorem 8.6.1 the identities

$$\begin{aligned}
Df(X) \bullet dX &= \sum_{i=1}^{n} D_i f(X) \bullet dX^i , \\
Df(X) \circ dX &= \sum_{i=1}^{n} D_i f(X) \circ dX^i , \\
DDf(X) \bullet (dX, dX) &= \sum_{i=1}^{n} D_i D_j f \bullet (dX^i, dX^j) .
\end{aligned}$$

If f is locally equal to a power series then

$$\begin{aligned}
\Big[D_i f(C)\Big](A) &= \Big[(D_i^* f(C))\Big]A, \\
\Big[D_i D_j f(C)\Big](A,B) &= \Big[(D_i^* D_j^* f(C))\Big]AB
\end{aligned}$$

for all $C \in U$ and $A, B \in \mathbf{E}$ where on the right-hand side we have usual matrix products. The assertion follows now from theorem 8.6.1. ■

Example. The exponential of a square matrix A is defined as the power series

$$\exp(A) = \sum_{i=0}^{\infty} \frac{1}{i!} A^i.$$

If X is an a.s. continuous semimartingale with values in a commutative algebra of matrices then theorem 8.6.2 yields

$$d \exp(X) = \exp(X)\bullet dX + \tfrac{1}{2} \exp(X)\bullet(dX\bullet dX) = \exp(X)\circ dX .$$

Of special interest are processes taking values in the complex plane $\mathbb{C}$. Complex numbers will be written in the form $z = x + i\,y$.

8.6.3 Definition. Let $\boldsymbol{S}_c^{\mathbb{C}}$ denote the spaces of those complex-valued processes whose real and imaginary parts are in $\boldsymbol{S}_c$ and let $d\boldsymbol{S}_c^{\mathbb{C}}$ denote the corresponding space of stochastic differentials.

Having identified $\mathbb{C}$ with the commutative algebra of 2×2-matrices of the form $\begin{pmatrix} x & y \\ -y & x \end{pmatrix}$, we may adopt the operations defined in definition 7.5.1 and get

8.6.4 Proposition.

(A) If dZ and dZ' are in $d\boldsymbol{S}_c^{\mathbb{C}}$ then

$$d(Z + Z')(I) = dZ(I) + dZ'(I).$$

(M) If $H = K + i\,L \in \boldsymbol{B}^{\mathbb{C}}$ and $Z = X + i\,Y \in \boldsymbol{S}_c^{\mathbb{C}}$ then

$$H\bullet dZ = K\bullet dX - L\bullet dY + i\,(L\bullet dX + K\bullet dY).$$

(P) If $dZ, dZ' \in d\boldsymbol{S}_c^{\mathbb{C}}$ then

$$dZ\bullet dZ' = d(ZZ') - Z\bullet dZ' - Z'\bullet dZ .$$

Complex mutual variation has the form

$$\langle Z,Z'\rangle_t = dZ\bullet dZ'(]0,t])$$

and **complex quadratic variation** is

$$\langle Z\rangle = \langle Z,Z\rangle .$$

Then we can simply state

8.6.5 Theorem. Let U be an open subset of $\mathbb{C}$ and $Z^1, \dots ,Z^n$ be a.s. continuous complex semimartingales such that $\mathbb{P}\big((Z^1, \dots ,Z^n) \in U\big) = 1$ for all $t \geq 0$.. Let further $f : U \longrightarrow \mathbb{C}$ be a holomorphic function. Then setting $Z = (Z^n, \dots ,Z^n)$ one has

$$df(Z) = \sum_{i=1}^{n} \frac{\partial f}{\partial z_k}(Z) \bullet dZ^k + ½ \sum_{i,j=1}^{n} \frac{\partial^2 f}{\partial z_i \, \partial z_i}(Z)\langle Z^i, Z^k \rangle$$

$$= \sum_{i=1}^{n} \frac{\partial f}{\partial z_k}(Z) \circ dZ^k .$$

Proof. Rewrite theorem 8.6.2. ∎

CHAPTER 9

THE SPECIAL ROLE OF BROWNIAN MOTION

Brownian motion plays a universally important role among the continuous martingales. This will be illustrated in the subsequent sections. The first one deals with what we view as the most important result on Brownian motion (besides Wiener's existence theorem) namely P. Lévy's characterization by means of quadratic variation. The theorem of K. Dambis, L.E. Dubins and G.Schwarz in section 9.2 tells us that every a.s. continuous local martingale is a time-changed Brownian motion. The next two sections contain aspects of Brownian motion which make it a very special Markov process. Section 9.5 presents Lévy's theorem on conformal invariance of complex Brownian motion. Section 9.6 introduces Hermite polynomials and indicates their role in integration w.r.t. Brownian motion and section 9.7 shows that every martingale for the natural filtration of a Brownian motion automatically is a.s. continuous.

Only Lévy's characterization of Brownian motion by its increasing process is essential for the remaining chapters. However, we invite you to have a look and to enjoy seeing stochastic calculus at work also in the other parts of this chapter.

From now on we are really interested in multidimensional semimartingales, in particular in multidimensional Brownian motions. One-dimensional Brownian motion was defined in section 1.1 and the concept of a $(\boldsymbol{F}_t)$-Brownian motion was introduced in example 3.0.2. A **d-dimensional Brownian motion** (resp. $(\boldsymbol{F}_t)$-Brownian motion) is a $\mathbb{R}^d$-valued process B whose components $B^1, \dots, B^d$ are independent Brownian motions ((resp. $(\boldsymbol{F}_t)$-Brownian motions with the *same* filtration $(\boldsymbol{F}_t)_{t \ge 0}$ for all components). A d-dimensional Brownian motion, provided all paths are at least right-continuous, is easily seen to be a $(\boldsymbol{F}_t)$-Brownian motion for the filtration $\boldsymbol{F}_t = \sigma(B^i_s : 1 \le i \le d, 0 \le s \le t)$. Up to now we considered only Brownian motions starting at zero. A ($(\boldsymbol{F}_t)$-) **Brownian motion starting at** $x \in \mathbb{R}^d$ is a process of the form $B^x = x + B$ with a d-dimensional ($(\boldsymbol{F}_t)$-) Brownian motion B (starting at zero).

Since we have developed the stochastic calculus only for right-continuous processes we always choose a right-continuous adapted version of Brownian motion (which exists by lemma 4.3.5).

9.1 Lévy's Characterization of Brownian Motion

We have already seen that $\langle B\rangle_t = t$ is the increasing process of Brownian motion. Paul Lévy's striking observation states that this property singles out Brownian motion from all a.s. continuous local martingales.

9.1.1 Theorem. Suppose that B is an a.s. continuous process starting at zero. Let $(\boldsymbol{F}_t)_{t\geq 0}$ be some filtration. Then the following are equivalent:

(a) B is an $(\boldsymbol{F}_t)$-Brownian motion,

(b) B and $(B_t^2 - t)_{t\geq 0}$ are local martingales w.r.t. $(\boldsymbol{F}_t)_{t\geq 0}$.

Remark. (a) The continuity assumption in the theorem is essential: for the Poisson martingale the process $(M_t^2 - t)_{t\geq 0}$ is a local martingale too.

(b) A straightforward calculation allows to deduce from the theorem for a > 0 that B and $(B_t^2 - at)_{t\geq 0}$ are local martingales if and only if $B_t = \sqrt{a}\, B'_t$ with a Brownian motion B'.

A simulation of 20 paths of the process $(B_t^2 - t)_{t\geq 0}$ between t = 0 and t = 1 is shown in figure 9.1.1.

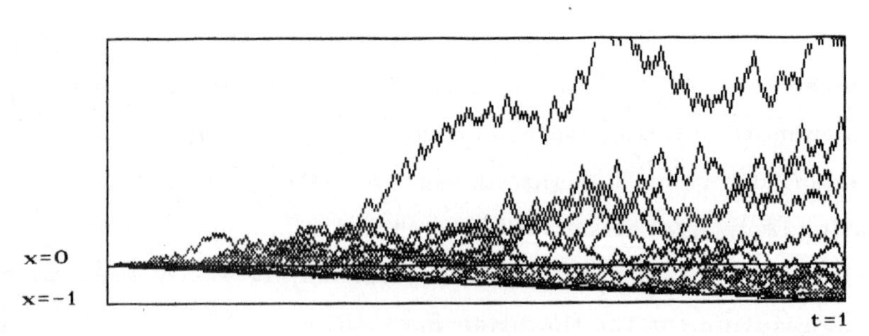

Fig. 9.1.1

We formulate part of the proof as two assertions which stand on their own. The first one characterizes independence by means of conditional characteristic functions.

9.1.2 Lemma. A random variable X on $(\Omega, \boldsymbol{F}, \mathbb{P})$ is independent of the sub-σ-field $\boldsymbol{G}$ of $\boldsymbol{F}$ if and only if for all $a \in \mathbb{R}$,

$$\mathbb{E}(e^{iaX} \mid \boldsymbol{G}) = \mathbb{E}(e^{iaX}).$$

Proof. If X is independent of $\boldsymbol{G}$ then so is e^{iaX} and hence its conditional expectation equals the expectation. Conversely, assume that the conditional and the unconditional characteristic functions coincide. We have to prove for each $G \in \boldsymbol{G}$ of positive probability that

$$\frac{1}{\mathbb{P}(G)} \mathbb{P}\left(\left\{X \in B\right\} \cap G\right) = \mathbb{P}\left(X \in B\right) \text{ for all } B \in \mathscr{B}(\mathbb{R}),$$

or equivalently, that the distributions $\mathbb{P}_X$ of X under $\mathbb{P}$ and $\mathbb{P}_{X|G}$ under $\mathbb{P}(\cdot|G)$ are equal. The computation

$$\int_{\mathbb{R}} e^{iat}\, d\mathbb{P}_{X|G} = \frac{1}{\mathbb{P}(G)} \int_G e^{iaX}\, d\mathbb{P} = \frac{1}{\mathbb{P}(G)} \int_G \mathbb{E}(e^{iaX} \mid \boldsymbol{G})\, d\mathbb{P}$$

$$= \frac{1}{\mathbb{P}(G)} \int_G \mathbb{E}(e^{iaX})\, d\mathbb{P} = \mathbb{E}(e^{iaX}) = \int_{\mathbb{R}} e^{iat\cdot}\, d\mathbb{P}_X$$

shows that the characteristic functions of these distributions and hence the distributions themselves coincide. This completes the proof. ■

The second assertion characterizes Brownian motion by means of exponential martingales (cf. section 8.2).

9.1.3 Proposition. Suppose that B is an a.s. continuous right-continuous process adapted to the filtration $(\boldsymbol{F}_t)_{t \geq 0}$ and starting at zero. Suppose further that for every real number a the exponential process

$$\left(e^{iaB_t + \frac{1}{2} a^2 t}\right)_{t \geq 0}$$

is a complex $(\boldsymbol{F}_t)$-martingale. Then B is a $(\boldsymbol{F}_t)$-Brownian motion.

Proposition 8.2.2 applied to M = B and the number a replaced by *ia* yields the converse of this result.

Proof. By assumption, the following identities hold a.s.:

$$\mathbb{E}(e^{ia(B_t - B_s)} \mid \boldsymbol{F}_s) = e^{-iaB_s} \cdot e^{-\frac{1}{2}a^2 t}\, \mathbb{E}(e^{iaB_t + \frac{1}{2}a^2 t} \mid \boldsymbol{F}_s)$$

$$= e^{-iaB_s - \frac{1}{2}a^2 t}\, e^{iaB_s}\, e^{\frac{1}{2}a^2 s} = e^{-\frac{1}{2}a^2(t-s)}\,.$$

Hence the conditional expectation does not depend on ω and therefore is equal to the expectation. By the preceding lemma the increment $B_t - B_s$ is independent of $\mathcal{F}_s$. Its Fourier transform is $\exp(-\frac{1}{2}a^2(t-s))$. This implies that $B_t - B_s$ has a centered normal distribution with variance $t - s$. Thus B is a $(\mathcal{F}_t)$-Brownian motion. ■

Proof of theorem 9.1.1. If B is a $(\mathcal{F}_t)$-Brownian motion then B and $(B_t^2 - t)$ are local martingales (examples 4.2.2 and 5.6.6).
Conversely, assume (b). We have $\langle B\rangle_t = t$ since for elements of $\mathcal{M}_c$ the increasing process is uniquely determined by the fact that $B^2 - \langle B\rangle$ is a local martingale according to theorem 5.6.2. For every real number a, the process

$$Y_t^a = \exp(iaB_t + \tfrac{1}{2}a^2\, t)$$

is a local martingale by proposition 8.2.2. Since

$$\sup_{s\le t} |\exp(iaB_s + \tfrac{1}{2}a^2 s)| = \sup_{s\le t} \exp(\tfrac{1}{2}a^2 s) = \exp(\tfrac{1}{2}a^2 t)$$

for any fixed $t > 0$, the processes Y^a are martingales by proposition 4.2.3. The above characterization finally shows that B is a $(\mathcal{F}_t)$-Brownian motion. ■

Lévy's theorem states that the law of Brownian motion can be identified from its increasing process. We cannot hope that in general the law of $M \in \mathcal{M}_c$ is uniquely determined by $\langle M\rangle$.

9.1.4 Example. Suppose that B is a $(\mathcal{F}_t)$-Brownian motion and T is a stopping time. Then $\langle B_{\cdot\wedge T}\rangle_t = \langle B\rangle_{t\wedge T} = t \wedge T$ according to proposition 7.3.11. In particular, for $T = \inf\left\{t \ge 0 : B_t \in \{-1,2\}\right\}$ the martingales

$$M_t = B_{t\wedge T} \text{ and } M'_t = -B_{t\wedge T} = -M_t$$

both have the increasing process $t \wedge T$ but for instance $\mathbb{P}(M_t = 2) > 0$ and $\mathbb{P}(M'_t = 2) = \mathbb{P}(M_t = -2) = 0$ for every $t > 0$.

This example leads also to the following question: What happens if the increasing process of a continuous martingale does not coincide with t everywhere but only

$$\langle M\rangle_t(\omega) = t \text{ for } 0 \le t \le T(\omega)$$

where T is a random time? Can we say that then M equals a Brownian motion up to time T ? Some pessimistic preliminary remarks are in order. Consider the extreme situation where M vanishes identically. Then the law of M is degenerate. How do we get a Brownian continuation? The underlying probability space may not carry a Brownian motion. Therefore we adjoin one. This leads to the following "local version". Lévy's theorem reads now:

9.1.5 Proposition. Suppose that M is an a.s. continuous local martingale with increasing process $\langle M\rangle_t = t\wedge T$ where T is a nonnegative random variable. Then T is a stopping time.

Choose now a $(\boldsymbol{G}_t)$-Brownian motion B on some probability space $(\Omega',\boldsymbol{G},\mathbb{P}')$ and set

$$W_t(\omega,\omega') = M_t(\omega) + B_t(\omega') - B_{t\wedge T(\omega)}(\omega').$$

Then $W = (W_t)_{t\geq 0}$ is an $(\boldsymbol{F}_t\otimes\boldsymbol{G}_t)$-Brownian motion on the probability space $(\Omega\times\Omega', \boldsymbol{F}\otimes\boldsymbol{G}, \mathbb{P}\otimes\mathbb{P}')$ and

$$W_t(\omega,\omega') = M_t(\omega) \text{ if } t < T(\omega).$$

This justifies:

9.1.6 Definition. Let $(\boldsymbol{F}_t)_{t\geq 0}$ be a filtration and T a nonnegative random variable. An a.s. continuous local $(\boldsymbol{F}_t)$-martingale W starting at zero is called a **Brownian motion stopped by** T if $\langle W\rangle_t = t\wedge T$.

Proof of proposition 9.1.5. For each $t > 0$, we have

$$\{T < t\} = \{\langle M\rangle_t < t\} \in \boldsymbol{F}_t$$

and hence T is a stopping time.

The three terms in the definition of W define canonically a.s. continuous local martingales on the product space. It should be sufficient to give a proof for the last term. To this end, we go back to the very definition of conditional expectation. Let $s < t$ and choose a rectangle $F_s\times G_s \in \boldsymbol{F}_s\otimes\boldsymbol{G}_s$. Then

$$\int_{F_s\times G_s} B_{t\wedge T(\omega)}(\omega')\, d\mathbb{P}\otimes\mathbb{P}'(\omega,\omega') = \int_{F_s}\int_{G_s} B_{t\wedge T(\omega)}(\omega')\, d\mathbb{P}'(\omega')d\mathbb{P}(\omega)$$

$$= \int_{F_s\times G_s} B_{s\wedge T(\omega)}(\omega')\, d\mathbb{P}\otimes\mathbb{P}'(\omega,\omega').$$

This proves that W is an a.s. continuous martingale. We compute now the increasing process of W:

$$\langle W\rangle_t = \lim_{|\rho|\to 0} \sum_{t_i\le t} (W_{t_i} - W_{t_{i-1}})^2$$
$$= \lim_{|\rho|\to 0} \sum_{t_i\le T\wedge t} (M_{t_i} - M_{t_{i-1}})^2 + \sum_{T\wedge t<t_i\le t} (B_{t_i} - B_{t_{i-1}})^2$$
$$= T\wedge t + (t - T\wedge t) = t$$

(the terms with $t_{i-1} \le T < t_i$ vanish in the limit). By Lévy's characterization, W is a Brownian motion. ■

We turn now to a multidimensional version of Lévy's theorem. Of course, we could have carried out the multidimensional proof straight away but we did not want to obscure the idea by additional notation. So we deduce it now from the linear case.

9.1.7 Proposition. Suppose that $B = (B^1, \dots, B^d)$ is a d-dimensional a.s. continuous local martingale starting at $x \in \mathbb{R}^d$. Then B is a d-dimensional Brownian motion starting in x if and only if $\langle B^j, B^k\rangle = \delta_{jk} t$ for $1 \le j, k \le d$ (δ_{jk} equals 1 if j = k and else is 0) .

Proof. We may suppose $x = 0 \in \mathbb{R}^d$. The "only if" part is example 7.3.17. Assume now $\langle B^j, B^k\rangle = \delta_{jk}t$. For real numbers $a^1, \dots, a^d$ the a.s. continuous process

$$B'_t = \sum_{k=1}^{d} a^k B^k$$

has quadratic variation

$$\langle B'\rangle_t = \left[\sum_{k=1}^{d} (a^k)^2\right] t.$$

Hence B' is by Lévy's theorem a linear Brownian motion where B'_1 has variance $(a^1)^2+\dots+(a^d)^2$. In order to show that the components of B are independent Brownian motions we write down the common characteristic function of the $n\cdot d$ increments

$$B^k_{t_{j+1}} - B^k_{t_j},\ 0 = t_0 \le t_1 \le \dots \le t_n,\ 0 \le k \le d$$

(we may choose the same partition for all k since we can consider a **common** refinement otherwise). Choose $(a_j^k) \in \mathbb{R}^{nd}$ and write $B'_j = \sum_{k=1}^d a_j^k B^k$. Then

$$\mathbb{E}\Big[\exp\Big(i\sum_{j=1}^{n-1}\sum_{k=0}^{d} a_j^k (B_{t_{j+1}}^k - B_{t_j}^k)\Big)\Big] = \mathbb{E}\Big[\exp\Big(i\sum_{j=0}^{n-1}(B'_{j,t_{j+1}} - B'_{j,t_j})\Big)\Big].$$

Conditioning inside the square brackets on $\boldsymbol{F}_{t_{n-1}}$ gives

$$\mathbb{E}\Big[\exp\Big(i\sum_{j=0}^{n-2}(B'_{j,t_{j+1}} - B'_{j,t_j})\Big)\Big]\cdot\exp\Big(-\tfrac{1}{2}\sum_{k=1}^{d}(a_n^k)^2(t_{n+1} - t_n)\Big)$$

and repeating this procedure shows that the increments have the desired characteristic function

$$\exp\Big[-\tfrac{1}{2}\sum_{j,k}(a_j^k)^2(t_{j+1} - t_j)\Big].$$

This completes the proof. ■

9.1.8 Example. If B is a d-dimensional Brownian motion and H is a predictable process with values in the orthogonal d × d-matrices then $\left(\int_0^t H\, dB\right)_{t\geq 0}$ is a d-dimensional Brownian motion. In fact, the last result applies since

$$\big(H\bullet dB\big)\bullet\big(H\bullet dB\big)^{\top} = H\bullet dB\bullet dB^{\top}\bullet H^{\top} = \big(HH^{\top}\big)\bullet dt = E_d\, dt$$

(where E_d is the unit matrix).

Using Lévy's theorem we prove the strong Markov property of Brownian motion. The key is the following result. Another proof of proposition 9.1.9 will be given in lemma 11.3.5.

9.1.9 Proposition. For every d-dimensional $(\boldsymbol{F}_t)$-Brownian motion B and every finite stopping time T the process $(B_{T+s} - B_T)_{s\geq 0}$ is a d-dimensional $(\boldsymbol{F}_{T+s}^{+})$-Brownian motion.

Proof. Compute the quadratic variation of the components of $M_s = B_{T+s} - B_T$:

$$\langle M^i\rangle_s = \lim_{|\rho|\to 0}\sum_{T<s_i<T+s}\Big(B_{T+s_{i+1}}^i - B_{T+s_i}^i\Big)^2 = \langle B^i\rangle_{T+s} - \langle B^i\rangle_T = s\,.$$

and similarly, for $i \neq j$,

$$\langle M^i, M^j\rangle_s = \langle B^i, B^j\rangle_{T+s} - \langle B^i, B^j\rangle_T = 0.$$

By the multidimensional form of Lévy's theorem it remains to see that M is a local $(\mathcal{F}^+_{T+s})$-martingale. This follows e.g. from the following abstract argument. ■

9.1.10 Lemma. For every local martingale N and every a.s. finite stopping time T the process $(N_{T+s})_{s\geq 0}$ is a local $(\mathcal{F}^+_{T+s})_{s\geq 0}$-martingale.

Proof. Let (T_k) be a localizing sequence such that for each k the process $N_{\cdot\wedge T_k} - N_0$ is a uniformly integrable martingale. Let $S_k = T \vee T_k - T$. Then

$$\{S_k < s\} = \{T + s > T \vee T_k\} \in \mathcal{F}^+_{T+s}.$$

Hence (S_k) is a localizing sequence w.r.t. $(\mathcal{F}^+_{T+s})$. Moreover,

$$N_{T+s\wedge S_k} - N_T = N_{(T+s)\wedge T_k} - N_{T\wedge T_k} \tag{9.1.1}$$

since on the set $\{T < T_k\}$ one has

$$T + s \wedge S_k = (T + s) \wedge T_k$$

and on the complement $\{T_k \leq T\}$ both sides in (9.1.1) vanish. The stopping theorem (corollary 3.2.8(c)) applied to the martingale $N_{\cdot\wedge T_k} - N_0$ shows that the right-hand side and therefore also the left-hand side of (9.1.1) is a uniformly integrable $(\mathcal{F}^+_{T+s})$-martingale. ■

Let $\mathbb{W}^{d,x}$ and $\mathbb{E}^x$ denote the distribution and expectation with respect to the distribution of a Brownian motion starting in $x \in \mathbb{R}^d$. Then the expectation of a Borel function $f : \mathscr{C}(\mathbb{R}_+, \mathbb{R}^d) \longrightarrow \mathbb{R}_+$ can be written as

$$\mathbb{E}^x(f) = \mathbb{E}^0(f(\cdot + x)) = \int_{\mathscr{C}(\mathbb{R}_+, \mathbb{R}^d)} f(\omega + x)\, d\mathbb{W}^{d,0}(\omega)$$

where $\omega + x \in \mathscr{C}(\mathbb{R}_+, \mathbb{R}^d)$ is given by $(\omega + x)_t = \omega_t + x$.

9.1.11 Corollary. (Strong Markov Property of Brownian motion). Let B be a d-dimensional $(\mathcal{F}_t)$-Brownian motion and T be an a.s. finite stopping time. Let further $f : \mathscr{C}(\mathbb{R}_+, \mathbb{R}^d) \longrightarrow \mathbb{R}_+$ be a Borel function. Then

$$\mathbb{E}\Big(f\big((B_{T+s})_{s\geq 0}\big) \;\Big|\; \mathcal{F}^+_T\Big) = \mathbb{E}^{B_T}(f).$$

Proof. Let $h : \mathscr{C}(\mathbb{R}_+, \mathbb{R}^d) \longrightarrow \mathbb{R}_+$ be a Borel function. Then

$$\mathbb{E}^0(h) = \mathbb{E}^x\big(h(\cdot - x)\big) = \mathbb{E}^{B_T}\big(h(\cdot - B_T)\big).$$

For every Borel function $g : \mathbb{R}^d \longrightarrow \mathbb{R}_+$ the variable $g(B_T)$ is $\boldsymbol{F}_T^+$-measurable and according to proposition 9.1.9,

$$\mathbb{E}\Big(g(B_T)\, h\big((B_{T+s} - B_T)_{s\geq 0}\big) \;\Big|\; \boldsymbol{F}_T^+\Big) = g(B_T)\; \mathbb{E}^0(h)$$

$$= g(B_T)\; \mathbb{E}^{B_T}\big(h(\cdot - B_T)\big) = \mathbb{E}^{B_T}\big(g(B_T) h(\cdot - B_T)\big).$$

Monotone convergence allows to extend this to the more general statement: For every product-measurable function $k : \mathbb{R}^d \times \mathscr{C}(\mathbb{R}_+, \mathbb{R}^d) \longrightarrow \mathbb{R}_+$ we have

$$\mathbb{E}\Big(k\big(B_T,(B_{T+s} - B_T)_{s\geq 0}\big) \;\Big|\; \boldsymbol{F}_T^+\Big) = \mathbb{E}^{B_T}\big(\, k(B_T, \cdot - B_T)\big),$$

and in particular, for $k(x,\omega) = f(\omega + x)$ this yields

$$\mathbb{E}\Big(f\big((B_{T+s})_{s\geq 0}\big) \;\Big|\; \boldsymbol{F}_T^+\Big) = \mathbb{E}^{B_T}(f)$$

which completes the proof. ■

9.2 Continuous Local Martingales are Time-Changed Brownian Motions

Lévy's result states that a continuous local martingale is a Brownian motion if its increasing process is t at time t. On the other hand, any increasing function $t \longmapsto f(t)$ can be transformed into the identity $t \longmapsto t$ on its range by a suitable transformation of the time-axis (namely the "right-inverse" of f). In particular, this applies to the paths of the increasing process $t \longmapsto \langle M \rangle_t$ of a continuous local martingale M. In this way M becomes a Brownian motion if each realization ω is endowed with its own appropriate clock, running individually with variable speed. This is made precise in theorem 9.2.3. The result was proved by K. Dambis (1965) and independently by L.E. Dubins and G. Schwarz (1965). We also give a multidimensional version of this result due to F. Knight and finally show how one gets a Cauchy process from a two-dimensional Brownian motion by a suitable time-change.

The intuitive idea of each ω having an own clock corresponds to the concept of time-change. By a **time-change** we simply understand an increasing net $(T_r)_{r\geq 0}$ of stopping times. Any process $M = (M_t)_{t\geq 0}$ may be transformed into the **time-changed** process $(M_{T_r})_{r\geq 0}$. Nasty things can happen if the time-change is not "reasonable".

9.2.1 Example. Consider a $(\boldsymbol{F}_t)$-Brownian motion $B = (B_{t\wedge 2})_{t\geq 0}$ (frozen at time 2 just to make sure that the stopping theorem applies). We transform B by the time-change $T_r = \inf\{t \geq 0 : |B_t| > r\}$, $r \geq 0$.

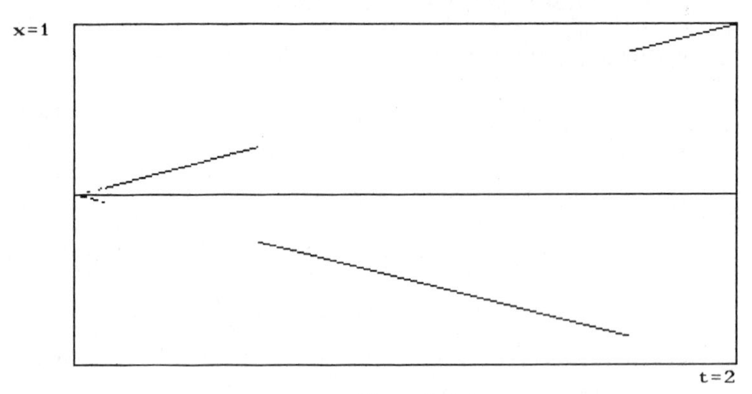

Fig. 9.2.1

A path of $(B_{T_r})_{r\geq 0}$ might look like that in figure 9.2.1. The time-changed process is a martingale by the stopping theorem. On the other hand, the stopping theorem did not promise that the time changed process is again continuous and in fact, we loose continuity here. Now consider a Brownian motion with unlimited lifetime and the stopping times $T_r = \{t \geq 0 : B_t > r\}$. If this time change is applied to Brownian motion the typical time-changed path will follow the diagonal forever. Of course, this time-changed Brownian motion is no longer a martingale. It is even no local martingale since it starts at zero and else has strictly positive values.

The appropriate time-change is given by the stopping times

$$T_r = \inf\left\{t \geq 0 : \langle M\rangle_t > r\right\}. \tag{9.2.1}$$

9.2.2 Lemma. Suppose that $M \in \boldsymbol{M}_c$, consider the just defined stopping times and set $\langle M\rangle_\infty = \sup_{t\geq 0} \langle M\rangle_t$. Then

$$\left\{\langle M\rangle_\infty < \infty\right\} = \bigcup_{r<\infty} \left\{T_r = \infty\right\} \tag{9.2.2}$$

and there is a random variable M_∞ such that $M_\infty = \lim_{t\to\infty} M_t$ a.s. on this set.

Proof. Equality of sets in (9.2.2) is clear. Moreover, $\langle M\rangle_{T_r} \leq r$ a.s. because of a.s. continuity. Thus the stopped process $M_{\cdot\wedge T_r} - M_0$ converges a.s. since it is a $\boldsymbol{L}^2$-bounded martingale (proposition 5.6.3). In particular,

$$\lim_{t\to\infty} M_t = \lim_{t\to\infty} M_{t\wedge T_r}$$

exists a.s. on $\{T_r = \infty\}$. Since the sets $\{T_r = \infty\}$ increase with r and

$$\bigcup_{r<\infty} \{T_r = \infty\} = \bigcup_{n\geq 1} \{T_n = \infty\}$$

the limit $\lim_{t\to\infty} M_t$ exists a.s. on $\bigcup_{r<\infty} \{T_r = \infty\}$. ■

We prove now the announced result.

9.2.3 Theorem. Suppose that $M \in \boldsymbol{M}_c$. Set for $r \geq 0$,

$$T_r = \inf\left\{t \geq 0 : \langle M\rangle_t > r\right\}, \quad B_r = M_{T_r} - M_0 \quad \text{and} \quad \boldsymbol{G}_r = \boldsymbol{F}^+_{T_r}.$$

Then the process B is a $(\boldsymbol{G}_r)$-Brownian motion if $\mathbb{P}(\langle M\rangle_\infty = \infty) = 1$. If $\mathbb{P}(\langle M\rangle_\infty < \infty) > 0$ then B is a Brownian motion stopped by $\langle M\rangle_\infty$. The process M can be reconstructed from B by the inverse time-change

$$B_{\langle M\rangle_t} = M_t - M_0 \text{ a.s. .}$$

This relation extends to

$$\int_0^t H_s \, dM_s = \int_0^{\langle M\rangle_t} H_{T_s} \, dB_s \text{ for every } t \geq 0 \qquad (9.2.3)$$

for every progressively measurable process H in the sense that one side is defined whenever the other is, and we accept the convention that H = 0 if $T_r = \infty$.

Comments. (a) The definition of B_r is justified by lemma 9.2.2.

(b) The theorem implies in particular that the stochastic integral for continuous local martingales is in a precise sense not more general than the integral w.r.t. Brownian motions. However, the increasing process $\langle M\rangle$ which establishes this connection is not so easy to construct without the apparently more general integral of chapter 5.4.

(c) Another remarkable feature of the result is that after the rescaling the dependence of the increments of M disappears.

(d) Here is a variation which differs from the theorem only if the "usual conditions" on the filtration are not assumed:

Assume that *all* paths of $\langle M\rangle$ are continuous and nondecreasing - e.g. if $M = \int_0^\cdot H \, dW$ for some $(\boldsymbol{F}_t)$-Brownian motion W then $\langle M\rangle_t$ has the version $\int_0^t H_s^2 \, ds$. Consider

$$S_r = \inf \{t : \langle M\rangle_t = r\} .$$

Then $S_r \le T_r$ and $\{\langle M\rangle = r\} = [S_r, T_r]$. By proposition 5.6.8 almost all paths of M are constant on this interval and $B'_r = M_{S_r}$ is a.s. equal to $B = M_{T_r}$. Moreover, S_r is a strict stopping time (proposition 2.3.5) and B'_r is $\boldsymbol{F}_{S_r}$-measurable. The process B' may not be right-continuous but it has a right-continuous $(\boldsymbol{F}_{S_r})$-adapted modification by 4.3.5.

Proof of theorem 9.2.3. Since $r_n \downarrow r$ implies $T_{r_n} \downarrow T_r$ by the right-continuity of $\langle M\rangle$, we conclude that the right-continuity of M implies the right-continuity of B. That B_r is $\boldsymbol{F}^+_{T_r}$-measurable follows from proposition 2.3.11.
For the a.s. statements choose a nullset N such that for $\omega \notin N$ the following hold: The paths $M.(\omega)$ and $\langle M\rangle.(\omega)$ are continuous, the latter is nondecreasing, and finally,

$$(9.2.4) \qquad \langle M\rangle_t(\omega) = \langle M\rangle_s(\omega) \quad \text{implies} \quad M_t(\omega) = M_s(\omega) .$$

This is possible by proposition 5.6.8. Outside N we have

$$(9.2.5) \qquad \langle M\rangle_{T_r} = r \wedge \langle M\rangle_\infty \text{ for all } r$$

and in particular,

$$\langle M\rangle_{T_{\langle M\rangle_t}} = \langle M\rangle_t .$$

Assuming for notational simplicity $M_0 = 0$, this identity and (9.2.4) imply

$$B_{\langle M\rangle_t} = M_{T_{\langle M\rangle_t}} = M_t .$$

For the continuity of B, it remains to verify the left-continuity. Suppose $r_n \uparrow r$ and set $s = \sup T_{r_n}$. Then using (9.2.5)

$$\langle M\rangle_s = \lim_{n\to\infty} \langle M\rangle_{T_{r_n}} = \langle M\rangle_{T_r} .$$

Hence by (9.2.4) and the continuity of M

$$B_r = M_{T_r} = M_s = \lim_{n\to\infty} M_{T_{r_n}} = \lim_{n\to\infty} B_{r_n} .$$

By the stopping theorem the processes B and $(B_r^2 - \langle M\rangle_{T_r})_{r\ge 0}$ are a.s. continuous local martingales w.r.t. $(\boldsymbol{G}_r)_{r\ge 0}$, hence $(\langle M\rangle_{T_r})_{r\ge 0}$ is the increasing process of B. Because of (9.2.5) this shows that B is a Brownian motion stopped by $\langle M\rangle_\infty$.
Suppose now that H is progressively measurable w.r.t. $(\boldsymbol{F}_t)_{t\ge 0}$. The transformation formula (5.1.8) for Lebesgue-Stieltjes integrals yields

$$(9.2.6) \qquad \int_0^{T_t} H_s^2 d\langle M\rangle_s = \int_0^t H_{T_s}^2 d\langle M\rangle_{T_s} = \int_0^t H_{T_s}^2 d\langle M\rangle_{\infty \wedge s}.$$

Hence the integrability condition in definition 6.3.4 is fulfilled either for both sides of the equation (9.2.3) or for neither of them.

To prove the equation (9.2.3), let us start with an elementary process $H = \mathbf{1}_{]0,R]}$, where R is a stopping time. Set

$$R' = \sup\{s \geq 0 : T_s < R\}.$$

Then

$$H_{T_t}(\omega) = \mathbf{1}_{]0,R']}(t,\omega) - \mathbf{1}_{N_t}(t,\omega) \text{ where } N_t(\omega) \subset \{t\} \times \Omega$$

and we have the following three observations:

(i) R' is a strict $(\boldsymbol{G}_t)$-stopping time. In fact:

$$\{R' \leq t\} = \bigcap_{s \in \mathbb{Q} \cap [0,t[} \{T_s < R\} \in \boldsymbol{F}^+_{T_t} = \boldsymbol{G}_t.$$

(ii) $\lambda \otimes \mathbb{P}(N_t) = 0$.

(iii) $\langle M\rangle_{T_s \wedge R} \longrightarrow t \wedge R'$ if $s \uparrow t$.

It follows that

$$\int_0^{T_t} H\, dM = M_{T_t \wedge R} - M_0 = \lim_{s \uparrow t} M_{T_s \wedge R} - M_0 = B_{t \wedge R'}$$

$$= \int_0^t \mathbf{1}_{]0,R']}\, dB = \int_0^t H_{T_s}\, dB_s.$$

The equality extends to elementary processes by linearity and to general H by approximation because of (9.2.6). This completes the proof. ■

We use the theorem of Dambis-Dubins-Schwarz to prove a crude criterion for a stopped exponential local martingale to be a martingale.

9.2.4 Proposition. Let $M \in \boldsymbol{M}_c$ start at zero and let T be a stopping time such that $\langle M\rangle_T \leq c$ a.s. for some finite number c. Then the exponential local martingale $\exp(M_{\cdot \wedge T} - \frac{1}{2}\langle M\rangle_{\cdot \wedge T})$ is a $\boldsymbol{L}^2$-bounded martingale.

Proof. Recall that the exponential process is a local martingale by proposition 8.2.2 . We use criterion 4.2.3. Hence it is sufficient to show that the family

$$\{\exp(M_S - \tfrac{1}{2}\langle M\rangle_S) : S \text{ is a stopping time with } S \leq T\}$$

is bounded in $\boldsymbol{L}^2$. The following relations are straightforward:

$$\mathbb{E}([\exp(M_S - \tfrac{1}{2}\langle M\rangle_S)]^2) \le \mathbb{E}(\exp(2M_S))$$

$$\le \mathbb{E}(\exp(2 \max_{0\le t\le S} M)) = \mathbb{E}(\exp(2 \max_{0\le t\le S} B_{\langle M\rangle_t}))$$

$$\le \mathbb{E}(\exp(2 \max_{0\le s\le c} B_s)) = \mathbb{E}((\max_{0\le s\le c} \exp(B_s))^2) .$$

Application of Doob's inequality 3.1.3 to the submartingale $(\exp(B_s))_{0\le s\le c}$ yields the bound $\mathbb{E}(\exp(2B_c))$. This completes the proof. ■

The multidimensional version of the theorem is due to F. Knight (1970) .

9.2.5 Theorem. Suppose that $M = (M^1,\dots,M^d) \in \boldsymbol{M}_c^d$ starts at $0 \in \mathbb{R}^d$ and fulfills
(i) $\langle M^k\rangle_\infty \equiv \infty$ a.s. for every $k = 1,\dots,d$,
(ii) $\langle M^j, M^k\rangle \equiv 0$ a.s. if $j \neq k$.
Let

$$T_r^k = \inf\{t : \langle M^k\rangle_t > r\} \text{ and } B_r^k = M_{T_r}^k .$$

Then $B = (B^1,\dots,B^d)$ is a d-dimensional Brownian motion starting at $0 \in \mathbb{R}^d$.

Proof. The idea is to compute the characteristic function of the increments like in proposition 9.1.7. Plainly, the proof is a bit more involved. Define the continuous martingale

$$X_t = \sum_{k=1}^{d} \sum_{j=1}^{n} a_j^k \left(B_{t\wedge t_{j+1}}^k - B_{t\wedge t_j}^k\right) .$$

For convenience of notation we write it in the form

$$X_t = \sum_{k=1}^{d} \int_0^t h_k \, dB^k \text{ where } h_k = \sum_{j=1}^{n} a_j^k \mathbf{1}_{]t_j,t_{j+1}]} .$$

We must compute the increasing process of X. To this end we note the transformation

$$\int_0^t h_k \, dB^k = \int_0^{T_t^k} h_k(\langle M^k\rangle) \, dM^k$$

which follows from (9.2.3), and

$$\int_0^t h_k^2(s) \, ds = \int_0^{T_t^k} h_k^2(\langle M^k\rangle) \, d\langle M^k\rangle$$

which corresponds to (9.2.6) . Differential calculus gives

$$\langle X\rangle_t = \sum_{j,k=1}^{d} \int_0^{T_t^j \wedge T_t^k} h_j(\langle M^j\rangle) \, h_k(\langle M^k\rangle) \, d\langle M^k, M^j\rangle$$

and by the orthogonality assumption (ii),

$$\langle X\rangle_t = \sum_{k=1}^{d} \int_0^{T_t^k} h_k^2\left(\langle M^k\rangle\right) d\langle M^k\rangle = \sum_{k=1}^{d} \int_0^t h_k^2(s)\, ds.$$

This quantity is deterministic and bounded by a constant c not depending on t. Hence the exponential local martingale $Y = \exp(iX + \frac{1}{2}\langle X\rangle)$ is uniformly bounded:

$$|Y_t| \le |e^{\langle X\rangle_t}| \le e^c.$$

Therefore Y is even a martingale and the characteristic function of the increments can be found as in proposition 9.1.7:

$$\mathbb{E}\left(e^{iX_t}\right) = \exp\left(-\frac{1}{2} \sum_{k=1}^{d} \sum_{j=1}^{n} (a_j^k)^2 \,(t_{j+1} - t_j)\right).$$

Thus B is a $(\boldsymbol{F}_t)$-Brownian motion where $(\boldsymbol{F}_t)$ is its natural filtration. ■

Remarks. (a) Knight's original proof is rather technical and hard to read. The above proof is due to C. Cocozza and M. Yor (1980).

(b) If one drops condition (i) one arrives at stopped Brownian motion rather than Brownian motions. As in proposition 9.1.5 the underlying probability space has to be enlarged.

(c) There is no counterpart for the filtrations $\boldsymbol{F}_{T_r}^+$ from theorem 9.2.3 since several time-changes are involved.

Here is another example of a time-change of Brownian motion.

9.2.6 Definition. Let $(\boldsymbol{G}_r)_{r\ge 0}$ be a filtration. A right-continuous $(\boldsymbol{G}_r)$-adapted process X is called a $(\boldsymbol{G}_r)$-**Cauchy process** if for every $r \ge 0$ and $s \ge 0$ the increment $X_{s+r} - X_s$ is independent of $\boldsymbol{G}_s$ with a Cauchy(r)-distribution, i.e.

$$\mathbb{P}\left(X_{s+r} - X_s \in A\right) = \int_A \frac{r}{\pi(x^2 + r^2)}\, dx.$$

The following result shows that the point on the vertical line $\{x\} \times \mathbb{R}$ at which a two-dimensional Brownian motion first crosses this line defines a Cauchy process $(X_r)_{r\ge 0}$.

9.2.7 Theorem. Let B^1 and B^2 be two independent $(\boldsymbol{F}_t)$-Brownian motions. Consider the stopping times

$$T_r = \inf\left\{t : B_t^1 > r\right\}.$$

Then the process $(B^2_{T_r})_{r \geq 0}$ is a $(\mathcal{F}^+_{T_r})$-Cauchy process.

The main step in the proof is to determine the distribution of T_r.

9.2.8 Lemma. Let B be a Brownian motion and $r > 0$. The distribution of the stopping time $T_r = \inf\{t : B_t > r\}$ has the density

$$f_{T_r}(s) = \begin{cases} \frac{r}{s} \frac{1}{\sqrt{2\pi s}} e^{-\frac{r^2}{2s}} & s > 0 \\ 0 & s \leq 0 \end{cases}$$

Proof. We have $B_{T_r} = r$ a.s. and thus for $s > 0$,

$$\begin{aligned} \mathbb{P}(T_r < s) &= \mathbb{P}\left(T_r < s,\ B_s - B_{T_r} > 0\right) + \mathbb{P}\left(T_r < s,\ B_s - B_{T_r} < 0\right) \\ &= 2\mathbb{P}\left(T_r < s,\ B_s - B_{T_r} > 0\right) = 2\mathbb{P}\left(B_s > r\right) \\ &= 2\int_r^\infty \frac{1}{\sqrt{2\pi s}} e^{-\frac{y^2}{2s}} dy\,. \end{aligned}$$

The density f_{T_r} at $s > 0$ is the derivative $\frac{d}{d\sigma}\mathbb{P}(T_r < \sigma)_{\sigma = s}$. We compute this by differentiating the Brownian transition function $p(s,y)$ under the integral sign and observing that p solves $\frac{\partial}{\partial s} p = ½ \frac{\partial^2}{\partial y^2} p$. Thus we get

$$f_{T_r}(s) = \frac{\partial}{\partial y} \frac{1}{\sqrt{2\pi s}} e^{-\frac{y^2}{2s}} \Big|_r^\infty = \frac{r}{s} \frac{1}{\sqrt{2\pi s}} e^{-\frac{r^2}{2s}}\,.$$

Hence f_{T_r} has the stated form. ■

Proof of the theorem. Since T_r is independent of B^2, the density of $B^2_{T_r}$ given $T_r = s$ is the standard density $p(s,y)$. Thus the lemma shows that the unconditional density of $B^2_{T_r}$ is given by

$$\begin{aligned} \int_0^\infty p(s,x)\, f_{T_r}(s)\, ds &= \int_0^\infty \frac{1}{\sqrt{2\pi s}} e^{-\frac{x^2}{2s}} \frac{1}{\sqrt{2\pi s}} \frac{r}{s} e^{-\frac{r^2}{2s}} ds \\ &= \int_0^\infty \frac{r}{2\pi s^2} e^{-\frac{x^2+r^2}{2s}} ds \\ &= \frac{r}{\pi(x^2+r^2)}\,. \end{aligned}$$

Thus $B^2_{T_r}$ has a Cauchy(r) distribution. Moreover, the increment $B^2_{T_{s+r}} - B^2_{T_s}$

is independent of $F^+_{T_s}$ and has the same distribution as $B^2_{T_r}$ since $T_{s+r} - T_s$ is independent of B^2 and has the same distribution as T_r. ■

Remark. As in the introductory example the family (T_r) is a.s. not left-continuous and thus in contrast to Brownian motion the paths of the Cauchy process a.s. have jumps.

Another proof of theorem 9.2.8 which avoids the above computations is given in Durrett (1984), p.33.

9.3 Harmonic Functions and Martingales

There is a close connection between classical potential theory and the theory of Brownian motion.This showed up already in Einstein's description of the Brownian transition function and also in example 8.1.3 for Itô's formula. Theorem 9.3.2 below is a probabilistic characterization of harmonic functions in terms of the associated functionals of Brownian motion. This implies a probabilistic interpretation of the Dirichlet problem which in turn is applied to the recurrence properties of Brownian motion. The interplay between Brownian motion and the Laplace operator will be extended in chapter 12 to diffusion processes and elliptic differential operators.

9.3.1 Definition. Let $U \subset \mathbb{R}^d$ be open. A realvalued function h of class $\mathscr{C}^2$ on U is called **harmonic** if $\Delta h = 0$ where $\Delta = \sum_{i=1}^{d} \frac{\partial^2}{\partial x_i^2}$ is the Laplacian.

The relation $\Delta h = 0$ translates into a martingale condition for the process h(B).

9.3.2 Theorem. Let U be an open subset of $\mathbb{R}^d$. A function h of class $\mathscr{C}^2$ on U is harmonic if and only if the following holds for every $x \in U$: Let B^x be a d-dimensional $(\boldsymbol{F}_t)$-Brownian motion starting at x and let $T = \inf\{t : B^x_t \notin U\}$. Then for every stopping time S such that $\mathbb{P}(S < T) = 1$ the process $h(B^x_{\cdot\wedge S})$ is a local $(\boldsymbol{F}_t)$-martingale.

Proof. Let $X_t = B^x_{t\wedge S}$. Then $\mathbb{P}(X_t \in U) = 1$ for all t. For every $\mathscr{C}^2$-function h on U, Itô's formula reads

$$dh(X) = \sum_{k=1}^{d} \partial_k h(X) \bullet dX + \tfrac{1}{2} \sum_{j,k=1}^{d} \partial_j \partial_k h(X) \bullet d\langle X^j, X^k \rangle .$$

Since by example, 7.3.17

$$\langle X^j, X^k \rangle_t = \langle B^{x,j}, B^{x,k} \rangle_{t \wedge S} = \delta_{jk} \cdot (t \wedge S)$$

this is equivalent to

$$dh(X) = \sum_{k=1}^{d} \partial_k h(X) \bullet dX + \tfrac{1}{2} \Delta h(X)\, d(t \wedge S). \tag{9.3.1}$$

Let now h be harmonic. Then the second term in (9.3.1) vanishes, i.e. $dh(X)$ is the differential of an integral with respect to a stopped Brownian motion. Thus $h(X)$ is a local martingale.

Conversely, let $h(B^x_{\cdot \wedge S})$ be a local martingale for every $x \in U$ and every stopping time S such that $\mathbb{P}(S < T) = 1$. Suppose that h were not harmonic. Then there is a $x_o \in U$ such that $\Delta h(x_o) \neq 0$. We may assume $\Delta h(x_o) > 0$. Let V be an open ball such that

$$x_o \in V \subset \overline{V} \subset U \cap \left\{ x : \Delta h(x) > \tfrac{1}{2} \Delta h(x_o) \right\} .$$

Let $S = \inf\{t : B_t^{x_o} \notin \overline{V}\}$. Then $\mathbb{P}(S < T) = 1$ and $h(B^{x_o}_{\cdot \wedge S})$ is uniformly bounded since h is bounded on $\overline{V}$. Therefore this process is a martingale according to criterion 4.2.3. The martingale property of $h(B^{x_o}_{\cdot \wedge S})$ and of the stochastic integrals (theorem 5.5.1) together with (9.3.1) imply

$$0 = \mathbb{E}\left(h(B_S^{x_o}) - h(B_o^{x_o})\right) = \mathbb{E}\left(\sum_{k=1}^{d} \int_o^S \partial_k\, h(B_t^{x_o})\, dB_t^{x_o,k} + \tfrac{1}{2} \int_o^S \Delta h(B_t^{x_o}) dt \right)$$

$$= \tfrac{1}{2}\, \mathbb{E}\left(\int_o^S \Delta h(B_t^{x_o})\, dt \right) > \tfrac{1}{4}\, \Delta h(x_o)\, \mathbb{E}(S) > 0 .$$

This is a contradiction and thus h is harmonic. This completes the proof of the theorem. ■

Solutions of the Dirichlet problem can now be represented as functionals of Brownian motion.

9.3.3 Corollary. Let $U \subset \mathbb{R}^d$ be open and bounded. Let $h \in \mathscr{C}^2(U) \cap \mathscr{C}(\overline{U})$ be a solution to the Dirichlet problem

$$\begin{cases} \Delta h = 0 & \text{in } U \\ h = \varphi & \text{on } \partial U . \end{cases}$$

Then for every $x_0 \in U$,

$$h(x_0) = \mathbb{E}\left(\varphi\left(B_T^{x_0}\right)\right)$$

where $(B_t^{x_0})_{t \geq 0}$ is a $(\boldsymbol{F}_t)$-Brownian motion starting in x_0 and T is its entrance time to the boundary ∂U of U.

This representation suggests a Monte-Carlo method for the actual evaluation of solutions h of the Dirichlet problem. To calculate $h(x_0)$, run a Brownian motion starting at x_0 and note where it leaves U. After several runs one has an approximation of the law of $B_T^{x_0}$ on ∂U. Integration of the boundary condition φ w.r.t. this measure yields then an approximation of $h(x_0)$.

Note that only that part of U contributes to the solution which can be reached from x_0 by a continuous path. Therefore we do not suppose U to be connected.

Proof of 9.3.3. Let

$$S_n = \inf\left\{t : \operatorname{dist}(B_t^{x_0}, \partial U) < \tfrac{1}{n}\right\}$$

be the entrance time into the $\frac{1}{n}$- neighbourhood of the boundary ∂U. Then $S_n \uparrow T$ and $h(B_{S_n}^{x_0}) \longrightarrow \varphi(B_T^{x_0})$ a.s. since the paths of B are a.s. continuous and since h is continuous at ∂U. By the theorem each process $h(B_{\cdot \wedge S_n}^{x_0})$ is a local martingale. Since h is bounded these processes are even martingales and thus

$$h(x_0) = \mathbb{E}\left(h(B_0^{x_0})\right) = \mathbb{E}\left(h(B_{S_n}^{x_0})\right).$$

Finally, these expectations converge to $\mathbb{E}\left(\varphi(B_T^{x_0})\right)$ by dominated convergence. Thus

$$h(x_0) = \mathbb{E}\left(\varphi(B_T^{x_0})\right)$$

which completes the proof. ■

9.3.4 Example. We consider for $d \geq 2$ a shell

$$U = \left\{x \in \mathbb{R}^d : a < |x| < b\right\}$$

and the boundary conditions

$$\varphi(x) = 0 \text{ for } |x| = a \text{ and } \varphi(x) = 1 \text{ for } |x| = b.$$

Denote by S(r) the sphere around the origin with radius r. Let $h_{a,b}$ denote the solution of the associated Dirichlet problem in U. The representation of $h_{a,b}$ in the corollary allows the following probabilistic interpretation:

$$h_{a,b}(x_0) = \mathbb{P}\left(B^{x_0} \text{ hits } S(b) \text{ before it possibly hits } S(a)\right). \tag{9.3.2}$$

The function $h_{a,b}$ can be determined analytically:

$$h_{a,b}(x) = \frac{f(|x|) - f(a)}{f(b) - f(a)} \quad \text{for } a < |x| < b \tag{9.3.3}$$

where

$$f(r) = \begin{cases} \log r & \text{if } d = 2 \\ r^{-(d-2)} & \text{if } d \geq 3 \end{cases}. \tag{9.3.4}$$

In fact, for a function $g(x) = f(|x|)$ one easily computes

$$\Delta g(x) = f''(|x|) + \frac{(d-1)f'(|x|)}{|x|}$$

and hence the condition $\Delta g = 0$ means that up to multiplicative and additive constants $f'(r) = r^{-(d-1)}$ and therefore (9.3.3) and (9.3.4) hold.

We establish now the basic recurrence and transience properties of Brownian motion in several dimensions.

9.3.5 Theorem. Let B^{x_0} be a d-dimensional Brownian motion starting at $x_0 \in \mathbb{R}^d$.
(a) For $d = 2$ the Brownian motion B^{x_0} visits with probability 1 every nonempty open set; in particular, almost all paths are dense in the plane. However, for every $x \neq x_0$ it visits x with probability 0.
(b) For $d \geq 3$ one has $\lim_{t\to\infty} |B_t^{x_0}| = \infty$ a.s. .

The first part of (a) means that Brownian motion in two dimensions is **neighbourhood recurrent**.

Proof. We note in advance two simple observations.
(1) From (9.3.3) we conclude that

$$\lim_{b\to\infty} h_{a,b}(x) = \begin{cases} 0 & \text{for } d = 2 \\ 1 - \left(\frac{|a|}{|x|}\right)^{d-2} & \text{for } d \geq 3 \end{cases}.$$

(2) Each random vector $B_t^{x_0}$ has density

$$\frac{1}{(2\pi t)^{d/2}} \cdot \exp\left(- \frac{|x - x_0|^2}{2t}\right).$$

Thus the probability that $B_t^{x_0}$ is in a fixed closed ball around the origin tends to zero as $t \longrightarrow \infty$. Hence B^{x_0} eventually leaves every such ball with radius greater than $|x_0|$ and thus hits the boundary in finite time a.s..

We consider now the case d = 2. Choose a starting point x_0 and assume that x_0 lies in the open ring between the spheres S(a) and S(b), i.e. $0 < a < |x_0| < b$. Since a continuous function on an interval $[0,t]$ is bounded a path of B^{x_0} cannot hit S(a) after having met *each* S(b) with $b > |x_0|$. Hence using first (2) and then (1) we get

$$\begin{aligned}
&\mathbb{P}\left(B^{x_0} \text{ hits } S(a)\right) \\
&= \mathbb{P}\left(\text{there is } b > |x_0| \text{ such that } B^{x_0} \text{ hits } S(a) \text{ before } S(b)\right) \\
&= 1 - \lim_{b\to\infty} \mathbb{P}\left(B^{x_0} \text{ does not hit } S(a) \text{ or it hits } S(b) \text{ before } S(a)\right) \\
&= 1 - \lim_{b\to\infty} h_{a,b}(x_0) = 1.
\end{aligned}$$

One easily concludes by translation that every open set is visited a.s.. Note that one has to check this only for a countable number of balls, hence almost every path visits *all* open sets.
For $d \geq 3$ it follows by a similar argument that

$$\begin{aligned}
&\mathbb{P}\left(B_s^{x_0} \in S(a) \text{ for some time } s \geq t\right) \\
&= \int_{\mathbb{R}^d} \mathbb{P}\left(B_s^{x} \in S(a) \text{ for some time } s \geq 0\right) d\nu_t(x) \\
&= \int_{\mathbb{R}^d} \left(\frac{a}{|x|}\right)^{d-2} d\nu_t(x)
\end{aligned}$$

where $\nu_t(x)$ is the distribution of $B_t^{x_0}$. Because of (2) the integral converges to 0 as $t \longrightarrow \infty$. Thus a visit of S(a) after t becomes arbitrarily unlikely. This implies that $|B_s^{x_0}| \longrightarrow \infty$ a.s. as $s \to \infty$. Finally, for d = 2,

$$\begin{aligned}
\mathbb{P}\left(B^{x_0} \text{ visits } 0\right) &= \mathbb{P}\left(B^{x_0} \text{ visits } 0 \text{ before } S(b) \text{ for some } b\right) \\
&= \lim_{b\to\infty} \mathbb{P}\left(B^{x_0} \text{ visits } 0 \text{ before } S(b)\right) \\
&= \lim_{b\to\infty} \lim_{a\to 0} \mathbb{P}\left(B^{x_0} \text{ visits } S(a) \text{ before } S(b)\right) \\
&= \lim_{b\to\infty} \lim_{a\to 0} \left(1 - h_{a,b}(x_0)\right) = 0
\end{aligned}$$

since $h_{a,b}(x) \xrightarrow[a\to 0]{} 1$. Similarly $\mathbb{P}(B^{x_0} \text{ visits } x) = 0$. This completes the proof. ■

As a typical application of the strong Markov property we would like to strengthen the recurrence statement in two dimensions as follows.

9.3.6 Corollary. Let B be a two-dimensional Brownian motion starting in x_0 and let U be a nonempty open set in $\mathbb{R}^2$. Then B^{x_0} spends a.s. an infinite amount of time inside U.

Proof. By translation we can assume that U contains a ball $B(0,a)$, $a > 0$. Define successively stopping times by $T_0 = 0$ and

$$\begin{aligned} S_n &= \inf\left\{t > T_{n-1} \ : \ |B_t| = \tfrac{a}{2}\right\} \\ T_n &= \inf\left\{t > S_n \ : \ |B_t| = a\right\}. \end{aligned}$$

Then the strong Markov property 9.1.11 implies that $T_n - S_n$ is independent of $\mathcal{F}^+_{S_n}$ with a distribution given by

$$\mathbb{P}\left(T_n - S_n \geq t\right) = \mathbb{E}^{B_{S_n}}\left(1_{\{|B_s|<a \text{ for } s<t\}}\right).$$

The right-hand side is constant a.s. since it depends only on $|B_{S_n}|$ which a.s. equals a. Therefore the sequence $(T_n - S_n)_{n\geq 1}$ is i.i.d. and nonnegative . The strong law of large numbers gives

$$\sum_{n=1}^{\infty} T_n - S_n = +\infty \text{ a.s.}\,.$$

In particular, the random set

$$\left\{t \ : \ |B_t^{x_0}| < a\right\}$$

which contains all the intervals $[S_n, T_n[$ has infinite length. ■

9.4 Characterization of Homogeneous Transition Functions

Here is another consequence of Lévy's theorem. The Brownian transition function

$$p(t,x,y) = \frac{1}{\sqrt{2\pi t}}\, e^{\frac{(y-x)^2}{2t}}$$

is **spatially homogeneous** i.e. $p(t,x+h,y+h) = p(t,x,y)$ for all $t > 0$ and $x, y, h \in \mathbb{R}$. In particular, Brownian motion is not only a Markov process but has also independent increments. So the transition density from x to y does only depend on the distance $y - x$ and not on the previous path which led the particle to x. Surprisingly, the converse is also true.

9.4.1 Theorem. Suppose that X is a Markov process with continuous paths and starting at zero. Assume further that X has a (stationary) continuous and spacially homogeneous transition function

$$q :]0,\infty[\times \mathbb{R} \times \mathbb{R} \longrightarrow [0,\infty[$$

and that

$$\mu = \int_{\mathbb{R}} y\, q(1,0,y)\, dy \text{ and } \sigma^2 = \int_{\mathbb{R}} (y-\mu)^2\, q(1,0,y)\, dy$$

exist. Then

$$q(t,x,y) = \frac{1}{\sqrt{2\pi\sigma^2 t}}\, e^{-\frac{1}{2\sigma^2 t}(y-x-\mu t)^2} .$$

Proof. Since the transition function of X is homogeneous in space and time, X has independent increments and so does $M_t = X_t - \mu t$. Denote the natural filtration of M by $(\boldsymbol{G}_t)_{t\geq 0}$. Because the increments are independent we have

$$\mathbb{E}(M_t \mid \boldsymbol{G}_s) = \mathbb{E}(M_t - M_s + M_s | \boldsymbol{G}_s) = \mathbb{E}(M_t - M_s) + M_s.$$

Since further

$$\mathbb{E}(M_t) = \mathbb{E}(X_t) - \mu t = \mu t - \mu t = 0 ,$$

M is a continuous martingale w.r.t. $(\boldsymbol{G}_t)_{t\geq 0}$ starting at zero. Each increment $M_t - M_s$ is independent of $\boldsymbol{G}_s$ hence so is $\langle M\rangle_t - \langle M\rangle_s$ by the very definition of quadratic variation . Set

$$f(t) = \mathbb{E}(\langle M\rangle_t) = \mathbb{E}(M_t^2)\ (< \infty)$$

(cf. (5.6.5)). Stationarity of the increments yields again by (5.6.5)

$$\mathbb{E}(\langle M\rangle_t - \langle M\rangle_s) = \mathbb{E}(\langle M\rangle_{t-s})$$

hence

$$f(s+t) = f(s) + f(t) \text{ for all } s, t \geq 0 .$$

Since $\langle M\rangle$ and therefore f is increasing and since $f(1) = \sigma^2$ we conclude that

$$f(t) = \sigma^2 t \text{ for every } t \geq 0 .$$

In summary, the process $\langle M\rangle - f(\cdot)$ is of locally finite variation and has independent increments and expectation zero. So it is a continuous martingale with start at zero and by theorem 5.3.2 the zero process. Hence $\langle M\rangle_t = \sigma^2 t$.

In summary, the process $\langle M \rangle - f(\cdot)$ is of locally finite variation and has independent increments and expectation zero. So it is a continuous martingale with start at zero and by theorem 5.3.2 the zero process. Hence $\langle M \rangle_t = \sigma^2 t$. The salient point is again Lévy's theorem . It tells us that $B_t = M_{t/\sigma^2}$ is a Brownian motion with $B_{\sigma^2 t} + \mu t = X_t$. This implies the assertion. ■

Continuity of the paths as well as the existence of the first two moments can be removed from the assumptions and the converse of the theorem holds also true. The proof is similar to the respective parts in the construction of Wiener measure. A corresponding result holds on Euclidean n-space and even on Lie groups.

The theorem may be rephrased as follows: q is the transition function of a Brownian motion with "diffusion constant" $D = \sigma^2$ and "drift" μ. In fact, processes X as above are special "diffusion processes" which are the topic of the chapter 12.

One can derive a Markov process with transition function q from a Brownian motion by a simple transformation on the paths. Let W denote the coordinate process on Wiener space (cf. section 1.1) and set

$$\varphi : \mathscr{C}(\mathbb{R}_+) \longrightarrow \mathscr{C}(\mathbb{R}_+),\ \varphi(\omega)(t) = \sigma^2 \omega(t) + \mu t \ .$$

Then

(a) on $(\mathscr{C}(\mathbb{R}_+), \mathscr{B}(\mathscr{C}(\mathbb{R}_+)), \mathbb{W})$ the process $(\varphi(W_t))_{t \geq 0}$,

(b) on $(\mathscr{C}(\mathbb{R}_+), \mathscr{B}(\mathscr{C}(\mathbb{R}_+)), \mathbb{W} \circ \varphi^{-1})$ the process $(W_t))_{t \geq 0}$

both have transition function q.

9.5 Conformal Invariance

The central result of this section is Paul Lévy's theorem on conformal invariance of Brownian motion: if $B = X + iY$ is a complex Brownian motion and if f is an entire nonconstant function then f(B) is again a Brownian motion, although running with a variable speed. This is not too suprising: the derivative of an analytic function at any given point is given by a positive multiple of an orthogonal matrix and we know from example 9.1.8 that the composition UB of a d-dimensional Brownian motion and an orthogonal matrix U is again a Brownian motion. Lévy's theorem reveals a fundamental connection between complex - or planar - Brownian motion and analytic functions. It is appealing for both, probabilists and people interested in function theory. Conformal invariance can be used to study Brownian motion via analytic functions and, conversely, to look at analytic functions from a probabilistic point of view. We will only illustrate these aspects and refer to B. Davis' exposition "Brownian motion and analytic functions" (1979).

9.5.1 Definition. A stochastic process $B = X + iY$ is called a **complex** or **planar** $(\mathcal{F}_t)$- **Brownian motion** if (X,Y) is a two-dimensional $(\mathcal{F}_t)$-Brownian motion. A complex-valued a.s. continuous local martingale B is a **complex Brownian motion stopped by** (the nonnegative random variable) **T** if both

$$\langle X\rangle_t = \langle Y\rangle_t = t \wedge T \text{ and } \langle X,Y\rangle_t = 0.$$

Quadratic variation of real-valued Brownian motion is the identity on $\mathbb{R}_+$. Complex quadratic variation was introduced in section 8.6.

9.5.2 Lemma. The complex quadratic variation of a complex Brownian motion vanishes.

Proof. Let $B = X + iY$ be a complex Brownian motion. Then $\langle X,Y\rangle = 0$ and thus

$$\langle B\rangle = \langle X + iY, X + iY\rangle = \langle X\rangle - \langle Y\rangle + 2i\langle X,Y\rangle = t - t = 0. \blacksquare$$

Complex a.s. continuous local martingales with vanishing complex quadratic variation are called **conformal local martingales.** The complex version of the Dubins-Schwarz theorem reads

9.5.3 Theorem. For each conformal local martingale $Z = X + iY$ there is a Brownian motion B (possibly stopped by $\langle X\rangle_\infty$) such that $Z_{T_r} - Z_0 = B_r$ for the natural time-change given by $T_r = \inf\{t \geq 0 : \langle X\rangle_t > r\}$.

Proof. Since

$$0 \equiv \langle Z\rangle = \langle X\rangle - \langle Y\rangle + 2i\,\langle X,Y\rangle,$$

we have

$$\langle X\rangle = \langle Y\rangle \text{ and } \langle X,Y\rangle \equiv 0.$$

By the theorem of Dubins and Schwarz , the processes X^* and Y^* defined by

$$X_r^* = (X_{T_r} - X_0) \text{ and } Y_r^* = (Y_{T_r} - Y_0)$$

are Brownian motions stopped by $\langle X\rangle_\infty$ and by proposition 7.3.11,

$$\langle X^*,Y^*\rangle_r = \langle X_{T_r}, Y_{T_r}\rangle = \langle X,Y\rangle_{T_r} = 0.$$

Hence $B = X^* + i\, Y^*$ is a Brownian motion stopped by $\langle X\rangle_\infty$. ■

Note that we have reproved a special version of the theorem of Knight (cf. remark 9.2.7(b). However, the two components are transformed by the same time-change.

We can now prove Lévy's theorem on conformal invariance of complex Brownian motion.

9.5.4 Theorem. Let B be a complex Brownian motion starting with $B_0 = z_0$. Let further $f : \mathbb{C} \longrightarrow \mathbb{C}$ be a nonconstant entire function. Set

$$T_r = \inf\left\{t \geq 0 : \int_0^t |f'(B_s)|^2 ds > r\right\}.$$

Then $(f(B_{T_r}))_{r\geq 0}$ is a complex Brownian motion starting from $f(z_0)$.

Proof. The complex Itô-formula 8.6.5 and lemma 9.5.2 yield

$$df(B) = f'(B)\bullet dB + \tfrac{1}{2} f''(B)\, d\langle B\rangle = f'(B)\bullet dB.$$

Hence f(B) is an a.s. continuous complex local martingale and

$$d\langle f(B)\rangle = f'(B)^2\bullet d\langle B\rangle = 0$$

by lemma 9.5.2. Write f in the form $f = u + i\, v$ and set

$$T_r = \inf\left\{t \geq 0 : \langle u(B)\rangle_t > r\right\}.$$

According to theorem 9.5.3, the process $f(B_{T_r}))_{r\geq 0}$ is a Brownian motion stopped by $\langle u(B)\rangle_\infty$. It remains to be shown that $\langle u(B)\rangle_\infty = \infty$ and that T_r has the form stated in the theorem.

Let us check the latter. By the Itô formula (9.3.1) and since u is harmonic we get

$$du(B) = \frac{\partial u}{\partial x}(B)\bullet dX + \frac{\partial u}{\partial y}(B)\bullet dY + \tfrac{1}{2}\cdot\Delta u(B)\,dt = \frac{\partial u}{\partial x}(B)\bullet dX + \frac{\partial u}{\partial y}(B)\bullet dY$$

where $B = X + iY$. In particular, $u(B) \in \boldsymbol{M}_c$. Furthermore,

$$d\langle u(B)\rangle = du(B)\bullet du(B) = \left[\frac{\partial u}{\partial x}(B)\right]^2\bullet dX\bullet dX + \left[\frac{\partial u}{\partial y}(B)\right]^2\bullet dY\bullet dY$$

$$= \left[\left[\frac{\partial u}{\partial x}(B)\right]^2 + \left[\frac{\partial u}{\partial y}(B)\right]^2\right]dt = |f'(B)|^2 dt$$

where we have again used $dX\bullet dY = 0$.
To verify that $\langle u(B)\rangle_\infty = \infty$ a.s. recall from lemma 9.2.2 that $\lim_{t\to\infty} u(B)_t$ exists a.s. on the set $\{\langle u(B)\rangle_\infty < \infty\}$. Since the same holds for the imaginary part we conclude that $\lim_{t\to\infty} f(B)_t$ exists a.s. on this set. On the other hand, f being non-constant takes at least two distinct values z_1 and z_2. There are disjoint balls U_1 and U_2 around z_1 and z_2 and B visits each of the open sets $f^{-1}(U_1)$ and $f^{-1}(U_2)$ a.s. arbitrarily late since Brownian motion is neighbourhood recurrent (cf. theorem 9.3.5). Hence $f(B)_t$ a.s. cannot converge as $t \uparrow \infty$. In summary, we have shown that $\langle u(B)\rangle_\infty = \infty$ a.s. which completes the proof. ■

P. Lévy's proof from 1948 (cf. Lévy (1965), VII. 5.6.1) runs roughly as follows: Locally, analytic functions are almost like $az + b$. These functions clearly transform Brownian motion into time changed Brownian motion. Q.E.D. On the intuitive level the gap between Lévy's argument and our presentation is not large. However, a heavy machinery is needed to fill this gap on the technical level. The approach via stochastic calculus is indicated in McKean (1969), p. 109.

A last remark concerns dimensions higher than two. It is natural to ask for those transformations mapping Brownian motion into a time-change of itself. A simple argument using Itô's formula and a theorem of Liouville shows that such mappings necessarily have to preserve angles. To single out these angle preserving maps which fulfill Lévy's theorem in dimension higher than two, one uses Kelvin transformations; cf. Durrett (1984).

As an illustration we give another proof of one part of theorem 9.3.5.

9.5.5 Corollary. The probability that a complex Brownian motion B ever hits a fixed point other than its starting point $0 \in \mathbb{C}$ is zero; i.e. for each point $b \neq 0$ in the plane one has

$$\mathbb{P}(B_t = b \text{ for some } t \geq 0) = 0.$$

Proof. By theorem 9.5.4 the process $b - b\exp(B)$ is a time-changed Brownian motion starting at 0. It never hits b since the exponential function never vanishes and the same is true for B itself since the set covered by a path does not care about the speed the path is went through. ■

Let us now use theorem 9.5.4 the other way round and see how Brownian motion may help to study analytic functions. The following (very) weak version of Picard's little theorem states that the range of a nonconstant analytic function on the complex plane meets every set of positive Lebesgue measure:

9.5.6 Corollary. Suppose that f is an analytic function. Then

$$\lambda^2(\mathbb{C}\backslash f(\mathbb{C})) = 0.$$

Proof. In the notation of theorem 9.5.4 each variable $f(B_{T_r})$, $r > 0$, has a two-dimensional normal distribution which implies the assertion. ■

Picard's little theorem is sharper; it states that the range of a nonconstant entire function meets each point of the complex plane, except perhaps one. The proof requires some work on tangling of Brownian paths. This and related results are collected in the aforementioned exposition of B. Davis. Recent results on the asymptotic distribution of winding and crossing numbers are presented as part of a larger framework of asymptotic laws for planar Brownian motion in the (rather advanced) exposition "Asymptotic laws of planar Brownian motion" by J. Pitman and M. Yor (1986).

9.6 Hermite Polynomials and Wiener Chaos

In classical analysis power series expansions play an important role. In this section we give a first introduction into an analogous series representation of general functionals of Brownian motion. The main idea dates back to N. Wiener's paper on "Homogeneous chaos" (1930). Of course, Wiener did not use the Itô integral but a multiple integral with deterministic integrands (cf. the comments preceding theorem 9.6.8). Partially, this section is also a preparation for the study of Brownian filtrations in section 9.7.

We start out by calculating some concrete integrals. For integration with respect to Brownian motion the (generalized) Hermite polynomials play a role similar to the role of the monomials $\frac{t^n}{n!}$ for the Lebesgue integral. In analogy to the relations

$$t^0 = 1 \,, \; \frac{t^n}{n!} = \int_0^t \frac{s^{n-1}}{(n-1)!}\, ds$$

the Hermite functions $H^{(n)}(t,x)$ can be characterized by

$$(9.6.1) \qquad H^{(0)} \equiv 1 \,, \quad H^{(n)}(t,B_t) = \int_0^t H^{(n-1)}(s,B_s)\, dB_s$$

where B is any $(\boldsymbol{F}_t)$-Brownian motion. In this section we first solve this iteration scheme and then use the Hermite polynomials for an orthogonal decomposition of the space $\boldsymbol{L}^2$ over the Wiener probability space.

We already know that $\int_0^t 1\, dB = B_t$ and $\int_0^t B\, dB = \frac{1}{2}(B_t^2 - t)$, which means that $H^{(1)}(t,x) = x$ and $H^{(2)}(t,x) = \frac{1}{2}(x^2 - t)$.

How can one arrive at an algorithm to get $H^{(3)}$, $H^{(4)}$, etc.? The clue must be again in Itô's formula. Using as in lemma 8.2.3 the index notation for partial derivatives we get

$$(9.6.2) \quad dH^{(n)}(t,B_t) = H_x^{(n)}(t,B_t) \bullet dB_t + \left[H_t^{(n)}(t,B_t) + \tfrac{1}{2}H_{xx}^{(n)}(t,B_t)\right] dt \,.$$

Comparing (9.6.1) and (9.6.2) we arrive at the conditions

$$(9.6.3) \qquad H_t^{(n)}(t,x) = -\tfrac{1}{2}\, H_{xx}^{(n)}(t,x)$$

and

$$(9.6.4) \qquad H_x^{(n)}(t,x) = H^{(n-1)}(t,x) \,.$$

Under the presence of (9.6.4) the relation (9.6.3) may also be read as

$$H_t^{(n)}(t,x) = -\tfrac{1}{2} H^{(n-2)}(t,x) , \; n \geq 2 . \tag{9.6.5}$$

Moreover, the passage from the differential version to the integral version in (9.6.1) requires the initial condition

$$H^{(n)}(0,0) = 0 \tag{9.6.6}$$

and as previously noted we have

$$H^{(0)}(t,x) = 1 , \; H^{(1)}(t,x) = x . \tag{9.6.7}$$

9.6.1 Lemma. The recursive equations (9.6.4) - (9.6.7) have a unique solution. $H^{(n)}$ is a polynomial function in t and x. Its leading term is $\frac{x^n}{n!}$.

9.6.2 Definition. The polynomial $H^{(n)}$ given by (9.6.4) - (9.6.7) is called the n-th **generalized Hermite polynomial**.

Proof of 9.6.1. Suppose that $n \geq 2$ and $H^{(0)}, H^{(1)}, \ldots, H^{(n-1)}$ are given. Then $H^{(n)}$ is uniquely determined since (9.6.4) and (9.6.5) prescribe its partial derivatives and (9.6.6) determines the constant of integration. For the existence one has to make sure that the $\mathscr{C}^1$-vector field $(-\tfrac{1}{2} H^{(n-2)}, H^{(n-1)})$ on $\mathbb{R}^2$ is integrable, i.e. we have to verify the integrability condition

$$-\tfrac{1}{2} H_x^{(n-2)} = H_t^{(n-1)} . \tag{9.6.8}$$

Both sides vanish for n = 2 and for n > 2 we can use by induction the equations (9.6.4) and (9.6.5) for n - 1 to see that in (9.6.8) both sides equal $-\tfrac{1}{2} H^{(n-3)}$. A primitive of a polynomial is a polynomial and the only term of degree n in $H^{(n)}$ is $\frac{x^n}{n!}$ as is again seen by induction. ■

For the actual calculation one first determines term by term a primitive F of $H^{(n-1)}$ in the variable x. Then one chooses the free "constant of integration" h(t) in

$$H^{(n)}(t,x) = F(t,x) + h(t)$$

in such a way that (9.6.5) and (9.6.6) are satisfied. In this way one gets $H^{(3)}(t,x) = \frac{1}{6} x^3 - \tfrac{1}{2} tx$ and $H^{(4)}(t,x) = \frac{1}{24} x^4 - \frac{1}{4} tx^2 + \frac{1}{8} t^2$.

Comment. Here are two other definitions of $H^{(n)}$. Even though they give closed representations they do not reduce considerably the effort to determine say $H^{(15)}$. We omit the verification of these formulas:

$$H^{(n)}(t,x) = \frac{(-t)^n}{n!} \exp\left(\frac{x^2}{2t}\right) \frac{d^n}{dx^n}\left[\exp\left(-\frac{x^2}{2t}\right)\right] \tag{9.6.9}$$

$$H^{(n)}(t,x) = \frac{1}{n!} \sum_{k=0}^{[\frac{n}{2}]} (-1)^k \frac{n!}{(n-2k)!2^k k!} t^k x^{n-2k} . \tag{9.6.10}$$

Note that (9.6.9) once more reveals the close connection with the normal distribution. Part (a) of the following result is another aspect of this connection. Our proof, however, uses the relation (9.6.1).

9.6.3 Proposition. (a) Let ν_t denote the normal distribution on $\mathbb{R}$ with mean 0 and variance t. For every $t > 0$ the sequence

$$\left[\left(\frac{n!}{t^n}\right)^{1/2} H^{(n)}(t,\cdot)\right]_{n\geq 0}$$

of polynomial functions is an orthonormal base of $\mathcal{L}^2(\mathbb{R},\nu_t)$.

(b) The process $\left(H^{(n)}(t,B_t)\right)_{t\geq 0}$ is a $\mathcal{L}^2$-martingale such that

$$\|H^{(n)}(t,B_t)\|_2^2 = \frac{t^n}{n!} . \tag{9.6.11}$$

Proof. We first note that (b) easily follows from (9.6.1) by induction: If $\|H^{(n-1)}(s,B_s)\|_2^2 = \frac{s^{n-1}}{(n-1)!}$ then

$$\|H^{(n)}(t,B_t)\|_2^2 = \int_0^t \mathbb{E}\left(H^{(n-1)}(s,B_s)\right)^2 d\langle B\rangle_s = \int_0^t \frac{s^{n-1}}{(n-1)!}\, ds = \frac{t^n}{n!} .$$

Since B_t has distribution ν_t this also shows that the sequence in part (a) consists of vectors of norm 1 in $\mathcal{L}^2(\mathbb{R},\nu_t)$. For the orthogonality it suffices to show that for $m < n$ the expectation

$$\mathbb{E}\left(H^{(n)}(t,B_t)\, H^{(m)}(t,B_t)\right)$$

vanishes. We evaluate this expectation using stochastic integration by parts.

$$\begin{aligned} &H^{(n)}(t,B_t)\, H^{(m)}(t,B_t) \\ &= \int_0^t H^{(n)}(s,B_s)\, dH^{(m)}(s,B_s) + \int_0^t H^{(m)}(s,B_s)\, dH^{(n)}(s,B_s) \\ &\quad + \int_0^t \left[dH^{(n)}(s,B_s) \bullet dH^{(m)}(s,B_s)\right] . \end{aligned}$$

According to (b) the processes $H^{(n)}(s,B)$ are $\mathcal{L}^2$-martingales and so are the first two stochastic integrals. Their expectation hence vanishes. The expectation of the third integral also vanishes for $n \neq m$, $m \geq 1$, by induction

$$\mathbb{E}\Big(\big(H^{(n-1)}(s,B) \bullet dB\big) \bullet \big(H^{(m-1)}(s,B) \bullet dB\big)\big([0,t]\big)\Big)$$
$$= \mathbb{E}\left(\int_0^t H^{(n-1)}(s,B)\ H^{(m-1)}(s,B)\ ds\right)$$
$$= \int_0^t \mathbb{E}\left(H^{(n-1)}(s,B)\ H^{(m-1)}(s,B)\right) ds = 0.$$

The rest of the assertion follows from the following lemma. ■

9.6.4 Lemma. The polynomial functions are dense in $\mathcal{L}^p(\nu_t)$ for every $p \in [1,\infty[$.

Proof. Since $\int_{-\infty}^{+\infty} e^{\alpha|x|}\, d\nu_t(x) < \infty$ for every $\alpha \in \mathbb{R}$, the power series expansion of the trigonometric functions sin αx and cos αx converge in $\mathcal{L}^p(\nu_t)$ for every α by dominated convergence. Hence these functions are in the $\mathcal{L}^p(\nu_t)$-closure of the polynomial functions. Now for a probability measure ν every smooth function of compact support f can be approximated in $\mathcal{L}^p(\nu)$ by trigonometric polynomials. In fact,

$$\left(\int_{-\infty}^{+\infty} |f - f_n|^p\, d\nu\right)^{1/p} \leq \sup_{|x|\leq c} \left|f(x) - f_n(x)\right| + \left\|f_n\right\|_\infty \Big(\nu\big(\mathbb{R}\setminus[-c,c]\big)\Big)$$

where c is chosen such that f has its support in $[-c,c]$ and f_n is the n-th partial sum of the Fourier series of f with respect to this interval. Thus the closure of the polynomials contains the continuous functions of compact support and hence all of $\mathcal{L}^p(\nu_t)$. ■

Next we want to extend the preceding proposition to the case of products of Hermite polynomials in the Brownian variables. We start by the orthogonality relations.

9.6.5 Lemma. Every polynomial function $p(X_1, \ldots ,X_m)$ of independent Gaussian centered random variables $X_1, \ldots ,X_m$ with variances $\sigma_1^2, \ldots ,\sigma_m^2$ has a representation

$$\sum_{k_1,\ldots,k_m=0}^{n} a_{k_1,\ldots,k_m} \prod_{i=1}^{m} H^{(k_i)}\left(\sigma_i^2, X_i\right) \tag{9.6.12}$$

and in this sum all terms are orthogonal.

Proof. The polynomial $p(X_1, \ldots, X_m)$ has the form

$$(9.6.13) \qquad \sum_{k_1,\ldots,k_m=0}^{n} b_{k_1,\ldots,k_m} \prod_{i=1}^{m} X_i^{k_i} .$$

For every i the monomials x^k are linear combinations of the Hermite polynomials $H^{(n)}(\sigma_i^2, x)$ and this transforms (9.6.13) into (9.6.12). The main point is the orthogonality statement.

Now let us consider two terms of the sum in (9.6.12) which correspond to different multiindices $(k_1,\ldots,k_m)$ and $(l_1,\ldots,l_m)$. We have to show

$$(9.6.14) \qquad \mathbb{E}\Big(\prod_{i=1}^{m} H^{(k_i)}(\sigma_i^2, X_i) \prod_{i=1}^{m} H^{(l_i)}(\sigma_i^2, X_i)\Big) = 0 .$$

Choose i_0 such that $k_{i_0} \neq l_{i_0}$. By proposition 9.6.3,

$$\mathbb{E}\Big(H^{(k_{i_0})}(\sigma_{i_0}^2, X_{i_0})\, H^{(l_{i_0})}(\sigma_{i_0}^2, X_{i_0})\Big) = 0 .$$

The remaining part of the product in (9.6.14) is a function of the X_i, $i \neq i_0$, and independent of this i_0-part. This proves (9.6.14) and hence the lemma. ■

The density statement in the preceding proposition can be extended as follows.

9.6.6 Lemma. Let **V** be the linear span of the random variables of the form

$$(9.6.15) \qquad p_1\big(B_{t_2} - B_{t_1}\big) \cdot p_2\big(B_{t_3} - B_{t_2}\big) \cdot \ldots\ldots \cdot p_k\big(B_{t_{k+1}} - B_{t_k}\big)$$

where $0 \le t_1 < t_2 < \ldots\ldots < t_{k+1}$ and $p_1, \ldots, p_k$ are polynomials. Then **V** is dense in the space $\boldsymbol{L}^2(\Omega, \boldsymbol{F}, \mathbb{P})$ where $\boldsymbol{F}$ is the σ-field generated by the B_t, $t \ge 0$.

Proof. Let Z be of the form

$$f_1\big(B_{t_2} - B_{t_1}\big) \cdot f_2\big(B_{t_3} - B_{t_2}\big) \cdot \ldots\ldots \cdot f_k\big(B_{t_{k+1}} - B_{t_k}\big)$$

where the f_i are bounded measurable functions. Using the last result and the independence of the factors it is easy to see that Z can be approximated by random variables of the form (9.6.15) in $\boldsymbol{L}^2(\mathbb{P})$. The variables Z of this form generate all bounded $\boldsymbol{F}$-measurable functions as a monotone class. This implies the assertion. ■

The following lemma is an analogue of the binomial identity

$$\frac{1}{n!}(x_1 + x_2)^n = \sum_{k=0}^{n} \frac{x_1^k}{k!} \frac{x_2^{(n-k)}}{(n-k)!} .$$

Of course one could give a deterministic proof of the lemma but we deduce it from the recursive relation (9.6.1).

9.6.7 Lemma. The following identity holds for the generalized Hermite polynomials

$$H^{(n)}(t_1 + t_2, x_1 + x_2) = \sum_{k=0}^{n} H^{(k)}(t_1, x_1)\, H^{(n-k)}(t_2, x_2) . \tag{9.6.16}$$

Proof. Since Gaussian distributions on $\mathbb{R}^2$ have dense support and this identity relates continuous functions it suffices to show that (9.6.16) holds a.s. if x_1 and x_2 are replaced by independent Gaussian variables X_1 and X_2 of variance t_1 and t_2, respectively. We choose $X_1 = B_{t_1}$ and $X_2 = \tilde{B}_{t_2}$ where $\tilde{B}_s$ is the increment $B_{t_1+s} - B_{t_1}$. Using (9.6.1) we get by induction

$$H^{(n)}\left(t_1 + t_2, B_{t_1} + \tilde{B}_{t_2}\right) = H^{(n)}\left(t_1 + t_2, B_{t_1+t_2}\right)$$

$$= \int_0^{t_1+t_2} H^{(n-1)}(s, B_s)\, dB_s$$

$$= \int_0^{t_1} H^{(n-1)}(s, B_s)\, dB_s + \int_0^{t_2} H^{(n-1)}\left(t_1 + s, B_{t_1} + \tilde{B}_s\right) d\tilde{B}_s$$

$$= H^{(n)}(t_1, B_{t_1}) + \int_0^{t_2} \sum_{k=0}^{n-1} H^{(k)}(t_1, B_{t_1})\, H^{(n-1-k)}(s, \tilde{B}_s)\, d\tilde{B}_s$$

$$= H^{(n)}(t_1, B_{t_1}) + \sum_{k=0}^{n-1} H^{(k)}(t_1, B_{t_1})\, H^{(n-k)}(t_2, \tilde{B}_{t_2})$$

$$= \sum_{k=0}^{n} H^{(k)}(t_1, B_{t_1})\, H^{(n-k)}(t_2, \tilde{B}_{t_2}) .$$

This completes the proof. ■

N. Wiener (1930) viewed the random variables $H^{(n)}(t, B_t)$ as "multiple integrals". The identity (9.6.1) can be rewritten as

$$H^{(n)}(t,B_t) = \int_0^t \left\{\int_0^{s_1} \left\{\int_0^{s_2} \cdots \left\{\int_0^{s_{n-1}} 1\, dB_{s_n}\right\} \cdots dB_{s_3}\right\} dB_{s_2}\right\} dB_{s_1}$$

or

$$H^{(n)}(t,B_t) = \int_0^t \cdots \int_0^{s_{n-1}} dB_{s_n} \cdots dB_{s_1} = \int \cdots \int_{0<s_1<\ldots<s_n<t} dB_{s_1} \cdots dB_{s_n} .$$

It can be shown that this integral defines a random additive set function over the set $\{(s_1,...,s_n) \in \mathbb{R}^n_+ : s_1 <...< s_n\}$. Then the formula (9.6.16) corresponds to the disjoint decomposition of the area of integration

$$\{s : 0 < s_1 <...< s_n < t_1 + t_2\}$$
$$= \{s : t_1 < s_1 <...< s_n < t_1 + t_2\} \cup \{s : s_1 < t_1 < s_2 <...< s_n < t_1 + t_2\}$$
$$\cup ...\cup \{s : s_1 <...< s_n < t_1\} .$$

The following result thus gives a decomposition of functionals of Brownian motion into multiple integrals. S. Kakutani (1961) noticed a more abstract version of these results for Gaussian processes which does not involve the linear structure of the time parameter. D. Engel (1982) gives a systematic presentation of the set function approach which works also for certain non-Gaussian processes.

9.6.8 Theorem. ("Wiener chaos") Let $\boldsymbol{F}$ be generated by the Brownian motion, $\boldsymbol{F} = \sigma\{ B_t : t \geq 0 \}$. Then one has the orthogonal Hilbert space decomposition

$$\boldsymbol{L}^2(\Omega,\boldsymbol{F},\mathbb{P}) = \sum_{n=0}^{\infty}{}^{\oplus} Q_n$$

where Q_n is the closed linear hull of all random variables of the form

$$(9.6.17) \qquad H^{(n_1)}\left(t_1 - t_0 , B_{t_1} - B_{t_0}\right) ... H^{(n_k)}\left(t_k - t_{k-1} , B_{t_k} - B_{t_{k-1}}\right)$$

such that $0 \leq n_i$ and $\sum_{i=1}^{k} n_i = n$.

Proof. From lemma 9.6.6 it follows that the linear hull of the union of the Q_n is dense in $\boldsymbol{L}^2(\Omega,\boldsymbol{F},\mathbb{P})$. It remains to show that these spaces are orthogonal. Let $Z_1 \in Q_n$ and $Z_2 \in Q_m$, $m < n$, have representations as in (9.6.17). Using lemma 9.6.7 one can assume that the associated time indices $t_0, ... ,t_k$ which appear in (9.6.17) are the same for both representations. Note that the degrees m and n, respectively, of the polynomials are not affected by an eventual application of lemma 9.6.7. Since $m < n$ the orthogonality of Z_1 and Z_2 results now from lemma 9.6.5. ■

9.7 Brownian Filtration

This section shows that the natural filtration of a Brownian motion has a couple of special properties. Maybe the most surprising one is the fact that every local martingale for this filtration is a stochastic integral with respect to the underlying Brownian motion (theorem 9.7.4). The proof of this is based on the explicit calculus of Hermite polynomials in the previous section. In particular, every "Brownian martingale" is a.s. continuous and all stopping times for a Brownian filtration are a.s. predictable.

We start with a consequence of the Markov property. Recall that for a σ-field $\boldsymbol{G}$ the symbol ${}^{0}\boldsymbol{G}$ denotes the nullset augmentation of $\boldsymbol{G}$.

9.7.1 Theorem. Let $(\boldsymbol{G}_t)$ be the natural filtration of a process X. Suppose that for every $t \ge 0$ and every bounded $\sigma(X_{t+s} : s \ge 0)$-measurable function g the conditional expectation $\mathbb{E}(g \mid \boldsymbol{G}_t^+)$ has a $\boldsymbol{G}_t$-measurable version. Then $\boldsymbol{G}_t^+$ is contained in ${}^{0}\boldsymbol{G}_t$ for all $t \ge 0$. In particular, the filtration $({}^{0}\boldsymbol{G}_t)$ is right-continuous.

Proof. Let f be bounded and $\boldsymbol{G}_t$-measurable and let g be as in the hypothesis. Then

$$\mathbb{E}(fg \mid \boldsymbol{G}_t^+) = f\,\mathbb{E}(g \mid \boldsymbol{G}_t^+)$$

has a $\boldsymbol{G}_t$-measurable version. These products generate the full σ-field $\boldsymbol{G} = \sigma(X_s : s \ge 0)$. Hence for every $\boldsymbol{G}$-measurable h there is a $\boldsymbol{G}_t$-measurable version of $\mathbb{E}(h \mid \boldsymbol{G}_t^+)$. In particular, every $\boldsymbol{G}_t^+$-measurable h is a.s. equal to a $\boldsymbol{G}_t$-measurable function, i.e. $\boldsymbol{G}_t^+ \subset {}^{0}\boldsymbol{G}_t$.
For the right-continuity of the filtration $({}^{0}\boldsymbol{G}_t)$, let $(t_n)_{n\ge1}$ decrease to t and let G be a set in $\bigcap {}^{0}\boldsymbol{G}_{t_n}$. We have to show that $G \in {}^{0}\boldsymbol{G}_t$. For every n there is a set G_n in $\boldsymbol{G}_{t_n}$ such that $G = G_n$ a.s. . Then the intersection $G' = \bigcap G_n$ is in $\boldsymbol{G}_t^+$ and hence in ${}^{0}\boldsymbol{G}_t$. Moreover, G and G' differ only on a $\mathbb{P}$-nullset. Thus $G \in {}^{0}\boldsymbol{G}_t$ which completes the proof. ■

This implies e.g. that the asymptotic behaviour of Brownian motion for small times is "trivial". Note that Brownian motion satisfies the hypothesis of theorem 9.7.1 according to theorem 9.1.11.

9.7.2 Corollary. (**Blumenthal's 0-1-law**) If X is as in the theorem and X_0 is a.s. constant then $\mathbb{P}(G) \in \{0,1\}$ for every $G \in \boldsymbol{G}_0^+$.

Proof. If X_0 is a.s. constant then $\mathbb{P}$ is trivial on $\boldsymbol{G}_0 = \sigma(X_0)$ and hence also on ${}^0\boldsymbol{G}_0$. Thus the assertion is a special case of the theorem. ∎

The following results are of a more specific nature.

9.7.3 Definition. A filtration $(\boldsymbol{F}_t)_{t\geq 0}$ is called a **Brownian filtration** if it lies between the natural filtration of a d-dimensional $(\boldsymbol{F}_t)$-Brownian motion B and its augmentation by nullsets, i.e.

$$\sigma\left(B_s \;:\; 0 \leq s \leq t\right) \subset \boldsymbol{F}_t \subset {}^{\circ}\sigma\left(B_s \;:\; 0 \leq s \leq t\right).$$

9.7.4 Theorem. For a Brownian filtration every local martingale M is a Brownian stochastic integral. More precisely, there is a process $H = (H^1, \ldots, H^d)$ in $\boldsymbol{L}^{2,loc}(\mathbb{R}_+ \times \Omega, \boldsymbol{P}, \lambda \otimes \mathbb{P})$ such that for the d-dimensional Brownian motion B inducing the filtration one has

$$M_t = \int_0^t H\, dB = \sum_{i=1}^{d} \int_0^t H^i\, dB^i \text{ a.s..} \tag{9.7.1}$$

Proof. The proof is divided into four steps. In the first three steps it is shown that for every $Z \in \boldsymbol{L}^2(\mathbb{P})$ the martingale $(\mathbb{E}(Z \mid \boldsymbol{F}_t))_{t\geq 0}$ of its conditional expectations can be written in the form

$$\mathbb{E}\left(Z \mid \boldsymbol{F}_t\right) = \int_0^t H\, dB \tag{9.7.2}$$

for some $H \in \boldsymbol{L}^2\,(\lambda \otimes \mathbb{P})$.

(1) We first assume that d = 1 and that Z is of the form

$$p_1\left(B_{t_2} - B_{t_1}\right) \cdot p_2\left(B_{t_3} - B_{t_2}\right) \cdot \ldots \cdot p_k\left(B_{t_{k+1}} - B_{t_k}\right) \tag{9.7.3}$$

for some polynomials p_i. For each t the Hermite polynomials $H^{(n)}(t,\cdot)$ span the linear space of all polynomials; this follows from the fact that the leading term is $\frac{x^n}{n!}$ (lemma 9.6.1). Hence we may also assume that the last factor in (9.7.3) is a Hermite polynomial; more precisely, we assume that

$$Z = G \cdot H^{(n)}\left(t_{k+1} - t_k\,,\, B_{t_{k+1}} - B_{t_k}\right)$$

with a $\boldsymbol{F}_{t_k}$-measurable G. We may also assume that this last factor is nonconstant, i.e. that $n > 0$. We argue that for $t \geq t_k$

$$(9.7.4) \qquad H^{(n)}\left(t - t_k, B_t - B_{t_k}\right) = \int_{t_k}^{t} H^{(n-1)}\left(s - t_k, B_s - B_{t_k}\right) dB_s .$$

The process $(B_{t_k+r} - B_{t_k})_{r\geq 0}$ is a $(\boldsymbol{F}^{+}_{t_k+r})_{r\geq 0}$-Brownian motion (cf. theorem 9.1.9). The identity (9.7.4) is the defining relation for the Hermite polynomials (9.6.1) applied to this process except that the integrating process is replaced by the old Brownian motion. But since these two integrators have the same increments this change is justified. Abbreviate now the left-hand side of (9.7.4) by $J(t)$ and the integrand on the right by $K(t)$, i.e. write (9.7.4) as

$$J(t) = \int_{t_k}^{t} K(s)\, dB_s \text{ for } t \geq t_k .$$

Then $Z = G \cdot J(t_{k+1})$ and since the stochastic integral is a martingale by proposition 9.6.3(b) we get for $t \geq t_k$ using corollary 7.1.13,

$$\begin{aligned} \mathbb{E}\left(Z \mid \boldsymbol{F}_t\right) &= G \cdot \mathbb{E}\left(J(t_{k+1}) \mid \boldsymbol{F}_t\right) = G \cdot \int_{t_k}^{t\wedge t_{k+1}} K(s)\, dB_s \\ &= \int_{t_k}^{t\wedge t_{k+1}} G \cdot K(s)\, dB_s = \int_0^t 1_{]t_k,t_{k+1}]}\, G \cdot K(s)\, dB_s . \end{aligned}$$

In particular, $\mathbb{E}(Z \mid \boldsymbol{F}_{t_k}) = 0$ and hence also for $t < t_k$,

$$\mathbb{E}(Z \mid \boldsymbol{F}_t) = \mathbb{E}\left(\mathbb{E}(Z \mid \boldsymbol{F}_{t_k}) \mid \boldsymbol{F}_t\right) = 0 = \int_0^t 1_{]t_k,t_{k+1}]}\, G \cdot K(s)\, dB_s .$$

This proves the desired representation for Z as in (9.7.2). Note that the integrand in this representation is in $\boldsymbol{L}^2(\lambda \otimes \mathbb{P})$ which can be seen either by direct inspection or by using proposition 6.2.12.

(2) Now assume $Z \in \boldsymbol{L}^2(\mathbb{P})$. In the case $d = 1$ the density statement 9.6.6 together with (1) implies that there is a sequence of random variables $(Z_n)_{n\geq 0}$ such that $\|Z_n - Z\|_2 \longrightarrow 0$ and for each Z_n the associated martingale has a representation

$$\mathbb{E}\left(Z_n \mid \boldsymbol{F}_t\right) = \int_0^t H_n\, dB .$$

Z_n closes this martingale and we have

$$\begin{aligned} \|Z_n - Z_m\|_2^2 &= \mathbb{E}\left(\left(\int_0^\infty H_n - H_m\, dB\right)^2\right) = \mathbb{E}\left(\int_0^\infty \left(H_n - H_m\right)_s^2 ds\right) \\ &= \|H_n - H_m\|_{2,\lambda\otimes\mathbb{P}}^2 . \end{aligned}$$

Therefore (H_n) is a Cauchy sequence in $\boldsymbol{L}^2(\lambda\otimes\mathbb{P} \mid \boldsymbol{P})$. Let H be its limit. Then

$$\mathbb{E}\left(Z \mid \boldsymbol{F}_t\right) = \lim \mathbb{E}\left(Z_n \mid \boldsymbol{F}_t\right) = \lim \int_0^t H_n\, dB = \int_0^t H\, dB .$$

(3) Now let $d > 1$. Consider $Z \in \mathcal{L}^2(\mathbb{P})$ of the form

$$Z = Z^1 \cdot \ldots \cdot Z^d \tag{9.7.5}$$

where Z^i is $\sigma(B^i)$-measurable for $1 \le i \le d$. Let $(\boldsymbol{F}_t^i)_{t\ge0}$ be the natural filtration of B^i. Then the product

$$N = N^1 \cdot \ldots \cdot N^d$$

of the independent martingales $N^i = (\mathbb{E}(Z^i \mid \boldsymbol{F}_t^i))_{t\ge0}$ is a martingale (cf. theorem 7.3.16) which is closed by Z, i.e. $N = (\mathbb{E}(Z \mid \boldsymbol{F}_t))_{t\ge0}$. We know from (2) that each N^i is a stochastic integral process $\int_0^\cdot K^i\, dB^i$; in particular, N^i is a.s. continuous. Then $\langle N^i,N^j\rangle = 0$ for $i \ne j$ and hence by Itô's formula

$$dN = \sum_{i=1}^d \Big(\prod_{j\ne i} N^j\Big)\bullet dN^i = \sum_{i=1}^d \Big(\prod_{j\ne i} N^j\Big)K^i\bullet dB^i$$

i.e. N is a stochastic integral with respect to the d-dimensional Brownian motion $B = (B^1, \ldots, B^d)$. The linear combinations of the random variables of the form (9.7.5) are dense in $\mathcal{L}^2(\mathbb{P})$ and thus as in (2) the representation (9.7.2) is proved also for $d > 1$.

(4) Finally, assume that M is a local martingale. Let (T_k) be a localizing sequence such that $M_{\cdot\wedge T_k}$ is a uniformly integrable martingale for each k. Choose a sequence (Y_k) in $\mathcal{L}^2(\mathbb{P})$ such that $\|M_{T_k} - Y_k\|_1 \longrightarrow 0$. Then

$$\sup_{t\ge0} \left|M_{t\wedge T_k} - \mathbb{E}\big(Y_k \mid \boldsymbol{F}_t\big)\right| \longrightarrow 0$$

in probability by Doob's maximal inequality. Hence the a.s. continuous martingales $(\mathbb{E}(Y_k \mid \boldsymbol{F}_t))_{t\ge0}$ converge l.u.p. to M which therefore also is a.s. continuous. Thus it is a local $\mathcal{L}^2$-martingale. We may assume $M_{\cdot\wedge T_k} \in \mathcal{L}^\square$ and

$$M_{t\wedge T_k} = \int_0^{t\wedge T_k} H_k\, dB$$

for some $H_k \in \mathcal{L}^2(\lambda\otimes\mathbb{P})$. Then $M = \int_0^\cdot H\, dB$ for

$$H = \sum_{k=1}^\infty H_k\, \mathbf{1}_{]T_{k-1},T_k]}\,.$$

This is the desired representation. ■

The following is an immediate consequence.

9.7.5 Corollary. For a Brownian filtration every local martingal has a right-continuous a.s. continuous modification.

Proof of the corollary. Every stochastic integral w.r.t. Brownian motion is right-continuous a.s. continuous according to theorem 6.2.2(c). ■

Comment. (a) In the theorem it is essential that the integrands can be general elements of $\boldsymbol{L}^{2,loc}(\lambda \otimes \mathbb{P} \mid \boldsymbol{P})$. In particular, we need the integral for predictable integrands discussed in section 6.2. However, corollary 9.7.5 can be proved using the above argument based on Doob's inequality without reference to chapter 6.
(b) The proof of the corollary easily extends to the filtration induced by an infinite number of independent Brownian motions. It does not seem to be known whether any filtration on a Polish space for which every martingale is a.s. continuous can be induced by independent Brownian motions.

The continuity of (local) martingales has interesting consequences.

9.7.6 Corollary. A Brownian filtration does not admit an adapted Poisson process.

Proof. If (N_t) is an adapted Poisson process then $(N_t - t)_{t \geq 0}$ is an adapted non-continuous local martingale which cannot exist for a Brownian filtration. ■

The following result implies in particular that the entrance time of a d-dimensional Brownian motion into any open set can be a.s. announced by smaller stopping times. This is far from obvious.

9.7.7 Theorem. Let T be a stopping time for a Brownian filtration $(\boldsymbol{F}_t)$. Then T is $({}^o\boldsymbol{F}_t)$-predictable and the three σ-fields ${}^o\boldsymbol{F}_T$, ${}^o\boldsymbol{F}_T^+$ and ${}^o\boldsymbol{F}_T^-$ coincide.

Proof. According to theorem 9.7.1 the filtration $({}^o\boldsymbol{F}_t)$ is right-continuous. Therefore the stopping time T is a strict stopping time and thus the right-continuous increasing process $A = \mathbf{1}_{[T,\infty[}$ is adapted since $\mathbf{1}_{[T,\infty[}(t) = \mathbf{1}_{\{T \leq t\}}$. According to Meyer's desintegration theorem in the form given by proposition 7.2.5 there is a predictable right-continuous a.s. increasing process A' such that $A'_0 = 0$ and A and A' differ only by a martingale M. Because this martingale is a.s. continuous the two processes A and A' have a.s. the same jumps. Thus

A' has a.s. a jump of size 1 at T and since A' is increasing we have a.s. $A' \geq 0 = A$ on $[0,T[$ and $A' \geq 1 = A$ on $[T,\infty[$. But $\mathbb{E}(A'_\infty) = \mathbb{E}(A_\infty)$ and hence these inequalities a.s. cannot be strict anywhere, i.e. $A = A'$ a.s. and A is a predictable process and $[T,\infty[$ is a predictable set and T is predictable.

Since $F_t^+ \subset {}^o\boldsymbol{F}_t$ for every $t \geq 0$ the equality of the σ-fields ${}^o\boldsymbol{F}_T$ and ${}^o\boldsymbol{F}_T^+$ is an easy consequence of the definition in 2.2.3. For the proof of $\boldsymbol{F}_T^- = {}^o\boldsymbol{F}_T$ let A be in ${}^o\boldsymbol{F}_T$ and consider the martingale $M_t = \mathbb{E}(\mathbf{1}_A | {}^o\boldsymbol{F}_t)$. Since the filtration does not care about nullsets we may choose a continuous version according to corollary 9.7.5. By the stopping theorem,

$$M_T = \mathbb{E}(\mathbf{1}_A | {}^o\boldsymbol{F}_T) = \mathbf{1}_A$$

and by the predictable stopping theorem 6.4.10,

$$M_T = M_T^- = \mathbb{E}(\mathbf{1}_A | {}^o\boldsymbol{F}_T^-).$$

Thus A is in ${}^o\boldsymbol{F}_T^-$. This completes the proof. ■

CHAPTER 10

CHANGE OF MEASURES

The class of semimartingales proved to be convenient for the development of stochastic calculus because it is stable under natural operations:

(i) if X and Y are semimartingales then the Itô integral process $\int_0^t Y\,dX$ is a semimartingale;

(ii) if X is a semimartingale and if f is a twice continuously differentiable function then $f \circ X$ is a semimartingale according to the Itô formula.

There is another important stability property:

(iii) The class of semimartingales remains unchanged if on the underlying measurable space $(\Omega, \boldsymbol{F}, \mathbb{P})$ the measure $\mathbb{P}$ is replaced by an equivalent measure Q.

Notice that in general a (local) $\mathbb{P}$-martingale will not be a (local) Q-martingale since expectations are taken w.r.t. a different measure. Therefore the statement (iii) really is remarkable.

We use this invariance property of semimartingales to develop two concepts of contrary flavour: The first one is the "change of drift technique" in section 10.2 which proves to be a useful technical tool in many concrete examples. The second one is a rather theoretical analogue of the Riesz representation theorem, characterizing semimartingales by means of the continuity of their elementary integrals.

We start by making a list of concepts which stay unchanged if we switch from $\mathbb{P}$ to an equivalent probability measure Q on $\boldsymbol{F}$ and back. Both measures define the same a.s.-properties, e.g. a member of the classes $\boldsymbol{A}$, $\boldsymbol{B}$ or $\boldsymbol{H}$ will remain in the respective class if $\mathbb{P}$ is replaced by Q because these classes are defined by a.s.-properties of the paths. Similarly, $\mathbb{P}$ and Q have the same a.s. convergent sequences and the same localizing sequences (T_k) of stopping times. Also convergence in probability is the same for both measures. This can be seen either by observing that $\mathbb{P}(A_n) \to 0$ if and only if $Q(A_n) \to 0$ or using the fact that a sequence of random variables converges in probability if and only if every subsequence has an a.s. convergent subsequence. Consequently, the l.u.p. convergence of processes (cf. definition 4.3.2) is the same for both $\mathbb{P}$ and Q.

If X is a semimartingale for both measures and $H \in \boldsymbol{H}$ then the stochastic integral $\int_0^t H\,dX$ is not affected by the change from $\mathbb{P}$ to Q. Indeed, by formula (7.3.3) the integral can be written as

$$\int_I H\,dX = \lim_{|\rho| \to 0} \sum_i H(T_i)dX(I_i) \text{ in probability}$$

and convergence in probability is the same for both measures. Similarly, if X and Y are semimartingales w.r.t. $\mathbb{P}$ and Q then (cf. theorem 7.3.7).

$$d[X,Y](I) = \lim_{|\rho| \to 0} \sum_i dX(I_i)dY(I_i) \text{ in probability}$$

and thus mutual and quadratic variation are the same for both measures.

10.1 Semimartingales under a Change of Measure

In this section we prove that equivalent probability measures $\mathbb{P}$ and Q have the same semimartingales. A key role is played by the martingale of conditional expectations of the Radon-Nikodym density of Q with respect to $\mathbb{P}$.

10.1.1 Definition. Let Q be a probability measure equivalent to $\mathbb{P}$. Let Z_Q denote the Radon-Nikodym derivative $dQ/d\mathbb{P}$ and $Z_t = \mathbb{E}(Z_Q \mid \boldsymbol{F}_t)$.

In order to apply our calculus to the process Z, we must have right-continuity. Hence this assumption arises in the corresponding assertions. In important special cases it is automatically fulfilled:
If the filtration is right-continuous then the martingale Z has a right-continuous version (cf. theorem 3.2.6(b)). Also if $(\boldsymbol{F}_t)_{t \geq 0}$ is the natural filtration of a Brownian motion then every martingale has a right-continuous and even a.s. continuous version (cf. corollary 9.7.5).

The martingale Z has a natural interpretation.

10.1.2 Lemma. Suppose that Z is right-continuous.Let S be a stopping time and denote by $\mathbb{Q}_S$ and $\mathbb{P}_S$ the restrictions of $\mathbb{P}$ and Q to the σ-field $\boldsymbol{F}_S$. Then

$$Z_S = \frac{dQ_S}{d\mathbb{P}_S} \text{ almost surely.}$$

In particular, we have $Z_t = \dfrac{dQ_t}{d\mathbb{P}_t}$ a.s. for every $t \geq 0$.

Proof. Choose a bounded $\boldsymbol{F}_S^+$-measurable function f. Then by the stopping theorem and the definition of conditional expectations

$$\int f \cdot Z_S \, d\mathbb{P}_S = \int f \cdot Z_S \, d\mathbb{P} = \int f\left(\frac{dQ}{d\mathbb{P}}\right) d\mathbb{P} = \int f \, dQ = \int f \, dQ_S = \int f\left(\frac{dQ_S}{d\mathbb{P}_S}\right) d\mathbb{P}$$

which shows that $Z_S = dQ_S \big/ d\mathbb{P}_S$ a.s. for both $\mathbb{P}$ and Q. ■

Since $\mathbb{P}$ is absolutely continuous w.r.t. Q, the density Z_Q is $\mathbb{P}$-a.s. strictly positive. The following lemma implies that the process Z has a.s. strictly positive paths.

10.1.3 Lemma. Let Z be a right-continuous martingale closed by the limit variable Z_∞. If $\mathbb{P}(Z_\infty > 0) = 1$ then almost all paths of Z are bounded away from zero.

Proof. Consider the stopping times $S_n = \inf\{s \geq 0 : Z_s < \frac{1}{n}\}$. Then we have

$$Z_{S_n} = \mathbb{E}(Z_\infty | \boldsymbol{F}_{S_n}^+), \text{ and } Z_{S_n} \leq \frac{1}{n} \text{ on } \{S_n < \infty\}$$

thus

$$\int_{\{S_n < \infty\}} Z_\infty \, d\mathbb{P} = \int_{\{S_n < \infty\}} Z_{S_n} \, d\mathbb{P} < \frac{1}{n}.$$

Therefore $\mathbb{P}(Z_\infty > 0) = 1$ implies $\mathbb{P}(S_n < \infty) \downarrow 0$ and hence $S_n \uparrow \infty$ $\mathbb{P}$-a.s.. By definition of S_n this means that almost every path of Z is bounded away from zero on each bounded interval. Since $\lim_{t \to \infty} Z_t = Z_\infty > 0$ this shows that the paths are bounded away from zero on $\mathbb{R}_+$. ■

The central result of this section will be proved in two versions:

(a) In theorem 10.1.7 we assume that the process Z is a.s. continuous. This allows to avoid reference to chapter 6. In particular, the process Z is a local $\boldsymbol{L}^2$-martingale and we can apply the calculus of stochastic differentials with-

out appealing to theorem 7.2.1. Also in this case the formulation becomes more symmetric because $[A,Z] = 0$ for all $A \in \boldsymbol{A}$ by proposition 7.3.12 .
(b) In theorem 10.1.8 we give a general version of the theorem. There we must indeed refer to chapter 6 and theorem 7.2.1.
We will keep track of these two distinct cases.

As a consequence of lemma 10.1.2 one can relate local Q-martingales to local $\mathbb{P}$-martingales in the following way.

10.1.4 Proposition. Consider a right-continuous adapted process M' and assume that Z is right-continuous. Then:
(a) M' is a local martingale for Q if and only if the process $M = M'\,Z$ is a local martingale for $\mathbb{P}$.
(b) If Z is a.s. continuous then (a) holds with "local martingale" replaced by "local $\boldsymbol{L}^2$-martingale."

Proof. We continue with notation from lemma 10.1.2. Choose any stopping time S and a $\boldsymbol{F}_S$-measurable bounded function f. Then lemma 10.1.2 yields

$$\int f\, M'_S \, dQ = \int f\, M'_S Z_S \, d\mathbb{P} = \int f\, (M'Z)_S \, d\mathbb{P}$$

provided that one of these integrals exists. From this follows for every localizing sequence $(T_k)_{k\geq 1}$ that $M'_{.\wedge T_k}$ is a uniformly integrable Q-martingale if and only if $(M'Z)_{.\wedge T_k}$ is a uniformly integrable $\mathbb{P}$-martingale. This proves the first assertion.
If Z is a.s. continuous then there is a localizing sequence $(T_k)_{k\geq 1}$ such that a.s. $\frac{1}{k} \leq Z_{T_k} \leq k$. Thus the $\boldsymbol{L}^2$-norms w.r.t. $\mathbb{P}_{T_k}$ and Q_{T_k} are equivalent for each k. This implies the second assertion. ■

We will need the following identity:

10.1.5 Lemma. Let M be a semimartingale. Then

$$dM + \frac{1}{Z^-} \bullet d[M,Z] = \frac{1}{Z^-} \bullet dMZ - \frac{M^-}{Z^-} \bullet dZ .$$

Proof. By (7.3.1) the left-hand expression equals

$$dM + \frac{1}{Z^-} \bullet \left(dMZ - M^- \bullet dZ - Z^- \bullet dM \right)$$

which in turn equals the right-hand side. ■

In proposition 10.1.4 we started from a local Q-martingale M' and found a multiplicative way of changing it into a local $\mathbb{P}$-martingale. Now we start from a local $\mathbb{P}$-martingale and show how to change it into a local Q-martingale by adding a suitable process from $\boldsymbol{A}$.

10.1.6 Proposition. Suppose that Z is a.s. continuous. A $\mathbb{P}$-semimartingale M is a local $\boldsymbol{L}^2$-martingale for $\mathbb{P}$ if and only if the process M' given by

$$M'_t = M_t - \int_0^t \frac{1}{Z} d[M,Z]$$

is a local $\boldsymbol{L}^2$-martingale for Q.

Proof. Plainly, the process

$$M - M' = \int_0^{\cdot} \frac{1}{Z} d[M,Z]$$

is in $\boldsymbol{A}$ (recall lemma 10.1.3). Since Z is a.s. continuous we have $d[M,Z] = d[M',Z]$ by proposition 7.3.12. The last lemma applied to M' yields

$$dM = dM' + 1/Z \bullet d[M,Z] = dM' + 1/Z \bullet d[M',Z] = 1/Z \bullet d(M'Z) - \frac{M'^{-}}{Z} \bullet dZ.$$

Because $M'^{-}/Z \bullet dZ \in d\boldsymbol{M}$, we have $dM \in d\boldsymbol{M}$ if and only if $1/Z \bullet d(M'Z) \in d\boldsymbol{M}$ or equivalently $d(M'Z) = Z \bullet (1/Z \bullet dM'Z) \in d\boldsymbol{M}$. The latter in turn is equivalent to M' being a local Q-martingale by proposition 10.1.4. ■

We are now ready to prove the main result. First we consider the case that Z is a.s. continuous.

10.1.7. Theorem. Suppose that Z is a.s. continuous. Then $\mathbb{P}$ and Q have the same semimartingales.

Moreover, if X is a semimartingale then $X = M + A$ is a decomposition of X into an $\boldsymbol{M}$-part and an $\boldsymbol{A}$-part w.r.t. $\mathbb{P}$ if and only if $X = M' + A'$ is such a decomposition of X w.r.t. Q where

$$(10.1.1) \qquad \begin{aligned} M'_t &= M_t - \int_0^t \frac{1}{Z} d[X,Z], \\ A'_t &= A_t + \int_0^t \frac{1}{Z} d[X,Z]. \end{aligned}$$

Proof. Let $X = M + A$ be a $\mathbb{P}$-semimartingale where $M \in \boldsymbol{M}$ and $A \in \boldsymbol{A}$. *Define* now M' and A' by the above formulae. Then $X = M' + A'$ and the definition (10.1.1) of M' coincides with the one in the preceding proposition since

$[X - M,Z] = [A,Z] = 0$ (by proposition 7.3.12). Therefore M' is a local $\boldsymbol{L}^2$-martingale for Q. Plainly, A' is of locally finite variation and hence X is a Q-semimartingale with decomposition $X = M' + A'$.

In particular, every $\mathbb{P}$-semimartingale is a Q-semimartingale. The process of the inverse densities $\frac{d\mathbb{P}_t}{dQ_t}$, $t \geq 0$, has the a.s. continuous version $\frac{1}{Z}$ and hence by symmetry, every Q-semimartingale is also a $\mathbb{P}$-semimartingale.

Finally, given a representation $X = M' + A'$ as a Q-semimartingale, define M and A by (10.1.1). Since M' is a $\mathbb{P}$-semimartingale the process M also is a $\mathbb{P}$-semimartingale. Therefore the previous proposition can be applied and M is a local $\boldsymbol{L}^2$-martingale for $\mathbb{P}$. Since $A \in \boldsymbol{A}$ this shows that $X = M + A$ is a representation of X as a $\mathbb{P}$-semimartingale. This completes the proof. ■

The general version reads as follows:

10.1.8 Theorem. Let $(\Omega, \boldsymbol{F})$ be a measurable space and let $(\boldsymbol{F}_t)_{t\geq 0}$ be a filtration of sub-σ-fields of $\boldsymbol{F}$. Let $\mathbb{P}$ and Q be two equivalent probability measures. Then $\mathbb{P}$ and Q have the same semimartingales.

Proof. Since the assumption on $\mathbb{P}$ and Q is symmetric and since the space $\boldsymbol{A}$ is the same for both measures it suffices to show that every local martingale for Q is a $\mathbb{P}$-semimartingale.

First assume that the filtration is right-continuous: Then the $\mathbb{P}$-martingale $Z_t = \mathbb{E}\left(\frac{dQ}{d\mathbb{P}} \mid \boldsymbol{F}_t\right)$ can be chosen to be right-continuous (theorem 3.2.6). Let a local Q-martingale M' be given and consider the process

$$M = M' + \int_0^{\cdot} \frac{1}{Z^-}\, d[M',Z].$$

Then by lemma 10.1.5 applied to M' one gets

$$dM = \frac{1}{Z^-} \bullet dM'Z - \frac{M'^-}{Z^-} \bullet dZ .$$

The processes Z and M'Z are local $\mathbb{P}$-martingales (cf. proposition 10.1.4). Stochastic integrals with respect to local martingales are local martingales (cf. proposition 7.2.4). Thus M is a local $\mathbb{P}$-martingale and M' is a $\mathbb{P}$-semimartingale.

Now let $(\boldsymbol{F}_t)$ be an arbitrary filtration. In order to show that a local Q-martingale M' is a $\mathbb{P}$-semimartingale we may assume that M' is locally bounded according to corollary 7.2.3. The first part of this proof implies that for the measure $\mathbb{P}$ and the filtration $(\boldsymbol{F}_t^+)$ the process M' is a semimartingale and by corollary 7.2.8 even a special semimartingale . Therefore there is a representation $M' = M + A$ where M is a local $(\boldsymbol{F}_t^+)$-martingale for $\mathbb{P}$ and A is $(\boldsymbol{F}_t^+)$-predictable and of locally finite variation. Then $\int_0^\cdot |dA|$ is also $(\boldsymbol{F}_t^+)$-predictable and lemma 6.5.5 shows that A is a.s. equal to a right-continuous $(\boldsymbol{F}_t)$-predictable process $\tilde{A}$. Then $\tilde{M} = M' - \tilde{A}$ is $(\boldsymbol{F}_t)$-adapted. The process $\tilde{M}$ is a local $(\boldsymbol{F}_t)$-martingale since $M = \tilde{M}$ a.s. and so $M' = \tilde{M} + \tilde{A}$ is a representation of M' as a $\mathbb{P}$-semimartingale w.r.t. the filtration $(\boldsymbol{F}_t)$. ■

10.2 The Formulas of Girsanov-Maruyama and Cameron-Martin

We now consider the important special case of a semimartingale $X = M + A$ for which the martingale part M is a (multiple of a stopped) Brownian motion and the bounded variation part is given by a pathwise Lebesgue integral. In the language of chapter 12 the process A is called the drift term. Frequently, it is convenient to remove the drift or to change it into a more amenable one. Such a "change of drift" can be obtained by replacing the original measure on the underlying space by a suitable equivalent measure. This proves to be a powerful tool in handling stochastic differential equations or diffusion processes. On the basis of the abstract results developed in the preceding section, we carry out now the details of this change of drift technique.

The following result is known as **Girsanov's theorem** (Girsanov (1960)). In a similar form it is already contained in Maruyama (1954). One of its merits is that it allows to compute explicitely the law of solutions of stochastic differential equations with constant diffusion coefficients.

10.2.1 Theorem. Let B be a $(\boldsymbol{F}_t)$-Brownian motion, let H be a process in $\boldsymbol{B}$ and consider the exponential of the local martingale $\int_0^\cdot H\,dB$:

(10.2.1) $$Z_t = \exp\Big(\int_0^t H_s\,dB_s - \tfrac{1}{2}\int_0^t H_s^2\,ds\Big).$$

Suppose that T is a stopping time such that $Z_{\cdot\wedge T}$ is a martingale closed by Z_T and let the probability measure Q be given by

$$dQ = Z_T\,d\mathbb{P}\,.$$

Then w.r.t. Q, the process B^* defined by

$$B_t^* = B_{t\wedge T} - \int_0^{t\wedge T} H_s\, ds$$

is a $(\boldsymbol{F}_t)$-Brownian motion stopped by T.

10.2.2 Remark. (a) Since $\int_0^{\cdot} H\, dB$ is a.s. continuous by corollary 7.1.12 (and right-continuous) we know from theorem 8.2.1 that the process Z is in $\boldsymbol{M}_c$.
(b) An equivalent condition on T for the conclusion to hold is $\mathbb{E}(Z_T) = 1$ (see proposition 10.2.3 below). A sufficient condition is

$$\int_0^T H_s^2\, ds \le c \text{ a.s.} \tag{10.2.2}$$

for some finite constant c. In fact, this means that the quadratic variation of $X = \int_0^{\cdot} H\, dB$ satisfies $\langle X\rangle_T \le c < \infty$ thus fulfilling the hypothesis of the martingale criterion 9.2.5 with M = X. Hence the process $Z_{\cdot\wedge T}$ is a $\boldsymbol{L}^2$-bounded martingale closed by Z_T. A more general sufficient condition is Novikov's criterion (cf. Liptser-Shiryayev (1977), theorem 6.1)

$$\mathbb{E}(\exp(½\int_0^T H_s^2\, ds)) < \infty. \tag{10.2.3}$$

Proof of the theorem. Suppose that the strictly positive a.s. continuous process $Z_{\cdot\wedge T}$ is a martingale closed by Z_T. Then it may play the role of Z in the previous section for the probability measure Q with density Z_T. Plainly, Q and $\mathbb{P}$ are equivalent. The process $B_{\cdot\wedge T}$ is in $\boldsymbol{M}_c$ for $\mathbb{P}$.
Let us compute its decomposition $B_{\cdot\wedge T} = M' + A'$ as a Q-semimartingale from theorem 10.1.7. With $Y = \int_0^{\cdot} H\, dB_{\cdot\wedge T}$ the process $Z_{\cdot\wedge T}$ satisfies the differential equation $dZ = Z \bullet dY$ from theorem 8.2.1 and hence

$$dZ_{\cdot\wedge T} = \left(Z_{\cdot\wedge T} H\right) \bullet dB_{\cdot\wedge T}.$$

Substitution yields now

$$d\langle B_{\cdot\wedge T}, Z_{\cdot\wedge T}\rangle = dB_{\cdot\wedge T} \bullet dZ_{\cdot\wedge T} = dB_{\cdot\wedge T} \bullet \left((Z_{\cdot\wedge T} H)\bullet dB_{\cdot\wedge T}\right)$$

$$= \left(Z_{\cdot\wedge T} H\right) \bullet dB_{\cdot\wedge T} \bullet dB_{\cdot\wedge T} = \left(Z_{\cdot\wedge T} H\right) \bullet d\cdot\wedge T$$

and we may proceed with

$$B_t^* = B_{t\wedge T} - \int_0^t H_s\, ds\wedge T = B_{t\wedge T} - \int_0^t \frac{1}{Z_s}\cdot\left(Z_s H_s\right) ds\wedge T$$

$$= B_{t\wedge T} - \int_0^t \frac{1}{Z_s}\, d\langle B_{\cdot\wedge T}, Z_{\cdot\wedge T}\rangle_s.$$

We recognize B^* as the local Q-martingale from theorem 10.1.7. Since the integral is in A and thus does not contribute to quadratic variation (corollary 7.3.14) and by proposition 7.3.11 its increasing process is given by

$$d\langle B^*\rangle_t = d\langle B\rangle_{t\wedge T} = d\, t\wedge T .$$

This proves the assertion. ∎

Let us add the assertion announced in part (b) of the remark.

10.2.3 Proposition. Let Z be a nonnegative right-continuous local martingale. Let further T be a stopping time such that $\mathbb{P}(Z_T > 0) = 1$ and $\mathbb{E}(Z_T) = \mathbb{E}(Z_0)$. Then the stopped process $Z_{\cdot\wedge T}$ is an a.s. strictly positive martingale closed by Z_T.

Proof. Choose a localizing sequence $(T_k)_{k\geq 1}$ such that $(Z_{t\wedge T_k})_{t\geq 0}$ is a uniformly integrable martingale for every k. Then

$$\mathbb{E}\big(Z_{T\wedge T_k}\big) = \mathbb{E}\big(\mathbb{E}(Z_{T_k}\,|\,\boldsymbol{F}_T^+)\big) = \mathbb{E}\big(Z_{T_k}\big) = \mathbb{E}\big(Z_0\big) = \mathbb{E}\big(Z_T\big).$$

Since $T_k \uparrow \infty$ a.s. one has $0 \leq Z_{T\wedge T_k} \longrightarrow Z_T$ a.s. and we conclude from Scheffé's lemma (cf. Billingsley (1979), p.184) that convergence takes place in $\boldsymbol{L}^1$. Hence

$$Z_S = \lim_{k\to\infty} Z_{S\wedge T_k} = \lim_{k\to\infty} \mathbb{E}\big(Z_{T\wedge T_k}\big|\boldsymbol{F}_S^+\big) = \mathbb{E}\big(Z_T\big|\boldsymbol{F}_S^+\big)$$

for every stopping time $S \leq T$. Thus $Z_{\cdot\wedge T}$ is a $(\boldsymbol{F}_t)$-adapted $(\boldsymbol{F}_t^+)$-martingale and hence a $(\boldsymbol{F}_t)$-martingale closed by Z_T. Strict positivity of the paths follows from lemma 10.1.3 . ∎

We will not prove the multidimensional version of theorem 10.2.1 and restrict ourselves to some comments.

10.2.4 Remark and example. Theorem 10.2.1 holds also if the processes H and B take values in Euclidean n-space (Liptser-Shiryayev (1977), page 234). One has just to interpret the products as scalar products. Let us rewrite the process Z in an important class of examples (this computation has been pointed out to us by A. Wakolbinger).

Suppose that B is a d-dimensional Brownian motion and $H_t(\omega) = \mu(t,B_t(\omega))$ where μ is of the form

$$\mu(t,y) = \text{grad} \log \varphi(t,y)$$

with some positive function $\varphi \in \mathscr{C}^{1,2}(\mathbb{R}\times\mathbb{R}^d)$. Write ∇ for grad, Δ for the Laplacian and $|\cdot|$ for Euclidean norm. Itô's formula yields

$$d \ln \varphi(\cdot,B.) = \nabla \ln \varphi(\cdot,B.)\bullet dB + \left(\partial_t \ln \varphi(t,B_t) + ½ \, \Delta \ln \varphi(t,B_t)\right)dt.$$

A straightforward calculation gives

$$\partial_t \ln \varphi + ½ \, \Delta \ln \varphi = \frac{1}{\varphi} \partial_t \varphi + ½\left(\frac{1}{\varphi} \Delta \varphi - \frac{1}{\varphi 2} |\nabla \varphi|^2\right)$$
$$= -½ \, |\nabla \ln \varphi|^2 + \frac{1}{\varphi}\left(\partial_t + ½ \, \Delta\right)\varphi .$$

Putting things together, we get

$$Z_t = \exp\left(\int_0^t \nabla \log \varphi(s,B_s) \, dB_s - ½ \int_0^t |\nabla \log \varphi(s,B_s)|^2 \, ds\right)$$

$$= \exp\left(\log \varphi(t,B_t) - \log \varphi(0,B_0) - \int_0^t \left[\frac{1}{\varphi} \left(\partial_t + ½ \, \Delta\right)\varphi\right](s,B_s) \, ds\right)$$

and hence

$$Z_t = \frac{\varphi(t,B_t)}{\varphi(0,B_0)} \exp\left(- \int_0^t \frac{(\partial_t + ½\Delta)\varphi}{\varphi}(s,B_s) \, ds\right). \tag{10.2.4}$$

If in particular φ is a space-time harmonic function - i.e. $\partial_t \varphi = - ½ \, \Delta \varphi$ - then

$$Z_t = \frac{\varphi(t,B_t)}{\varphi(0,B_0)} .$$

In practice, we are looking for *Brownian motions* rather than *stopped* Brownian motions. The following result for the setup of standard Brownian motion will be sufficient for our needs. Subsequent to Wiener's theorem in section 1.1, we showed that on the Polish space $\Omega = \mathscr{C}(\mathbb{R}_+)$ the Borel-σ-field coincides with the σ-field generated by the coordinate process W. Hence it is also the smallest σ-field containing the natural filtration of W.

10.2.5 Theorem. Let the underlying probability space be Wiener space $(\Omega, \boldsymbol{F}, \mathbb{W})$ and let $(\boldsymbol{F}_t)_{t\geq 0}$ be the natural filtration of standard Brownian motion W. Suppose that $H \in \boldsymbol{B}$ is such that

$$Z = \exp\left[\int_0^\cdot H_s \, dW_s - ½ \int_0^\cdot H_s^2 \, ds\right]$$

is a martingale. Then there is a unique probability measure Q on $\boldsymbol{F}$ with

$$dQ = Z_t d\mathbb{W} \text{ on } \boldsymbol{F}_t \text{ for every } t \geq 0$$

and under Q the process

$$B^* = W - \int_0^{\cdot} H_s \, ds \tag{10.2.5}$$

is a $(\boldsymbol{F}_t)$-Brownian motion. In particular, the conclusion holds if H is bounded.

Proof. Suppose that Z is a martingale. We apply the last theorem separately to each space $(\Omega, \boldsymbol{F}_t)$ and the constant stopping time $T \equiv t$. We may do so because $\mathbb{E}(Z_T) = \mathbb{E}(Z_0) = 1$. Thus we get measures $dQ_t = Z_t \, d\mathbb{W}$ under which $(B_s^*)_{0 \leq s \leq t}$ is a Brownian motion (up to time t). It remains the problem to extend the consistent measures Q_t to all of $\boldsymbol{F}$. But this is a routine measure theoretic procedure. The latter is the only argument where the special choice of Ω comes in.
Forget now about the martingale assumption on Z and assume that H is bounded. Then the exponential process Z is a martingale up to every time t by remark 10.2.2(b) and therefore a martingale. This reduces the proof to the first part. ∎

10.2.6 Example. Given the setting of the theorem consider a semimartingale which fulfills

$$dX_t = \mu(t)dt + \sigma(t) \bullet dW_t \tag{10.2.6}$$

for processes μ and σ in $\boldsymbol{B}$ and the Brownian motion W. Such processes are frequently called **Itô processes**. If we switch from $\mathbb{W}$ to Q then the coordinate process W is in general no longer a Brownian motion; but we get a Q-Brownian motion by the "change of drift" (10.2.5). A representation of X in terms of B^* is easily found. Since $dW = dB^* + H \, dt$ we compute (still on $(\Omega, \boldsymbol{F}, \mathbb{W})$)

$$dX_t = \Big(\sigma(t)H(t) + \mu(t)\Big)dt + \sigma(s) \bullet dB_s^* = \mu^*(t)dt + \sigma(t) \bullet dB_t^*. \tag{10.2.7}$$

Since $\mathbb{W}$ and Q have the same stochastic differentials, we have a representation of X as an Itô process w.r.t Q and the new Brownian motion B^*. This is a useful trick to reduce the solution of a stochastic differential equation to the solution of a simpler one.

Under the original measure $\mathbb{W}$ the process B^* is not a "pure" Brownian motion but rather a Brownian motion dislocated by some outer influence by $\int_0^t H_s\, ds$. Let us see what we can say about the distribution of this "Brownian motion with a drift". By (10.2.5) we get

$$\int_0^\cdot H_s\, dW_s = \int_0^\cdot H_s\, dB^*_s + \int_0^\cdot H_s^2\, ds.$$

We compute the law $\mathbb{W}\circ(B^*)^{-1}$ on $\boldsymbol{F}_t$ for $t \geq 0$. For $A \in \boldsymbol{F}_t$ we have

$$\begin{aligned}\mathbb{W}\circ B^{*\,-1}(A) &= \int_{\{B^*\in A\}} Z_t^{-1} Z_t\, d\mathbb{W}\\ &= \int_{\{B^*\in A\}} \exp\left(-\int_0^t H_s\, dW_s + \tfrac{1}{2}\int_0^t H_s^2\, ds\right) dQ,\end{aligned}$$

i.e.

$$\mathbb{W}\circ B^{*\,-1}(A) = \int_{\{B^*\in A\}} \exp\left(-\int_0^t H_s\, dB^*_s - \tfrac{1}{2}\int_0^t H_s^2\, ds\right) dQ. \tag{10.2.8}$$

If the integrand exp (..) in this formula is a function of B^* the right-hand side is an integral w.r.t. $\mathbb{W}$.

10.2.7 Proposition. Assume that there is a map $\psi = (\psi_s)_{s\geq 0}$ of $\mathscr{C}(\mathbb{R}_+)$ into itself such that $\omega_s = \psi_s(B^*(\omega))$ $\mathbb{W}$-a.s. and ψ_s is $\boldsymbol{F}_s$-measurable for every $s \geq 0$. Then there is a $\boldsymbol{F}_t$-measurable function φ_t on $\mathscr{C}(\mathbb{R}_+)$ such that

$$\varphi_t(B^*) = \exp\left(-\int_0^t H_s(\psi_s(B^*))\, dB^* - \tfrac{1}{2}\int_0^t (H_s(\psi_s(B^*)))^2\, ds\right) \ \mathbb{W}\text{-a.s.}.$$

On $\boldsymbol{F}_t$ the function φ_t is the density of the law of B^* w.r.t. Wiener measure $\mathbb{W}$:

$$\mathbb{W}\circ B^{*\,-1}(A) = \int_A \varphi_t\, d\mathbb{W}. \tag{10.2.9}$$

Proof. If the map ψ exists then the integrand on the right-hand side in (10.2.8) can be written as $\varphi_t(B^*)$ where φ_t is $\boldsymbol{F}_t$-measurable since the integrals inside the exp are adapted processes for the natural filtration of B^*. Hence equation (10.2.8) reads

$$\mathbb{W}\circ B^{*\,-1}(A) = \int_{\{B^*\in A\}} \varphi_t(B^*)\, dQ.$$

Since $\{B^* \in A\} \in \boldsymbol{F}_t$ and $\mathbb{W}$ is the distribution of B^* under Q this yields (10.2.9). ■

In example 12.4.2 we give a sufficient criterion which implies the hypothesis of this proposition. Here we restrict ourselves to deterministic functions H. If $H_s(\omega) = h(s)$ with a (Lebesgue-) square integrable function h then

$$\omega_t = B^*_t(\omega) + \int_0^t h(s)\, ds = \psi_t(B^*(\omega)).$$

In this case (10.2.9) gives the **Cameron-Martin formula**:

10.2.8 Corollary. For standard Brownian motion W and a square integrable function h on [0,t], the process B^* given by

$$B^*_r = W_r - \int_0^r h(s)\, ds$$

has on $\boldsymbol{F}_t$ a distribution with density

$$\exp\left(- \int_0^t h(s)\, d\, W_s - ½ \int_0^t h(s)^2\, ds\right)$$

with respect to Wiener measure.

If in particular the drift is a constant $h \equiv \mu$ this reduces to

10.2.9 Corollary. For standard Brownian motion W and $\mu \in \mathbb{R}$ the process B^* given by

$$B^*_r = W_r - r \cdot \mu$$

has on $\boldsymbol{F}_t$ a distribution with density

$$\exp\left(-\mu W_t - ½\, \mu^2 t\right)$$

with respect to Wiener measure.

This is the Brownian motion with constant drift of section 9.4.

10.3 An A Posteriori Justification of Semimartingales

This section contains the characterization of semimartingales by a continuity property of the associated elementary integrals. Roughly it says that only semimartingales admit a reasonable stochastic integral. It is possible to develop an axiomatic stochastic integration using only this basic continuity of the integral. This is carried out e.g. by P. Protter (1986).

10.3.1 Theorem. Let X be a right-continuous adapted process. Then X is a semimartingale if and only if for every $t \geq 0$ and every sequence $(H^n)_{n\geq 1}$ of elementary processes converging uniformly to 0 the sequence of random variables $\left(\int_0^t H^n \, dX\right)_{n\geq 1}$ converges in probability to 0.

That every semimartingale X has this property is a consequence of proposition 7.1.7. Basically, it follows from the two inequalities (5.1.4) for the Stieltjes integral and (5.4.2) for the martingale integral. The proof of the converse is divided into several steps:

(a) First of all it is easy to reduce the proof to bounded X.

(b) Due to theorem 10.1.8 one is free to pass from the underlying probability measure $\mathbb{P}$ to any other equivalent measure Q. A key step is a functional analytic argument implying that one may choose Q in such a way that the sets

$$\left\{\mathbb{E}_Q\left(\int_0^t H \, dX\right) \; : \; |H| \leq 1\right\}$$

are bounded (proposition 10.3.2).

(c) Then one shows that the set function

$$m \; : \;]S,T] \longmapsto \mathbb{E}_Q(X_T - X_S)$$

extends to a difference of two bounded nonnegative measures on the predictable σ-field (proposition 10.3.3).

(d) Applying the desintegration result of section 6.5 to these measures one finds a process A in $\boldsymbol{A}$ such that $X - A$ is a martingale, i.e. $X = M + A$ is a semimartingale.

We first state and prove the two propositions mentioned in (b) and (c) and then collect the loose ends to complete the proof of theorem 10.3.1.

10.3.2 Proposition. Let $(\Omega, \boldsymbol{F}, \mathbb{P})$ be a measure space with $0 < \mathbb{P}(\Omega) < \infty$ and let B be a convex subset of $\boldsymbol{L}^1(\mathbb{P})$ such that

$$\sup_{f\in B} \mathbb{P}\left(f > r\right) \xrightarrow[r\to\infty]{} 0 \,. \tag{10.3.1}$$

Then there is a bounded $\boldsymbol{F}$-measurable function g such that $g > 0$ $\mathbb{P}$-a.s. and

$$\sup_{f\in B} \int f \, g \, d\mathbb{P} < \infty \,. \tag{10.3.2}$$

Proof. (1) We first prove that there is a bounded measurable function $g \geq 0$ such that $\mathbb{P}(g > 0) > 0$ and for which (10.3.2) holds. We may assume $B \neq \emptyset$. Let C be the $\boldsymbol{L}^1(\mathbb{P})$-closure of the convex set

$$\left\{f - h : f \in B,\ h \geq 0,\ h \in \boldsymbol{L}^1(\mathbb{P})\right\}.$$

According to (10.3.1) there is a number r such that the constant function $r \cdot \mathbf{1}_\Omega$ does not belong to C. The Hahn-Banach separation theorem and the fact that $\boldsymbol{L}^\infty(\mathbb{P})$ is the dual space of $\boldsymbol{L}^1(\mathbb{P})$ under the natural bilinear form $\langle\cdot,\cdot\rangle$ imply that there is a nonzero element g of $\boldsymbol{L}^\infty(\mathbb{P})$ for which $\langle\cdot,g\rangle$ is bounded above on C, i.e.

$$\sup\left\{\int (f - h)\, g\, d\mathbb{P} : f \in B,\ h \geq 0,\ h \in \boldsymbol{L}^1(\mathbb{P})\right\} < \infty .$$

This is only possible if $\int h\, g\, d\mathbb{P} \geq 0$ for every $h \geq 0$, $h \in \boldsymbol{L}^1(\mathbb{P})$. Therefore $g \geq 0$ a.s.. Since $\mathbb{P}(g \neq 0) > 0$ the function g has the desired properties.

(2) We want to choose g such that $\mathbb{P}(g = 0) = 0$. That is, we have to show that the infimum

$$d = \inf\left\{\mathbb{P}(g = 0) : g \in \boldsymbol{L}^\infty(\mathbb{P}),\ g \geq 0,\ g \text{ satisfies } (10.3.2)\right\}$$

is attained and equals 0. To see that it is attained, choose a sequence in $\boldsymbol{L}^\infty_+(\mathbb{P})$ such that $\mathbb{P}(g_n = 0) \downarrow d$, $g_n \leq 1$ and

$$\sup_{f \in B} \int f g_n\, d\mathbb{P} = c_n < \infty .$$

Choose positive numbers a_n with $\sum a_n c_n < \infty$ and $\sum a_n < \infty$. Then $g = \sum a_n g_n$ is bounded, $\{g = 0\} = \bigcap \{g_n = 0\}$, hence $\mathbb{P}(g = 0) = d$ and $\int f g\, d\mathbb{P} \leq \sum a_n c_n$ for every $f \in B$ by dominated convergence. Thus the infimum is attained.

To see that $d = 0$ assume $d = \mathbb{P}\{g > 0\} > 0$. Apply the statement proved in (1) to the restricted measure $\mathbb{P}(\cdot \cap \{g = 0\})$ to get a function g' on $\{g = 0\}$ such that $\mathbb{P}(\{g' > 0\} \cap \{g = 0\}) > 0$. Then $g' \cdot \mathbf{1}_{\{g=0\}} + g$ defines a new function with all desired properties which vanishes on a set of strictly smaller measure than g in contradiction to the choice of g. ■

The following proposition generalizes the measure extension result 6.6.2 from nonnegative set functions to bounded ones.

10.3.3 Proposition. Let $\boldsymbol{G}$ be the field of sets generated by the stochastic intervals $]S,T]$. Let $m : \boldsymbol{G} \longrightarrow \mathbb{R}$ be finitely additive such that

$$\sup\left\{m(G) : G \in \boldsymbol{G}\right\} < \infty, \tag{10.3.3}$$

Moreover assume for all stopping times T_n and T that

(10.3.4) $$m\big(]T,T_n]\big) \longrightarrow 0 \text{ if } T_n \downarrow T,$$

and that for every localizing sequence (T_k) and every stopping time T

(10.3.5) $$m\big(]T_k \wedge T,T]\big) \longrightarrow 0 .$$

Then m extends to a difference of two nonnegative admissible measures on the predictable σ-field $\boldsymbol{P}$.

Proof. Consider the set function m^+ on $\boldsymbol{G}$ defined by

(10.3.6) $$m^+(G) = \sup\{m(F) : F \in \boldsymbol{G}, F \subset G\} .$$

Then m^+ is nonnegative and finite valued because of (10.3.3) and it is also finitely additive. Let us show that it satisfies

(10.3.7) $$m^+\big(]T_k, \infty]\big) \longrightarrow 0 \text{ for every localizing sequence } (T_k).$$

For this assume that there are a number $\varepsilon > 0$ and sets $F_k \in \boldsymbol{G}$ such that $F_k \subset]T_k, \infty]$ and $m(F_k) \geq \varepsilon$ for each k. Recall from the beginning of the proof of 6.6.2 that each F_k has a representation as a disjoint union of stochastic intervals

$$F_k = \bigcup_{i=1}^{i_k}]R_i^k, S_i^k] .$$

The relation (10.3.5) may be reformulated as $m(]0,T] \cap]T_j, \infty]) \xrightarrow[j\to\infty]{} 0$ for every stopping time T. This extends from $]0,T]$ to general stochastic intervals $]R,S]$ and therefore for fixed k one has $m(F_k \cap]T_j, \infty]) \xrightarrow[j\to\infty]{} 0$. Thus we can choose a subsequence (k_j) such that

$$\left|m\big(F_{k_j} \cap]T_{k_{j+1}}, \infty]\big)\right| < \frac{\varepsilon}{2}$$

and hence

$$m\big(F_{k_j} \cap]0, T_{k_{j+1}}]\big) \geq \frac{\varepsilon}{2}$$

for every j. The sets $G_j = F_{k_j} \cap]0,T_{k_{j+1}}]$ are disjoint. Thus

$$m\Big(\bigcup_{j=1}^{n} G_j\Big) \geq \frac{n\varepsilon}{2}$$

for every $n \in \mathbb{N}$ in contradiction to (10.3.3), and (10.3.7) is established. By a similar argument (10.3.4) implies

(10.3.8) $$m^+\big(]T,T_n]\big) \longrightarrow 0 \text{ for } T_n \downarrow T.$$

Thus according to proposition 6.6.2 there are finite nonnegative measures μ^+ and μ^- on $\boldsymbol{P}$ which extend m^+ and $m^- = m^+ - m$, respectively. By (10.3.7) and (10.3.5) they are both admissible in the sense of definition 6.5.1 which completes the proof. ■

Proof of theorem 10.3.1. Let X be a right-continuous adapted process whose elementary integral is continuous as required in the theorem. In order to show that X is a semimartingale it suffices to show that $X_{\cdot \wedge T_k}$ is a semimartingale for the stopping time

$$T_k = \inf\{t : |X_t| > k\} \wedge k .$$

for every k. Now

$$X_{\cdot \wedge T_k} = X\, 1_{[0,T_k[} + X_{T_k}\, 1_{[T_k,\infty[} .$$

The second term in this representation is in $\boldsymbol{A}$ and therefore it is sufficient to show that the first term is a semimartingale. Call it Y.
This process Y is right-continuous and bounded and it vanishes on $[k,\infty[\times \Omega$. Hence for every elementary process $H = \sum_{i=1}^{n} H_i\, 1_{]T_i,T_{i+1}]}$ the random variable

$$\int_0^\infty H\, dY = \sum_{i=1}^{n} H_i \left(Y_{T_{i+1}} - Y_{T_i}\right)$$

exists and is bounded. Moreover, the elementary integral with respect to Y inherits the continuity property of the elementary integral with respect to X. Let $\boldsymbol{E}$ denote the space of elementary processes. Let B be the set of these random variables where H runs through $\{H \in \boldsymbol{E} : |H| \le 1\}$. Then B is convex and

$$\begin{aligned} \sup_{f\in B} \mathbb{P}\left(f > r\right) &= \sup\left(\mathbb{P}\left\{\int_0^\infty H\, dY > r\right) : H \in \boldsymbol{E},\ |H| \le 1\right\} \\ &= \sup\left\{\mathbb{P}\left(\int_0^\infty H\, dY > 1\right) : H \in \boldsymbol{E},\ |H| \le \frac{1}{r}\right\} \xrightarrow[r\to\infty]{} 0 \end{aligned}$$

by the required continuity of the stochastic integral.
Thus proposition 10.3.2 shows that there is a probability measure Q (with $\frac{dQ}{d\mathbb{P}} = g$) equivalent to $\mathbb{P}$ such that

$$\sup\left\{\mathbb{E}_Q\left(\int_0^\infty H\, dY\right) : H \in \boldsymbol{E},\ |H| \le 1\right\} = \sup\left\{\int_\Omega f\, \frac{dQ}{d\mathbb{P}}\, d\mathbb{P} : f \in B\right\} < \infty .$$

Hence on the field $\boldsymbol{G}$ which was introduced in proposition 10.3.3 the additive set function m defined by

$$m : G \longmapsto \mathbb{E}_Q\left(\int_0^\infty 1_G\, dY\right)$$

satisfies (10.3.3). We verify the other hypotheses in proposition 10.3.3:

If T_n are stopping times such that $T_n \downarrow T$ then by the right-continuity and boundedness of Y one has

$$m\big(\,]T,T_n]\big) = \mathbb{E}_Q\big(Y_{T_n} - Y_T\big) \longrightarrow 0$$

and similarly, for every localizing sequence (T_k) and every stopping time T

$$m\big(\,]T_k \wedge T,T]\big) = \mathbb{E}_Q\big(Y_T - Y_{T_k \wedge T}\big) \longrightarrow 0$$

since Y is bounded $Y_t = 0$ for $t \geq k$. Therefore, according to proposition 10.3.3 there are two finite admissible measures μ^+ and μ^- on $\boldsymbol{P}$ such that $m(G) = \mu^+(G) - \mu^-(G)$ for all $G \in \boldsymbol{G}$. Then by theorem 6.5.4 there is a process A in $\boldsymbol{A}$, bounded in $\boldsymbol{L}^1(Q)$, such that for every stopping time T,

$$\mathbb{E}_Q(Y_T) = m([0,T]) = \mathbb{E}_Q(A_T)\ .$$

The difference $M = Y - A$ then satisfies $\mathbb{E}_Q(M_T) = 0$ for every stopping time. Thus it is a martingale and Y is a semimartingale for Q and hence also for $\mathbb{P}$. ∎

CHAPTER 11

STOCHASTIC DIFFERENTIAL EQUATIONS

A stochastic (ordinary) differential equation (SDE) usually looks like this

$$(11.0.1)\qquad dX^i(t) = \mu_i(t,X(t))dt + \sum_{j=1}^{d} \sigma_{ij}(t,X(t))\bullet dB^j(t),\ 1 \le i \le n .$$

Here X is an unknown $\mathbb{R}^n$-valued semimartingale and $B = (B^1, \dots, B^d)$ a d-dimensional Brownian motion. The coefficients μ_i and σ_{ij} are functions on $\mathbb{R}_+ \times \mathbb{R}^n$. This contains as a special case for $\sigma_{ij} = 0$ and $\mu = (\mu_1, \dots, \mu_n)$ the situation of a vector-valued deterministic ordinary differential equation

$$(11.0.2)\qquad \dot{x} = f(t,x) .$$

It is a pleasant fact that some basic techniques and results like the Picard-Lindelöf method of successive approximation carry over from equation (11.0.2) to equation (11.0.1). In demonstrating this, we consider instead of (11.0.1) the more abstract form

$$(11.03)\qquad dX = F(X)\bullet dY$$

which is notationally much simpler and also will show the analogy to (11.0.2) more clearly. Here Y is a (known) $\mathbb{R}^d$-valued semimartingale. A solution is a $\mathbb{R}^n$-valued semimartingale X whose differential satisfies equation (11.0.3). The symbol F(X) denotes a process in $\boldsymbol{B}^{n\times d}$ which depends on the (unknown) semimartingale X. Thus for each X the right-hand side of (11.0.3) really is in $d\boldsymbol{S}^n$ making the equation meaningful in our setting.

The equation (11.0.1) is a special case of (11.0.3). Take for Y the d+1-dimensional semimartingale

$$\left(t,\ B^1(t),\ \dots\ , B^d(t)\right)^{\top}$$

and for F(X) the matrix process

$$\begin{bmatrix} \mu_1(t,X(t)) & \sigma_{11}(t,X(t)) & \cdots\cdots\cdots & \sigma_{1d}(t,X(t)) \\ \vdots & \vdots & & \vdots \\ \mu_n(t,X(t)) & \sigma_{n1}(t,X(t)) & \cdots\cdots\cdots & \sigma_{nd}(t,X(t)) \end{bmatrix}.$$

Also included in (11.0.3) are stochastic **functional** differential equations like

$$dX(t) = -X(t-1)dt + \left(\int_0^1 X(t-s)ds\right) \bullet dB(t)$$

in which the process F(X) - in the example the process

$$\left(-X(t-1), \int_0^1 X(t-s)\,ds\right)_{t\geq 0}$$

uses at time t not only the value X(t) but the behaviour of X at earlier times. The theory of stochastic functional differential equations is a fairly young and interesting field. The only systematic account of it is Mohammed (1984).

Note the following asymmetry of SDEs with respect to the direction of time. For the solution of the *deterministic* equation (11.0.2) in the time interval $[t_0, t_1]$, it does not really matter whether $x(t_0)$ or $x(t_1)$ is given in advance, the two problems being equivalent in principle. The natural initial conditions for the stochastic equations (11.0.1) or (11.0.3) is given by a $\boldsymbol{F}_{t_0}$-measurable random vector $X(t_0)$. Then one can solve the equation at least approximately by careful forward steps taking up the information of the approximate solution in the past. The analoguous idea of "backward" solution of (11.0.1) with "initial condition" $X(t_1)$ breaks down since in addition to the infinitesimal restriction of the differential equation the measurability conditions on the solution get more stringent at each descending time step.

Let us mention that the process F(X) may depend on ω not only via the actual behaviour of the solution X but also in some explicit way like e.g. in

$$dX(t)(\omega) = A(t,\omega)\ X(t,\omega)\ dt$$

where A is another - e.g. stationary - process. Equations of this type are also of much interest in current research. We, however, shall restrict our attention in the next chapter to equations of type (11.0.1). So for us the main profit of the form (11.0.3) lies in the optical simplicity of the formulas.

In applied fields a SDE is frequently written in the form

$$(11.0.4) \qquad \dot{X} = \mu(t,X) + \sigma(t,X)\xi \, ,$$

where ξ is white noise $\frac{dB(t)}{dt}$ (which does not exist in the classical set up, cf. chapter 1). The reason for this notation is that it is more close to the deterministic form (11.0.2): one thinks of an ordinary differential equation disturbed by white noise.

Another remark concerns modeling of real phenomena by SDE's. Imagine that a scientist has good reasons to describe the time evolution of a given system by a SDE of the form (11.0.4). To give it a mathematically rigorous meaning he might rewrite it as the Itô equation

$$(11.0.5) \qquad dX_t = \mu(t,X)dt + \sigma(t,X)\bullet dB_t \, .$$

But he might as well choose the Stratonovich interpretation

$$(11.0.6) \qquad dX_t = \mu(t,X)dt + \sigma(t,X)\circ dB_t \, .$$

This Stratonovich equation is easily transformed into the Itô equation

$$(11.0.7) \qquad dX_t = \mu(t,X)dt + \sigma(t,X)\bullet dB_t + \tfrac{1}{2}d\sigma(t,X)\bullet dB_t \, ,$$

and conversely. In general, the SDEs (11.0.5) and (11.0.7) have different solutions for a given initial condition and this causes irritation about the choice of the model. Usually, an equation (11.0.4) arises as an approximation of less tractable classical equations. A simulation of these on a computer might produce solutions which support the belief in (11.0.5) or in (11.0.6). The only way to get out of this dilemma is a careful inspection of the limiting procedure which leads to the SDE. Results by Wong and Zakai (1969) and Sussman (1978) show: If one approximates white noise ξ as the "derivative" of Brownian motion by processes ξ_n with smooth paths, then the solutions X_n of

$$\dot{X}_n = \mu(\cdot,X_n) + \sigma(\cdot,X_n)\xi_n$$

tend to the solution X of (11.0.6), i.e. the Stratonovich model is the correct one. On the other hand, in biological sciences (11.0.4) frequently arises as a limit of difference equations and then somtimes it can be argued that the Itô model is the right interpretation (cf. Turelli (1977)). (Sometimes it even turns out that neither is appropriate). The controversy of Itô versus Stratonovich models is the topic of a variety of books and papers; for some references and a readable introduction we refer to Gard (1988), chapter 6.1.

A recent study of more advanced topics in the theory of SDEs by one of its pioneers is Skorokhod (1989). The theory of SDEs for noncontinuous semimartingales is not yet as well standardized as the continuous theory. As an example of research in this direction we refer to Blankenship and Li (1986) and the references therein.

Here is the plan for this chapter: In section 11.1 we develop a stochastic analogue of the classical method of successive approximation. We incorporate the noncontinuous case (Therefore the proof is more involved than the standard proof for SDEs corresponding to Itô processes for which the reader may consult e.g. Arnold (1974)). In section 11.2 we prove for the continuous case some estimates for the moments of the solutions. They allow to show in chapter 12 that certain functionals of the solution process are not only local martingales but martingales. Section 11.3 shows that the solutions of certain SDEs are Markov processes.

11.1 A Stochastic Picard-Lindelöf Theorem

We give now a stochastic version of the Picard-Lindelöf method of successive approximation. It yields an existence and uniqueness theorem for solutions of stochastic initial value problems provided the operator F satisfies some Lipschitz conditions. Moreover, we show that the solutions depend continuously on the initial conditions. In the last chapter we need only SDE's for continuous semimartingales. But we treat the more general case because there are no easily available presentations of it and the noncontinuous SDEs are getting more and more important in the literature.

The initial values will be random, i.e. a solution should be equal to some prescribed process on a stochastic interval $[0,S]$. Hence the "interesting" part of the solution will live on an interval $]S,T]$ and we adopt the following way of speaking: Two processes X and X' are **a.s.equal on a stochastic interval** $]S,T]$ if $1_{]S,T]}X$ is a.s. equal to $1_{]S,T]}X'$. Similarly, **the relation** $dX = Z \bullet dY$ **is satisfied on** $]S,T]$ if $dX(I) = Z \bullet dY(I)$ for every stochastic interval $I \subset]S,T]$.

Suppose now that an initial condition $Z \in \boldsymbol{S}^n$, a "driving" process $Y \in \boldsymbol{S}^d$, a map $F : \boldsymbol{H}^n \longrightarrow \boldsymbol{H}^{n\times d}$ and stopping times S and T are given. Then the process X solves the **initial value problem** corresponding to these data if

$$(11.1.1) \qquad \begin{aligned} dX &= F(X^-) \bullet dY \quad \text{on} \quad]S,T] \text{ and} \\ X &= Z \qquad\qquad\quad \text{on} \quad [0,S] . \end{aligned}$$

Lipschitz conditions are characteristic for Picard-Lindelöf type arguments. Most naturally, they will also be random. We say that F **satisfies a Lipschitz condition** on $]S,T]$ with the random Lipschitz "constant" $K \in \boldsymbol{A}$, if for all H and H' in $\boldsymbol{H}^n$

$$(11.1.2) \qquad \begin{aligned} &|F(H)_t(\omega) - F(H')_t(\omega)| \le K_t(\omega) \sup_{s\le t} |H_s(\omega) - H'_s(\omega)| \\ &\text{for} \quad S(\omega) \le t \le T(\omega). \end{aligned}$$

Like in the classical theory, we shall use Banach's fixed point principle to produce a solution of (11.1.1) as the fixed point of an integral operator. Given an initial condition Z this operator J_Z will be defined on $\boldsymbol{S}^n$ by

$$(11.1.3) \qquad (J_Z X)_t = \begin{cases} Z_t & \text{for } t \in [0,S[\\ Z_S + \int_S^t F(X^-)dY & \text{for } t \in [S,T] \\ \text{constant} & \text{on } \quad [T,\infty[\end{cases}$$

or in more compact notation by

$$J_Z X = Z_{\cdot\wedge S} + \int_{\cdot\wedge S}^{\cdot\wedge T} F(X^-)\, dY.$$

Let $J_Z^{(1)} = J_Z$ and $J_Z^{(k+1)} = J_Z \circ J_Z^{(k)}$; the symbol $|\cdot|$ denotes any (e.g. Euclidean) norm on the underlying finite dimensional vector spaces.

11.1.1 Theorem. Suppose that the Lipschitz condition (11.1.2) is fulfilled. Then the following holds:

(a) For every $X^* \in \boldsymbol{S}^n$ the processes $J_Z^{(k)}X^*$ converge l.u.p. to a solution of the initial value problem (11.1.1).

(b) The solution is a.s. unique on $[0,T]$; more precisely, if Z and Z' have the solutions X and X', respectively, and if $1_G Z = 1_G Z'$ for a set $G \in \boldsymbol{F}$ then

$1_G X = 1_G X'$ a.s. on $[0,T]$.

(c) The solution depends continuously on Z, i.e. if $(Z^{(m)})_{m \geq 1}$ is a sequence in $\boldsymbol{S}^n$ such that $Z^{(m)}_{.\wedge S}$ converges l.u.p. to $Z_{.\wedge S}$ and if $X^{(m)}$ is the solution for the initial condition $Z^{(m)}$ then $X^{(m)}_{.\wedge T}$ converges l.u.p. to $X_{.\wedge T}$.

Like in the deterministic case the idea is to divide the interval [S,T] into small parts on which a contraction argument will work. For this we need a couple of preparations. One crucial tool in the proof is the following estimate which is also useful in other contexts.

11.1.2 Proposition. Let $Y = M + A$ be a one-dimensional semimartingale and let S and T be stopping times. Then for every $H \in \boldsymbol{B}$ and in particular for every $H \in \boldsymbol{H}$ the following estimates hold

$$\mathbb{E}\Big(\sup_{S \leq s \leq t \leq T} \Big(\int_s^t H\, dY\Big)^2 \Big) \tag{11.1.4}$$
$$\leq 2\, \mathbb{E}\Big(4 \int_S^T H^2\, d\langle M\rangle + \Big(\int_S^T H^2\, |dA|\Big)\Big(\int_S^T |dA|\Big)\Big).$$

Proof. The inequality $(a+b)^2 \leq 2(a^2 + b^2)$ implies

$$\mathbb{E}\Big(\tfrac{1}{2} \sup_{S \leq s \leq t \leq T} \Big(\int_s^t H\, dY\Big)^2 \Big)$$
$$\leq \mathbb{E}\Big(\sup_{S \leq s \leq t \leq T} \Big(\int_s^t H\, dM\Big)^2 \Big) + \mathbb{E}\Big(\sup_{S \leq s \leq t \leq T} \Big(\int_s^t H\, dA\Big)^2 \Big).$$

The first term is at most

$$\mathbb{E}\Big(4 \int_S^T H^2\, d[M] \Big).$$

For $H \in \boldsymbol{H}$ this follows from corollary 5.6.5 and proposition 5.6.9. The more general case $H \in \boldsymbol{B}$ (which is not needed in the proof of the theorem) follows from (6.2.3). Here we may replace $[M]$ by $\langle M\rangle$: For $M \in \boldsymbol{M}_c$ by definition and for general $M \in \boldsymbol{M}$ by proposition 6.6.5. The second term can be dominated using the pathwise Cauchy-Schwarz estimate

$$\int_S^T |H|\, |dA| \leq \Big(\int_S^T |H|^2\, |dA|\Big)^{1/2} \Big(\int_S^T |dA|\Big)^{1/2}$$

proving the proposition. ■

We need the following consequence of this proposition.

11.1.3 Lemma. Let $Y = M + A$ be a d-dimensional semimartingale and S and T be stopping times such that $S \le T$ and for some $\delta > 0$,

$$\sup_{S \le t < T} D(Y,S,t) \le \delta \quad \text{a.s.} \tag{11.1.5}$$

where

$$D(Y,s,t) = \sum_{i=1}^{d} \left[4\left(\langle M^i\rangle_t - \langle M^i\rangle_s\right) + \left(\int_s^t |dA^i|\right)^2\right].$$

Then for every $H \in \boldsymbol{H}^{n\times d}$ one has the inequality

$$\mathbb{E}\Big(\sup_{S \le t < T} \Big|\int_S^t H\,dY\Big|^2\Big) \le c\,\delta\,\mathbb{E}\Big(\sup_{S \le t < T} |H_t|^2\Big) \tag{11.1.6}$$

with a constant c which depends only on the dimensions and on the particular choice of the norms but not on δ,H,S,T,M or A.

Proof. The proposition implies

$$\begin{aligned}
&\mathbb{E}\Big(\sup_{S \le t \le T} \Big|\int_S^t H\,dY\Big|^2\Big)\\
&\le \text{const} \max_{i,j} \mathbb{E}\Big(\sup_{S \le t < T} \Big|\int_S^t H^{i,j}\left(dM^i + dA^i\right)\Big|^2\Big)\\
&\le 2\,\text{const}\,\mathbb{E}\Big(\sum_{i=1}^{d} 4\int_S^T |H|^2 d\langle M^i\rangle + \Big[\int_S^T |H|^2 |dA^i|\Big]\Big[\int_S^T |dA^i|\Big]\Big)\\
&\le c \cdot \mathbb{E}\Big(\sup_{S \le t < T} |H_t|^2\Big) \cdot D(Y,S,T)
\end{aligned}$$

for some constant c. This implies the assertion (11.1.6) if $D(Y,S,T) \le \delta$ a.s.. The latter condition is a little stronger than (11.1.5) since Y may jump at time T. But for a.s. continuous M and A the lemma is proved. The reader may want to skip the proof for the general case in the first reading. It uses the theory of predictable stopping. First we modify the $\boldsymbol{A}$-part of Y setting

$$\tilde{A} = \begin{cases} A_t & \text{for } t < T \\ A_T^- & \text{for } t \ge T \end{cases}$$

and let $\tilde{Y} = M + \tilde{A}$. Moreover, we introduce

$$\tilde{D}_t = \sum_{i=1}^{d} 4\left(\langle M\rangle_{t\vee S} - \langle M\rangle_S\right) + \Big[\Big(\int_S^{\cdot\vee S} |dA|\Big)^-\Big]_t^2.$$

The process $\tilde{D}$ is predictable and $\tilde{D}_T = D(\tilde{Y}, S, T)$ a.s. since the **left** limit in the definition of D_T does not reflect the difference between A **and** $\tilde{A}$. Moreover, the predictable set $\tilde{G} = \{\tilde{D} > \delta\}$ a.s. does not meet $[0,T[$. Therefore according to corollary 6.4.7 there is a sequence (R_j) of stopping times such that $R_j \uparrow T$ and $\tilde{D}_{R_j} \leq \delta$ a.s. for each j and hence also

$$D(\tilde{Y}, S, R_j \vee S) \leq \delta \quad \text{a.s..}$$

Thus the first part of the proof applied to $R_j \vee S$ instead of T gives

$$\mathbb{E}\Big(\sup_{S \leq t < T} \Big|\int_S^t H\, dY\Big|^2\Big) = \mathbb{E}\Big(\sup_{S \leq t < T} \Big|\int_S^t H\, d\tilde{Y}\Big|^2\Big)$$

$$\leq \sup_j \mathbb{E}\Big(\sup_{S \leq t \leq R_j} \Big|\int_S^t H\, dY\Big|^2\Big) \leq c \cdot \delta \cdot \mathbb{E}\Big(\sup_{S \leq t < T} |H|^2\Big).$$

This completes the proof. ■

For a n-dimensional process X and a stopping time T we set

$$\|X\|_{2,T} = \mathbb{E}\Big(\sup_{0 \leq t \leq T} |X_t|^2\Big)^{1/2}$$

provided the supremum is measurable. This is true e.g. if X is separable. The contraction argument will take place in the following space:

$$\boldsymbol{H}_2 = \Big\{ H \in \boldsymbol{H}^n : H \text{ satisfies } \|H^-\|_{2,\infty} < \infty \Big\}.$$

11.1.4 Lemma. The space $\boldsymbol{H}_2$ with the (semi-) norm $\|H^-\|_{2,\infty}$ is complete. If $X \in \boldsymbol{S}^n$ and $\|X\|_{2,\infty} < \infty$ then $X^- \in \boldsymbol{H}_2$.

Proof. The proof of the completeness is straightforward using the arguments in proposition 5.4.3 and the fact that for $H \in \boldsymbol{H}$ the process H^- is separable and a.s. equal to H. The second statement in the lemma is trivial. ■

The following lemma proves an $\boldsymbol{H}_2$-version of the theorem in the case of a small stochastic interval]S,T]. Here the contraction argument comes in.

11.1.5 Lemma. Assume the hypothesis of theorem 11.1.1. Suppose in addition that for some positive and finite number $\varkappa$,

$$|F(0)| \leq \varkappa \quad \text{and} \quad K \leq \varkappa \quad \text{a.s. on } [0,T[. \tag{11.1.6}$$

Moreover, assume that there is $q < 1$ such that

$$\mathcal{O}D(Y,S,t) \leq \frac{q^2}{\varkappa^2 \cdot c} \quad \text{a.s. on } [0,T[\tag{11.1.7}$$

where c is the constant of lemma 11.1.3. Then

(a) For all initial conditions $Z \in \boldsymbol{S}^n$ the solution X is a.s. unique on [S,T].

(b) For every $X^* \in \boldsymbol{S}^n$ and $Z \in \boldsymbol{S}^n$ with $\|Z\|_{2,S} < \infty$ the sequence $(J_Z^{(k)} X^*)$ converges in $\|\cdot\|_{2,\infty}$-norm to a solution X. The solution satisfies

(11.1.8) $$\| X^- \|_{2,T} \le (1-q)^{-1}\big(\| Z \|_{2,S} + q\big).$$

If, similarly, $\|Z'\|_{2,S} < \infty$ and X' solves (11.1.1) for the initial condition Z' then

(11.1.9) $$\| (X - X')^- \|_{2,T} \le (1-q)^{-1} \| Z - Z' \|_{2,S} .$$

Proof. We proceed in three steps:

(1) For all X and X' we have

(11.1.10) $$\mathbb{E}\Big(\sup_{S \le t < T} \Big|\int_S^t (F(X^-) - F(X'^-))\, dY \Big|^2\Big) \le q^2 \cdot \mathbb{E}\Big(\sup_{0 \le t < T} \big| X_t - X'_t \big|^2\Big).$$

Indeed, the left-hand side is at most

$$c\cdot\delta\cdot \mathbb{E}\Big(\sup_{S \le t < T} \big| F(X^-)_t - F(X'^-)_t\big|^2\Big)$$

by lemma 11.1.3 where δ is the right-hand side in (11.1.7). By the definiton of the Lipschitz-coefficient process K and the assumption (11.1.6) this expectation does not exceed the right-hand side of (11.1.10).

(2) The estimate (11.1.10) implies that for every $Z \in \boldsymbol{S}^n$ the solution is a.s. unique on [S,T]: Let X and X' be two solutions for the initial condition Z. Then

(11.1.11) $$X_t - X'_t = \int_{t\wedge S}^t F(X^-) - F(X'^-)\, dY \quad \text{on } [0,T].$$

Choose a localizing sequence $(S_k)_{k\ge 1}$ such that X and X' are uniformly bounded on $[0,S_k[$. Since $D(Y,S,t) \le \delta$ a.s. on $[S,S\vee S_k\wedge T[$ the inequality (11.1.10) is applicable to this interval and the stochastic integral in (11.1.11). Hence

$$\mathbb{E}\Big(\sup_{0 \le t < S_k\wedge T} \big| X_t - X'_t \big|^2\Big) \le q^2\, \mathbb{E}\Big(\sup_{0 \le t < S_k\wedge T} \big| X_t - X'_t \big|^2\Big) < \infty .$$

As $q^2 < 1$ this is possible only if the expectation vanishes for each k and thus X = X' a.s. on [0,T[. Then in (11.1.11) the integrand vanishes and hence one has also $X_T = X'_T$. This proves the a.s. uniqueness.

(3) For the existence of the solution and the convergence of the iteration we introduce the "left" version I_Z of the operator J_Z by

$$I_Z H = \Big(Z_{\cdot\wedge S} + \int_{\cdot\wedge S}^{\cdot\wedge T} F(H)\, dY\Big)^-.$$

Assume first $\|Z\|_{2,S} < \infty$. Because of (11.1.7) we may apply lemma 11.1.3 with $\delta = q^2 \varkappa^{-2} c^{-1}$. Since F(0) is bounded by $\varkappa$ on $[0,T[$ the estimate (11.1.6) yields

$$\mathbb{E}\Big(\sup_{S\le t<T}\Big|\int_S^t F(0)\, dY\Big|^2\Big) \le c\cdot\delta\cdot\varkappa^2 \le q^2.$$

This together with $\|Z\|_{2,S} < \infty$ implies

$$\|I_Z 0\|_{2,\infty} \le \|Z_{\cdot\wedge S}\|_{2,T} + \Big\|\Big(\int_{\cdot\wedge S}^{\cdot\wedge T} F(0)\, dY\Big)^-\Big\|_{2,T} \le \|Z\|_{2,S} + q. \tag{11.1.12}$$

Therefore the process $I_Z 0$ belongs to $\boldsymbol{H}_2$. Letting $X' = 0$ in the estimate (11.1.10) shows

$$\|I_Z H - I_Z 0\|_{2,\infty} \le q\cdot\|H\|_{2,\infty}$$

and hence I_Z maps $\boldsymbol{H}_2$ into itself. Again by (11.1.10) one gets for any other pair (Z',H') the estimate

$$\begin{aligned}\|I_Z H - I_{Z'} H'\|_{2,\infty} &\le \|Z - Z'\|_{2,S} + \Big\|\int_{\cdot\wedge S}^{\cdot\wedge T} F(H) - F(H')\, dY\Big\|_{2,\infty} \\ &\le \|Z - Z'\|_{2,S} + q\,\|H - H'\|_{2,\infty}.\end{aligned} \tag{11.1.13}$$

In particular, for Z = Z' we infer that I_Z is a contraction of the space $\boldsymbol{H}_2$. This implies by Banach's contraction principle that for any $H^* \in \boldsymbol{H}_2$ the sequence $(I_Z^{(k)} H^*)$ converges (with speed q^k) to the unique fixed point H^Z of I_Z in $\boldsymbol{H}_2$. The relation $I_Z H^Z = H^Z$ shows that for the semimartingale

$$X^Z = Z_{\cdot\wedge S} + \int_{\cdot\wedge S}^{\cdot\wedge T} F(H)\, dY$$

one has $H^Z = (X^Z)^-$ (in particular $\|H^Z\|_{2,\infty} = \|X^Z\|_{2,\infty}$!). Moreover, $J_Z X^Z = X^Z$ and hence X^Z solves (11.1.1). If $H^{Z'}$ corresponds in the same way to Z' then similarly $I_{Z'} H^{Z'} = H^{Z'}$. Inequality (11.1.13) reads for H^Z and $H^{Z'}$ by this fixed point property:

$$(1-q)\|H^Z - H^{Z'}\|_{2,\infty} \le \|Z - Z'\|_{2,S}.$$

This proves the inequality (11.1.9). By (11.1.12) we have $\|I_0 0\|_{2,\infty} \le q$. Thus the estimate (11.1.13) proves for Z' = 0 and H' = 0 in particular

$$\begin{aligned}\|H^Z\|_{2,\infty} &= \|I_Z H^Z\|_{2,\infty} \\ &\le \|I_Z H^Z - I_0 0\|_{2,\infty} + \|I_0 0\|_{2,\infty} \le \|Z\|_{2,S} + q\cdot\|H^Z\|_{2,\infty} + q\end{aligned}$$

which implies (11.1.8). Note that this proof also shows that for an initial condition Z with $\|Z\|_{2,S} < \infty$ and any X the sequence $(J_Z^{(k)} X)_{k\ge1}$ converges in

$\|.\|_{2,\infty}$-norm to the solution X^Z since $(J_Z X)^- = I_Z(X^-)$ for every X. This completes the proof of the lemma. ∎

Now we drop the condition $\|Z\|_{2,S} < \infty$ The idea is to replace Z by a bounded semimartingale which agrees with Z on a large subset of Ω.

11.1.6 Lemma. The theorem is valid under the additional assumptions (11.1.6) and (11.1.7) of lemma 11.1.5.

Proof. Let us start with the following observation. Suppose that two initial conditions Z and Z' agree a.s. on a set $G \in \boldsymbol{F}$, i.e. $\mathbf{1}_G Z_{.\wedge S} = \mathbf{1}_G Z'_{.\wedge S}$ a.s.. If there is a process $\tilde{X}$ with

$$\sup_{s \le t} \mathbf{1}_G \left| (J_Z^{(k)} X^*)_s - \tilde{X}_s \right| \xrightarrow[k \to \infty]{} 0 \quad \text{in probability for all } t$$

i.e. if $J_Z^{(k)} X^*$ "converges l.u.p. on G" to a process $\tilde{X}$ then the sequence $(J_{Z'}^{(k)} X^*)_{k \ge 1}$ converges l.u.p. on G to $\tilde{X}$, too.
In fact, if $Z_{.\wedge S} = Z'_{.\wedge S}$ a.s. and $X = X'$ a.s. on a measurable set G then the Lipschitz condition (11.1.2) implies $F(X^-) = F(X'^-)$ a.s. on G (set $R = \infty \cdot \mathbf{1}_G$ in proposition 5.4.7(c)). Hence $J_Z X = J_{Z'} X'$ a.s. on G. Therefore we see that $J_Z^{(k)} X^* = J_{Z'}^{(k)} X^*$ a.s. on G for every k by induction and so the iteration has the same l.u.p.-limit on G.
Next we prove that for every $Z \in \boldsymbol{S}^n$ the iteration converges l.u.p. to a solution. This corresponds to part (a) of the theorem. Together with the uniqueness statement in lemma 11.1.5 and the observation above this also proves part (b) of the theorem.
For $\alpha > 0$ let $f_\alpha : \mathbb{R}^n \longrightarrow \mathbb{R}^n$ be a bounded $\mathscr{C}^2$-function such that $f_\alpha(x) = x$ for $|x| \le \alpha$. On the set

$$B_m = \left\{ \sup_{t \le S} |Z_t| \le m \right\} \in \boldsymbol{F}$$

the bounded semimartingale $Z^{(m)} = f_m(Z)$ satisfies $Z^{(m)}_{.\wedge S} = Z_{.\wedge S}$. The iteration for the initial condition $Z^{(m)}$ converges for the norm $\|\cdot\|_{2,\infty}$ and hence also l.u.p. to a process $X^{(m)}$. Since the initial conditions $Z^{(m)}$ and Z coincide on B_m we conclude from lemma 11.1.5 that $J_Z^{(k)} X^*$ converges l.u.p. on B_m to $X^{(m)}$. In particular,

$$X^{(m)} = X^{(m')} \text{ a.s. on } B_m \text{ for } m' \ge m.$$

As $\mathbb{P}(B_m) \longrightarrow 1$, the theorem 4.3.3 yields a right-continuous adapted process X which agrees a.s. with $X^{(m)}$ on B_m for every m. Therefore X is the l.u.p. limit of the sequence $(J_Z^{(k)}X^*)_{k\geq 1}$. By the Lipschitz condition we have also $F(X^-) = F(X^{(m)-})$ a.s. on B_m and therefore X is a solution on [S,T] for the initial condition Z.

In order to prove the continuity assertion (c) let $Z^{(m)}$ converge l.u.p. to Z and let X and $X^{(m)}$ be the corresponding solutions. There are a nullsequence (ε_m) and a sequence (G_m) of measurable sets such that $\mathbb{P}(G_m) \longrightarrow 1$ and

$$\sup_{t\leq S} \left| Z_t - Z_t^{(m)} \right| < \varepsilon_m$$

on G_m. Again we use the transformations f_α. We may assume that

$$|f_\alpha(x) - f_\alpha(x)| \leq |x - x'| \text{ and } |f_\alpha| \leq 2\alpha \text{ for each } \alpha.$$

Choose a sequence (α_m) such that $\alpha_m \longrightarrow \infty$ and $\alpha_m(1 - \mathbb{P}(G_m))^{1/2} \longrightarrow 0$. The semimartingales $f_{\alpha_m}(Z)$ and $f_{\alpha_m}(Z^{(m)})$ satisfy

$$|f_{\alpha_m}(Z) - f_{\alpha_m}(Z^{(m)})| \leq |Z - Z^{(m)}|\cdot \mathbf{1}_{G_m} + 4\cdot\alpha_m\cdot\mathbf{1}_{\Omega\setminus G_m}$$

and thus

$$\left\| f_{\alpha_m}(Z) - f_{\alpha_m}(Z^{(m)})\right\|_{2,S} \leq \varepsilon_m + 4\alpha_m\Big(1 - \mathbb{P}(G_m)\Big)^{1/2} \longrightarrow 0.$$

If $X'^{(m)}$ is the solution for $f_{\alpha_m}(Z)$ and $X''^{(m)}$ is the solution for $f_{\alpha_m}(Z^{(m)})$ we get from (11.1.9) that

$$\left\|X'^{(m)} - X''^{(m)}\right\|_{2,T} \longrightarrow 0. \tag{11.1.14}$$

Let

$$C_m = \Big\{ |Z|_{.\wedge S} \leq \alpha_m - \varepsilon_m \Big\}.$$

Then $\mathbb{P}(C_m) \longrightarrow 1$ and hence $\mathbb{P}(G_m \cap C_m) \rightarrow 1$. On $G_m \cap C_m$ the processes $Z_{.\wedge S}$ and $Z^{(m)}_{.\wedge S}$ are equal to $f_{\alpha_m}(Z)$ and $f_{\alpha_m}(Z^{(m)})$, respectively. Thus by (11.1.14)

$$\left\|\mathbf{1}_{G_m\cap B_m}(X - X^{(m)})\right\|_{2,T} \longrightarrow 0.$$

Since $\mathbb{P}(G_m \cap C_m) \longrightarrow 1$ this implies that $X^{(m)}_{.\wedge T}$ converges l.u.p. to $X_{.\wedge T}$. The proof is now complete. ■

Proof of the theorem. Choose q such that $0 < q < 1$. Let $T_0 = S$ and

$$T_n = \inf\Big\{t > T_{n-1} : |F(0)_t| > n \text{ or } K_t > n \text{ or } c\cdot n^2\cdot D(Y, T_{n-1}, t) > q^2\Big\}.$$

We want to prove by induction that for every n the assertion of the theorem holds for $T_n \wedge T$ instead of T. We have

$$|F(0)_t| > n \text{ and } K \le n \quad \text{on } [0,T_n[\text{ and}$$

$$D(Y,T_{n-1},t) \le \frac{q^2}{c \cdot n^2} \quad \text{on } [T_{n-1},T_n[.$$

Hence the assumptions of lemma 11.1.5 are satisfied with $\varkappa = n$. Thus the lemma shows that the theorem holds for the interval $[T_{n-1} \wedge T, T_n \wedge T]$. By induction hypothesis it holds for $[S, T_{n-1} \wedge T]$. These two statements together imply that the theorem holds for $[S,T_n]$. Since $T_n \uparrow \infty$ a.s. this proves the theorem. ∎

Remarks. (a) The crucial part of the proof of the theorem is the estimate (11.1.10). A similar estimate and therefore a similar result can also be proved if the operator F does not satisfy a pathwise Lipschitz condition but rather a mean square Lipschitz condition of the form

$$\left\| F(X^-) - F(X'^-) \right\|_{2,T} \le K \left\| X - X' \right\|_{2,T} .$$

For a Brownian motion M and dA = dt this has been done in Mohammed (1984).

(b) In the case of $Y \in \boldsymbol{S}_c^d$ the solutions of the equation are of course also a.s. continuous on the interval [S,T] even if the initial condition is not continuous on [0,S]. In this case the SDE (11.1.2) can be rewritten as

$$dX = F(X) \bullet dY.$$

(c) In many applications, e.g. in diffusion theory, the operator F is induced by a map $f : \mathbb{R}_+ \times \mathbb{R}^n \longrightarrow \mathbb{R}^{n \times d}$ in the form $F(X)_t = f(t, X_t)$. This case is studied in section 11.3.

11.2 Moments of the Solutions

This section does not use the results of the previous section.

For the study of the growth and the stability of the solutions of SDEs information about the moments of these solutions is helpful. In this section we want to estimate the moments of order $p \ge 2$ of the solutions of the equation

$$dX = F(X) \bullet dY. \tag{11.2.1}$$

We restrict ourselves to the continuous case. The main tool is the following estimate which is a special case of the inequality of Burkholder-Davis-Gundy. Its proof is a nice application of Itô's formula. For p = 2 it is a direct consequence of Doob's inequality. For the general BDG-inequality cf. Rogers-Williams(1987), theorem IV. 42.1.

11.2.1 Theorem. For every $p \geq 2$ there is a constant c_p such that for every a.s. continuous local martingale N starting at zero one has

$$\mathbb{E}\Big(\sup_{s\geq 0} |N_s|^p\Big) \leq c_p \mathbb{E}\Big((\langle N\rangle_\infty)^{p/2}\Big).$$

Proof. By localization it is easy to reduce the proof to the case of a uniformly bounded martingale N. Let $q = p/2$. Since $p \geq 2$ the function $x \longmapsto |x|^q$ is convex and thus $(|N|_s^q)_{0\leq s\leq\infty}$ is a nonnegative submartingale (cf. lemma 3.0.3 and the convergence theorem 3.2.4). Therefore, Doob's $\boldsymbol{L}^2$-inequality yields

$$\mathbb{E}\Big(\sup_{s\geq 0} |N_s|^p\Big) = \mathbb{E}\Big(\sup_{s\geq 0} (|N_s|^q)^2 \Big) \leq 4\ \mathbb{E}\Big(|N_\infty|^q)^2\Big) = 4\ \mathbb{E}\Big(|N_\infty|^p\Big). \tag{11.2.2}$$

In order to estimate the right-hand side note first that by Itô's formula, $dN^2 = 2N\bullet dN + d\langle N\rangle$ and, because of corollary 7.4.3 and $d\boldsymbol{A}_c\bullet d\boldsymbol{A}_c = \{0\}$, $d\langle N^2\rangle = 4N^2\bullet d\langle N\rangle$. Thus, writing down the Itô formula for the function $f(x) = x^q$ and the semimartingale N^2 gives

$$\begin{aligned} d|N|^p &= q(N^2)^{q-1}\bullet dN^2 + \tfrac{1}{2}q\,(q-1)(N^2)^{q-2}\bullet d\langle N^2\rangle . \\ &= q|N|^{p-2}\bullet\big(2N\bullet dN + d\langle N\rangle\big) + 2q\,(q-1)|N|^{p-4}N^2\bullet d\langle N\rangle \\ &= 2q\,|N|^{p-2}\bullet dN + \big(2q(q-1) + q\big)|N|^{p-2}\bullet d\langle N\rangle. \end{aligned}$$

The first term in this expression is the differential of a martingale. Hence

$$\mathbb{E}\Big(|N_\infty|^p\Big) = c\ \mathbb{E}\Big(\int_0^\infty |N|^{p-2} d\langle N\rangle\Big) \leq c\ \mathbb{E}\Big((\sup_{s\geq 0} |N_s|^{p-2})\langle N\rangle_\infty\Big).$$

Now $\frac{p}{2}$ and $\frac{p}{(p-2)}$ are conjugate exponents, in other words

$$\frac{2}{p} = 1 - \frac{(p-2)}{p}. \tag{11.2.3}$$

Thus by Hölder's inequality

$$\mathbb{E}\Big(|N_\infty|^p\Big) \leq c\bullet\mathbb{E}\Big(\sup_{s\geq 0}|N_s|^p\Big)^{(p-2)/p}\ \mathbb{E}\Big((\langle N\rangle_\infty)^{p/2}\Big)^{2/p}.$$

Using (11.2.2) and once again (11.2.3) yields

$$\Big(\mathbb{E}\Big(\sup_{s\geq 0} |N_s|^p\Big)\Big)^{2/p} \leq 4c\Big(\mathbb{E}\Big((\langle N\rangle_\infty)^{p/2}\Big)\Big)^{2/p} .$$

This is equivalent to the desired result. ■

As a consequence we can extend the estimate in proposition 11.1.2 from $p = 2$ to $p \geq 2$.

11.2.2 Corollary. Let $M \in \boldsymbol{M}_c$, $A \in \boldsymbol{A}$, $G \in \boldsymbol{B}$ and $H \in \boldsymbol{B}$ (or $G \in \boldsymbol{H}$ and $H \in \boldsymbol{H}$). Then for every stochastic interval $]S,T]$ and for every $p \geq 2$ the estimate

$$\mathbb{E}\Big(\sup_{S \leq t \leq T} \Big| \int_S^t G\, dM + \int_S^t H\, dA \Big|^p \Big)$$
$$\leq c_p \mathbb{E}\Big[\Big(\int_S^T d\langle M\rangle \Big)^{p/2-1} \int_S^T |G|^p d\langle M\rangle + \int_S^T |H|^p\, |dA|\, \Big(\int_S^T |dA| \Big)^{p-1} \Big]$$

holds with a constant c_p which depends only on p.

Proof. Since $|a + b|^p \leq 2^p(|a|^p + |b|^p)$ it suffices to estimate the two terms

$$\sup_{S \leq t \leq T} \Big| \int_S^t G\, dM \Big|^p \quad \text{and} \quad \sup_{S \leq t \leq T} \Big| \int_S^t H\, dA \Big|^p \tag{11.2.4}$$

separately. Because of the general Hölder estimate

$$\Big| \int_A f\, d\mu \Big|^p \leq \mu(A)^{p-1} \int |f|^p d\mu \tag{11.2.5}$$

we have the following inequality for the second term in (11.2.4):

$$\sup_{S \leq t \leq T} \Big| \int_S^t H\, dA \Big|^p \leq \Big(\int_S^T |dA| \Big)^{p-1} \int_S^T |H|\, |dA| .$$

The main point is that Burkholder's inequality gives a similar estimate for the first (i.e. the martingale) part of (11.2.4). Let N be the local martingale $\int_{\cdot \wedge S}^{\cdot \wedge T} G\, dM$. Then the increasing process $\langle N\rangle$ of N satisfies

$$\langle N\rangle_\infty = \int_S^T G^2\, d\langle M\rangle .$$

Therefore, theorem 11.2.1 and (11.2.5) for $p/2$ yield

$$\mathbb{E}\Big(\sup_{S \leq t \leq T} \Big| \int_S^t G\, dM \Big|^p \Big) \leq c_p\, \mathbb{E}\Big((\langle N\rangle_\infty)^{p/2} \Big)$$
$$= c_p\, \mathbb{E}\Big(\Big(\int_S^T G^2\, d\langle M\rangle \Big)^{p/2} \Big) \leq c_p\, \mathbb{E}\Big(\Big[\int_S^T d\langle M\rangle \Big]^{p/2-1} \int_S^T |G|^p d\langle M\rangle \Big).$$

This completes the proof. ■

Another important tool in our context is Gronwall's lemma: For a function $f : \mathbb{R}_+ \longrightarrow \mathbb{R}_+$ the inequality

$$f(t) \leq a + b \int_0^t f(s)\, ds \text{ for all } t \geq 0$$

implies the estimate

$$f(t) \leq a\, e^{bt} \text{ for all } t \geq 0.$$

11.2.3 Theorem. Let the constants $\varkappa < \infty$, $\tau < \infty$, $p \geq 2$ and dimensions d and n be given. There are constants γ and γ' such that the following is true:
Suppose that in the SDE (11.2.1) the driving semimartingale Y has a representation $Y = M + A$ with $M \in \boldsymbol{M}_c^d$ and $A \in \boldsymbol{A}^d$ and that for two stopping times S, T with $S \leq T$ one has

$$\sum_{i=1}^{d} \int_S^T \left(d\langle M^i\rangle + |dA^i|\right) \leq \tau. \tag{11.2.6}$$

Let further X and X' be solutions of (11.2.1) on $]S,T]$. Then:
(a) If F satisfies on $]S,T]$ the growth condition

$$|F(H)_t(\omega)| \leq \varkappa \cdot \left(1 + \sup_{s\leq t} |H_s(\omega)|\right) \tag{11.2.7}$$

then the p-th moments of the solution X fulfils

$$\mathbb{E}\left(\sup_{s\leq T} |X|_s^p\right) \leq \gamma\left(1 + \mathbb{E}\left(\sup_{s\leq S} |X|_s^p\right)\right). \tag{11.2.8}$$

(b) Similarly, if $\varkappa$ is a Lipschitz constant for F on $]S,T]$, i.e.

$$|F(H)_t(\omega) - F(H')_t(\omega)| \leq \varkappa \cdot \sup_{s\leq t} |H_s(\omega) - H'_s(\omega)|$$

then

$$\mathbb{E}\left(\sup_{s\leq T} |X - X'|_s^p\right) \leq \gamma' \mathbb{E}\left(\sup_{s\leq S} |X - X'|_s^p\right).$$

Proof. In this proof we use the inequality $|a+b|^p \leq 2^p(|a|^p+|b|^p)$ without any further comment. Set

$$X_t^* = \sup_{s\leq t} |X_s|$$

and write

$$dV = \sum_{i=1}^{d} \left(d\langle M^i\rangle + |dA^i|\right).$$

Let R be any stopping time with $S \leq R \leq T$. Then

$$X_R^* \leq X_S^* + \sup_{S\leq t\leq R} \left| \int_S^t F(X)dM + \int_S^t F(X)dA \right|.$$

We apply corollary 11.2.2 to the components of the stochastic integral vector

$$\mathbb{E}\left((X_R^*)^p\right) \leq c_1\left[\mathbb{E}\left((X_S^*)^p\right) + \int_S^R |F(X)|^p \, dV\right]$$

where the constant c_1 depends only on τ, d and n. The growth condition on F then implies

$$\begin{aligned} \mathbb{E}\left((X_R^*)^p\right) &\leq c_1\left[\mathbb{E}\left((X_S^*)^p\right) + c_2\, \mathbb{E}\left(\int_S^R (1 + X^*)^p \, dV\right)\right] \\ &\leq c_3\left[1 + \mathbb{E}\left((X_S^*)^p\right) + \mathbb{E}\left(\int_S^R (X^*)^p \, dV\right)\right]. \end{aligned} \tag{11.2.9}$$

This is applied for $t \geq 0$ to the stopping time

$$R_t = \inf\left\{s : \int_S^s dV > t\right\} \wedge T.$$

By pathwise change of variables,

$$\int_S^{R_t} (X^*)^p \, dV \leq \int_0^t (X^*_{R_s})^p \, ds \qquad \text{a.s.}$$

and thus by Fubini's theorem the estimate (11.2.9) implies

$$\mathbb{E}\left((X^*_{R_t})^p\right) \leq c_3 \left[1 + \mathbb{E}\left((X^*_S)^p\right) + \int_0^t \mathbb{E}\left((X^*_{R_s})^p\right) ds\right].$$

Gronwall's lemma gives for the function $f(t) = \mathbb{E}\left((X^*_{R_t})^p\right)$ the inequality

$$\mathbb{E}\left((X^*_{R_\tau})^p\right) \leq \gamma \left[1 + \mathbb{E}\left((X^*_S)^p\right)\right]$$

where $\gamma = c_3 e^{c_3 \tau}$. Since $R_\tau = T$ this is the assertion of part (a). The proof of (b) can be given along the same lines replacing X^*_t by $\sup_{s \leq t} |X_s - X'_s|$. ■

Consider the exponential of a one-dimensional continuous local martingale M, i.e. the solution of the equation $dX = X \bullet dM$ with initial condition $X_0 = 1$. Then $F(X) = X$ and the growth-condition is satisfied and we recover from the theorem the integrability criterion of proposition 9.2.4. Note that something like the time change of the Dambis-Dubins-Schwarz theorem occured also in the above proof.

11.3 Markovian Solutions

There is a close interplay between the theory of stochastic differential equations and the theory of Markov processes. In this section we show that the solutions of certain SDE's have the (strong) Markov property. First we rewrite the existence and uniqueness theorem for operators F which are induced by a Borel function $f : \mathbb{R}_+ \times \mathbb{R}^n \longrightarrow \mathbb{R}^{n \times d}$. We study the equation

$$\text{(11.3.1)} \qquad \begin{aligned} dX_t &= f(t, X_t^-) \bullet dY_t && \text{on }]S,T], \\ X &= Z && \text{on } [0,S] \end{aligned}$$

for semimartingales $Z \in \boldsymbol{S}^n$ and $Y \in \boldsymbol{S}^d$ and stopping times S and T. The Lipschitz condition (11.1.2) now takes the following simple form: For every $t^* < \infty$ there is a constant $K \in \mathbb{R}$ such that

$$\text{(11.3.2)} \qquad |f(t,z) - f(t,z')| \leq K|z - z'| \text{ for all } t \in [0,t^*].$$

Specializing from theorem 11.1.1 we get

11.3.1 Theorem. Suppose that the Borel function $f : \mathbb{R}_+ \times \mathbb{R}^n \longrightarrow \mathbb{R}^{n\times d}$ satisfies the Lipschitz condition (11.3.2). Then for every $Z \in \boldsymbol{S}^n$ the SDE (11.3.1) has an a.s. unique solution. It is given as the l.u.p. limit of the sequence $X^{(k)}$ where $X_t^{(0)} = Z_{t\wedge S}$ and

(11.3.3) $$X_t^{(k+1)} = Z_{t\wedge S} + \int_{t\wedge S}^{t\wedge T} f\left(s, (X^{(k)})_s^-\right) dY_s \ .$$

Note that even in this case theorem 11.1.1 gives some additional information like continuous dependence on the initial condition. Now we turn to the (strong) Markov property. One way to define it is as follows. The perhaps more suggestive version of the Markov property which was used in corollary 9.1.9 for Brownian motion will be proved in section 12.5.

11.3.2 Definition. An $(\boldsymbol{F}_t^+)$-adapted stochastic process X has the **strong Markov property** if for all finite stopping times S and all bounded $\sigma(X_{S+s} : s \geq 0)$-measurable functions Φ the conditional expectations w.r.t. the σ-fields $\boldsymbol{F}_S^+$ and $\sigma(X_S)$ coincide:

$$\mathbb{E}(\Phi \mid \boldsymbol{F}_S^+) = \mathbb{E}(\Phi \mid X_S) \ .$$

Recall that X_S is $\boldsymbol{F}_S^+$-measurable if X is right-continuous.

In order to prove that the solution of the SDE (11.3.1) enjoys the Markov property we need that the driving semimartingale Y has independent and stationary increments. This requirement will be fulfilled for the applications of chapter 12: in the diffusion SDEs there the driving process will be of the form $(t, B_t)_{t\geq 0}$. However, theorem 11.3.3 below also applies if B is replaced by the Poisson process or by the Cauchy process.

11.3.3 Theorem. Suppose that the function $f : \mathbb{R}_+ \times \mathbb{R}^d \longrightarrow \mathbb{R}^{n\times d}$ is Borel measurable in the first and Lipschitz continuous in the second component such that the Lipschitz constant is bounded on bounded time intervals. Suppose further that for all $s < t$ the increment $Y_{t+s} - Y_t$ of the semimartingale $Y \in \boldsymbol{S}^d$ is independent of $\boldsymbol{F}_t$ with a distribution depending only on s. Then each solution $X = (X_t)_{t\geq 0}$ of the SDE

$$dX_t = f(t, X_t^-) \bullet dY$$

has the strong Markov property.

The following simple measure theoretic observation will be useful in the proof. The arrows $\leftarrow$ and $\rightarrow$ indicate "past" and "pure future". The index P stands for "present".

11.3.4 Lemma. Let a probability space $(\Omega, \boldsymbol{F}, \mathbb{P})$ and three sub-σ-fields $\boldsymbol{G}_{\leftarrow}$, $\boldsymbol{G}_{\mathrm{P}}$ and $\boldsymbol{G}_{\rightarrow}$ of $\boldsymbol{F}$ be given such that $\boldsymbol{G}_{\mathrm{P}} \subset \boldsymbol{G}_{\leftarrow}$ and moreover $\boldsymbol{G}_{\leftarrow}$ and $\boldsymbol{G}_{\rightarrow}$ are independent. Then for each bounded and $\sigma(\boldsymbol{G}_{\mathrm{P}} \cup \boldsymbol{G}_{\rightarrow})$-measurable function Φ the conditional expectation $\mathbb{E}(\Phi | \boldsymbol{G}_{\leftarrow})$ is $\boldsymbol{G}_{\mathrm{P}}$-measurable.

Proof. Consider first special functions $\Phi = \Phi_{\mathrm{P}} \cdot \Phi_{\rightarrow}$ where Φ_{P} is $\boldsymbol{G}_{\mathrm{P}}$- and $\Phi_{\rightarrow}$ is $\boldsymbol{G}_{\rightarrow}$-measurable. Then

$$\mathbb{E}(\Phi | \boldsymbol{G}_{\leftarrow}) = \mathbb{E}(\Phi_{\mathrm{P}} \cdot \Phi_{\rightarrow} | \boldsymbol{G}_{\leftarrow}) = \Phi_{\mathrm{P}} \cdot \mathbb{E}(\Phi_{\rightarrow}) .$$

Since $\sigma(\boldsymbol{G}_{\mathrm{P}} \cup \boldsymbol{G}_{\rightarrow})$ is generated by variables of such a product form the assertion holds for general Φ. ■

The second observation is the fact that the independence of the increments extends from the σ-fields $\boldsymbol{F}_s$ to the σ-fields $\boldsymbol{F}_S^+$.

11.3.5 Lemma. Let Y be a right-continuous process with values in $\mathbb{R}^d$. Suppose that for all s, t > 0 the increment $Y_{t+s} - Y_t$ of Y is independent of $\boldsymbol{F}_t$ and has a law ρ_s on $\mathbb{R}^d$ depending only on s. Then for every finite stopping time S the increment $Y_{S+s} - Y_S$ is independent of $\boldsymbol{F}_S^+$ and has distribution ρ_s.

Proof. Let $A \in \boldsymbol{F}_S^+$ and consider

$$S_n(\omega) = \min \left\{ \frac{j}{2^n} : j \in \mathbb{N} , \frac{j}{2^n} > S(\omega) \right\} .$$

Then

$$A_{j,n} = A \cap \left\{ S_n = \frac{j}{2^n} \right\} \in \boldsymbol{F}_{\frac{j}{2^n}} .$$

For any bounded continuous function h on $\mathbb{R}^d$ the independence of the increments of Y implies

$$\mathbb{E}\left(1_{A_{j,n}} h\left(Y_{S_n+s} - Y_{S_n}\right)\right) = \mathbb{E}\left(1_{A_{j,n}} h\left(Y_{\frac{j}{2^n}+s} - Y_{\frac{j}{2^n}}\right)\right) = \mathbb{P}\left(A_{j,n}\right) \mathbb{E}\left(h(Y_s - Y_0)\right).$$

Sum out j to get

$$\mathbb{E}\left(\mathbf{1}_A\, h\left(Y_{S_n+s} - Y_{S_n}\right)\right) = \mathbb{P}(A)\, \mathbb{E}\left(h(Y_s - Y_0)\right).$$

As n tends to infinity we get by dominated convergence

$$\mathbb{E}\left(\mathbf{1}_A\, h\left(Y_{S+s} - Y_S\right)\right) = \mathbb{P}(A)\, \mathbb{E}\left(h(Y_s - Y_0)\right).$$

This completes the proof. ■

Note that this lemma gives another and in some sense more natural proof of proposition 9.1.9.

Proof of the theorem. Let X be a solution of the SDE and let S be a finite stopping time. We apply lemma 11.3.4 with the following choice of the σ-fields:

$$\mathcal{G}_{\leftarrow} = \mathcal{F}_S^+,\ \mathcal{G}_P = \sigma\left(X_S\right) \text{ and } \mathcal{G}_{\rightarrow} = \sigma\left(Y_{S+s} - Y_S : s \geq 0\right).$$

Then according to lemma 11.3.5 the σ-fields $\mathcal{G}_{\leftarrow}$ and $\mathcal{G}_{\rightarrow}$ are independent and therefore lemma 11.3.4 shows that

$$\mathbb{E}\left(\Phi \mid \mathcal{F}_S^+\right) = \mathbb{E}\left(\Phi \mid X_S\right)$$

for every bounded function Φ which is measurable w.r.t. the σ-field

$$\sigma\left(\left\{X_S\right\} \cup \left\{Y_{S+s} - Y_S : s \geq 0\right\}\right).$$

Let $\mathcal{G}$ be the nullset augmentation of this σ-field. It remains to be shown that X_{S+s} is $\mathcal{G}$-measurable for every $s \geq 0$. For this we consider X as the solution of the SDE (11.3.1) on the stochastic interval $]S,\infty[$ with the initial condition $X_{\cdot\wedge S}$. We prove by induction that for every k the random variables $X^{(k)}_{S+s}$ are $\mathcal{G}$-measurable. For k = 0 the random variable $X^{(0)}_{S+s} = Z_S$ is $\mathcal{G}$-measurable by definition of $\mathcal{G}$. For the induction step the point is that the stochastic integral (11.3.3) can be approximated in probability by Riemann sums $\Sigma H_{S+s_i}(Y_{S+s_{i+1}} - Y_{S+s_i})$. Here by induction and the measurability of the function f the factors H_{T+s_i} are $\mathcal{G}$-measurable and the Y-increments are $\mathcal{G}$-measurable by definition of $\mathcal{G}$. Since $\mathcal{G}$ contains the nullsets the limit of the variables $X^{(k)}_{S+s}$ is also $\mathcal{G}$-measurable. This finishes the induction argument. Letting $k \to \infty$ shows that $X_{S+s} = \lim X^{(k)}_{S+s}$ is $\mathcal{G}$-measurable. This completes the proof of the theorem. ■

CHAPTER 12

TOWARDS DIFFUSIONS

In chapter 1 we started from the Brownian motion and the Ornstein-Uhlenbeck process, explaining some of their classical context in molecular physics and economy. These two processes are generalized by diffusion processes.

12.0 What is a Diffusion?

We quote from Karlin and Taylor (1981): »Apart from their intrinsic interest, diffusion processes are of value for a manifold of purposes.
(1) Many physical, biological, economic and social phenomena are either well approximated or reasonably modelled by diffusion processes. These include examples from molecular motions of enumerable particles subject to interactions, security price fluctuations in a perfect market, some communication systems with noise, neurophysiological activity with disturbances, variations of population growth, changes in species numbers subject to competition and other community relationships, gene substitutions in evolutionary development, etc.
(2) Many functionals, including first-passage probabilities, mean absorbtion times, occupation time distributions, boundary behaviour properties, and stationary distributions, can be calculated explicitely ... . The calculations largely reduce to solving second-order differential equations with simple boundary conditions ... .
(3) By transforming the time scale and renormalizing the state variable, many Markov processes can be approximated by diffusion processes ... .«

A diffusion process behaves in small time intervals like a Brownian motion subject to an affine transformation where the transformation may vary in time and space . Diffusions can also be considered as stochastically perturbed dynamical systems.

The approximation of Markov processes mentioned in (3) is a generalization of the approximation of random walks by Brownian motion (Donsker's invariance principle). This idea is of course also important in the numerical simulation of diffusion processes where the diffusion process in question is replaced by an appropriate discrete Markov process.

There are essentially three approaches to the mathematical structure of a diffusion process:

(a) Kolmogorov's analytical approach via transition functions;

(b) Itô's approach via stochastic differential equations;

(c) Stroock and Varadhan's approach via the martingale problem.

In this chapter we give a first introduction to (b) and (c). We do not describe Kolmogorov's approach (a). It uses the theory of partial differential equations and yields the easiest connection between diffusion theory and the theory of operator semigroups. (In our framework this connection is sketched in section 12.5.) Kolmogorov's approach still is valuable for explicit calculations as mentioned under (2) in the above quotation. We have used its more elementary countable state space version in the example 4.2.6(b). The interested reader may consult e.g. Karlin and Taylor (1981), Wentzell (1980), Williams (1979) or Feller (1966), volume II. He may also want to have a look at Kolmogorov's original articles (1931) and (1933).

In Kolmogorov's theory the Brownian motion and its transition function are not more than one particularly simple member of the more general class of diffusion processes described by their transition functions. In Itô's approach it turns out that for *every* diffusion process the purely probabilistic part can be viewed as being generated by the Brownian motion in the associated SDE (cf. (12.1.1)). Hence Brownian motion is ubiquitous in Itô theory.

The main advantages of the approaches (b) and (c) are: Firstly, a better probabilistic understanding of the underlying processes. This allows sometimes to avoid heavy calculations. Secondly, they lend themselves much more easily to generalizations to the "infinite dimensional" case which arises from either non-Markovian situations (cf. Mohammed (1983)) or large interacting systems (where the number of particles or system components tends to infinity, cf. e.g. Dawson and Gärtner (1989).)

Let us shortly discuss the mathematical concept of a diffusion. By $\mathbf{S}^n_+$ we denote the cone of all symmetric nonnegative definite n×n-matrices $\mathbf{a}$ (i.e. $x^T \mathbf{a} x \geq 0$ for every $x \in \mathbb{R}^n$). In nearly all practical cases, an $\mathbf{a} \in \mathbf{S}^n_+$ will be given in the form $\sigma\sigma^T$ with an $n\times d$-matrix σ. In lemma 12.2.1 we show how to find in a smooth way for a given $\mathbf{a} \in \mathbf{S}^n_+$ a matrix σ such that $\mathbf{a} = \sigma\sigma^T$. Let

$$(12.0.1) \qquad \mu : \mathbb{R}_+\times\mathbb{R}^n \longrightarrow \mathbb{R}^n, \; \mathbf{a} : \mathbb{R}_+\times\mathbb{R}^n \longrightarrow \mathbf{S}^n_+$$

be measurable and bounded on every compact subset of $\mathbb{R}_+\times\mathbb{R}^n$. To the vector μ and the matrix $\mathbf{a}$ associate the differential operators L_t given by

$$(12.0.2) \qquad L_t\varphi(x) = \sum_{i=1}^{n} \mu_i(t,x) \frac{\partial\varphi}{\partial x_i}(x) + \tfrac{1}{2}\sum_{i,j=1}^{n} \mathbf{a}_{ij}(t,x) \frac{\partial\varphi^2}{\partial x_i\partial x_j}(x).$$

These operators may be defined on every linear space $\mathcal{T}$ of twice continuously differentiable functions on $\mathbb{R}^n$. Fix for the time being such a space $\mathcal{T}$ of **test functions** and call an a.s. continuous $\mathbb{R}^n$-valued and $(\mathcal{F}_t)$-adapted process X a **diffusion process** if

$$(12.0.3) \qquad \frac{1}{h}\, \mathbb{E}\Big(\varphi(X_{t+h}) - \varphi(X_t) \;\Big|\; \mathcal{F}_t\Big) \longrightarrow L_t\varphi(X_t)$$

for every $t \geq 0$ and every test function $\varphi \in \mathcal{T}$ a.s. as $h \to 0$

and if moreover X has the strong Markov property. The vector μ is called the **drift vector** and the matrix $\mathbf{a}$ is called the **diffusion matrix** of X.

This preliminary definition seems to depend on the choice of $\mathcal{T}$. In particular, we have tacitly assumed that the conditional expectations above exist. This always is true if $\mathcal{T} \subset \mathcal{C}^2_c(\mathbb{R}^n)$, the space of twice continuously differentiable functions φ which vanish outside some compact set. Frequently, one chooses $\mathcal{T} = \mathcal{C}^\infty_c(\mathbb{R}^n)$. The conditional expectations will also exist for more general test-functions if the law of X has higher moments.

The intuitive meaning of our current "definition" will become clear by way of a simple example. Assume that $n = 1$ and that μ and $\mathbf{a}$ are constant. Then

$$L_t\varphi(x) = \mu\varphi'(x) + \tfrac{1}{2}\,\mathbf{a}\varphi''(x).$$

Assume further that $\varphi(x) = x$ and $\psi(x) = x^2$ are admissible test functions. Then $L_t\varphi(x) = \mu$ and $L_t\psi(x) = 2\mu x + \mathbf{a}$ and we conclude for a diffusion X associated to L_t that

$$(12.0.4) \qquad \begin{aligned} &\lim_{h\to 0} \frac{1}{h}\, \mathbb{E}\Big(X_{t+h} - X_t \Big| \mathcal{F}_t\Big) = \mu, \\ &\lim_{h\to 0} \frac{1}{h}\, \mathbb{E}\Big((X_{t+h} - X_t)^2 \;\Big|\; \mathcal{F}_t\Big) = \mathbf{a}. \end{aligned}$$

The first identity is clear; the second one requires a little computation:

$$\frac{1}{h}\mathbb{E}\left((X_{t+h} - X_t)^2 \mid \mathcal{F}_t\right) = \frac{1}{h}\mathbb{E}\left(X_{t+h}^2 - X_t^2 \mid \mathcal{F}_t\right) - 2\,X_t\,\frac{1}{h}\mathbb{E}\left(X_{t+h} - X_t \mid \mathcal{F}_t\right).$$

This expression converges by assumption (12.0.3) to

$$2\mu X_t + \mathbf{a} - 2X_t\mu = \mathbf{a}.$$

We may rephrase this observation as

$$\begin{aligned} \mathbb{E}\left(X_{t+h} - X_t \mid \mathcal{F}_t\right) &= \mu h + \text{terms of lower order}, \\ \mathbb{E}\left((X_{t+h} - X_t)^2 \mid \mathcal{F}_t\right) &= \mathbf{a}h + \text{terms of lower order} \end{aligned}$$

which motivates the names **infinitesimal mean** and **infinitesmal variance** for μ and $\mathbf{a}$. They are also called **infinitesimal parameters**. To be more concrete think of a Brownian particle modeled by Brownian motion B. Imagine that the temperature is raised which increases the amplitudes of zigzagging by a factor σ and that a wind blows on the particle such that it drifts away with a constant velocity μ. Then we would describe this motion by the process

$$X_t = \sigma B_t + \mu t.$$

A computation like the one above shows that X is a diffusion process with drift μ and diffusion $\mathbf{a} = \sigma^2$. In fact, this is the most simple diffusion process. It reads in differential form

$$dX = \mu\, dt + \sigma\, dB.$$

The Ornstein-Uhlenbeck velocity process v of section 1.2 given by

$$dv = -\beta v\, dt + dB$$

is also a diffusion process. Here the drift $\mu = -\beta v$ is not constant but proportional to v.

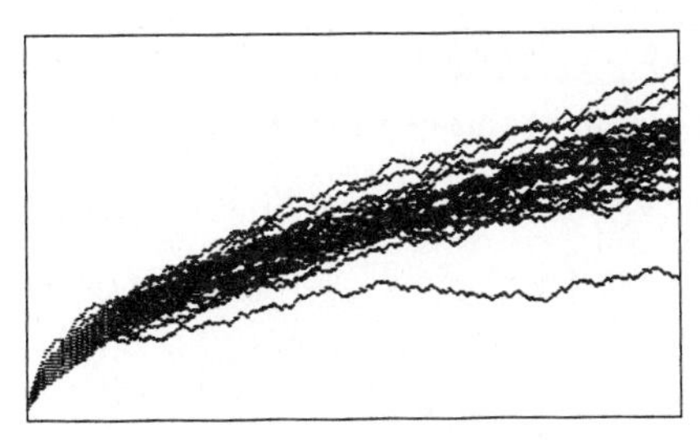

To be definite, we fix now the space of test functions.

12.0.1 Definition. In the sequel, we will call a process X a **diffusion** or a **diffusion process** associated to μ and $\mathbf{a}$ if it satisfies condition (12.0.3) with the space $\boldsymbol{T} = \mathscr{C}_c^\infty(\mathbb{R}^n)$ of test functions and if it has the strong Markov property.

It should be noted here that as soon as the space of test functions contains $\mathscr{C}_c^\infty(\mathbb{R}^n)$ and we are looking for "reasonable" processes the operators L_t *necessarily have the elliptic form* we use here (cf. section 12.2).

There is a variety of different definitions of diffusion processes in use and there are even texts dealing with diffusions without any precise definition. The most general one as a continuous (strong) Markov process is too wide for our purposes. It is useful if one is interested in general properties, but embraces rather exotic examples (cf. Williams (1979), III.38.12). In contrast, our program is to *construct* diffusion processes for given infinitesimal parameters. More appropriate for this purpose is a definition via these parameters using the limits of moments in the spirit of (12.0.4). We prefer the closely related definition by means of the differential operators L_t for various reasons:

(a) It is simultaneously a correct definition for the martingale problem. (Definitions even more adapted to this approach like that in Rogers-Williams (1987) are in our opinion less intuitive for the beginner.)
(b) It is open for the semigroup approach.
(c) The aforementioned definition via limits of moments becomes rather clumsy in more dimensions and also if the second moments of the variables X^i do not exist.

The monograph of Karlin and Taylor (1981) contains a wealth of concrete examples. Many explicit and deep results are in Itô and Mc Kean (1974). A systematic treatment from a more abstract point of view is Ikeda and Watanabe (1981).

12.1 Constructing Diffusions as Solutions of SDE's

One of the main uses of SDEs and their theory is to construct and study diffusion processes. In fact, K. Itô developed his stochastic calculus mainly for that purpose.

We continue with the notation from 12.0. We assume that the infinitesimal variance is of the form $\mathbf{a} = \sigma\sigma^{\top}$ where the infinitesimal deviation $\sigma : \mathbb{R}_{+}\times\mathbb{R}^{n} \longrightarrow \mathbb{R}^{n\times d}$ is measurable and bounded on every compact subset of $\mathbb{R}_{+}\times\mathbb{R}^{n}$. Usually σ will be the unique symmetric and nonnegative square-root of $\mathbf{a}$ and thus n will equal d. Also the map $\mu : \mathbb{R}_{+} \times \mathbb{R}^{n} \longrightarrow \mathbb{R}^{n}$ is required to be measurable and bounded on compact sets. To μ and σ we associate the SDE

$$dX_t = \mu(t,X_t)dt + \sigma(t,X_t)\bullet dB_t \tag{12.1.1}$$

with a d-dimensional driving Brownian motion $B = (B^1, \dots, B^d)^{\top}$. The fact that stochastic integrals w.r.t. Brownian motion are local martingales will be used in order to verify the infinitesimal condition (12.0.3) for the solutions of (12.1.1). So they are diffusion processes if in addition they have the strong Markov property.

We require the noncontinuous paths of the driving Brownian motion - if there are any - to be right-continuous. The reason is that in this text the integrating processes are right-continuous.

We first compute the mutual variation of the components of the solution.

12.1.1 Lemma. The components X^i of a solution X of the SDE (12.1.1) have mutual variation

$$\langle X^i, X^j\rangle = \int_0^{\cdot} \mathbf{a}_{ij}(s,X_s)\, ds.$$

Proof. We have because of $dB^k\bullet dB^m = \delta_{km}dt$ and $dB^k\bullet dt = 0$

$$dX^i\bullet dX^j = \Big(\sum_{k=1}^{d} \sigma_{ik}(\cdot\,,X)\bullet dB^k\Big)\bullet\Big(\sum_{k=1}^{d} \sigma_{jk}(\cdot\,,X)\bullet dB^k\Big)$$

$$= \Big(\sum_{k=1}^{d} \sigma_{ik}(\cdot\,,X)\,\sigma_{jk}(\cdot\,,X)\Big)\, dt = \mathbf{a}_{ij}(s,X_s)\, ds$$

which proves the identity. ∎

Most of the subsequent results do not only concern test functions $\varphi \in \mathscr{C}_c^\infty(\mathbb{R}^n)$ but also smooth "space-time test functions" which may depend explicitely on the time parameter t. We will as usual write $\mathscr{C}^{i,j}(\mathbb{R}_+\times U)$ for the space of functions $\varphi(t,x)$ which are i times continuously differentiable in t and j times in x. The following theorem is the Itô formula for smooth functionals of the solution X of (12.1.1).

12.1.2 Theorem. Suppose that the n-dimensional semimartingale X solves (12.1.1). If $\mathbb{P}(X_t \in U) = 1$ for some open subset U of $\mathbb{R}^n$ and all t and if $\varphi \in \mathscr{C}^{1,2}(\mathbb{R}_+\times U)$ then

$$\begin{aligned}&\varphi(t,X_t) - \varphi(0,X_0) - \int_0^t \Big[\frac{\partial}{\partial s} + L_s\Big]\varphi(s,X_s)\,ds \\ (12.1.2)\qquad &= \sum_{i=1}^n \sum_{j=1}^d \int_0^t \frac{\partial \varphi}{\partial x_i}(s,X_s)\sigma_{ij}(s,X_s)\,dB_s^j\,.\end{aligned}$$

In particular, the expression on the left-hand side is an a.s. continuous local martingale.

Proof. Apply the Itô formula and then plug in the right-hand sides of the of (12.1.1) and of the identity in the lemma:

$$\begin{aligned}d\varphi(\cdot,X) &= \partial_t\varphi(\cdot,X)dt + \sum_{i=1}^n \partial_i\varphi(\cdot,X)\bullet dX^i + \tfrac{1}{2}\sum_{i,j=1}^n \partial_i\partial_j\,\varphi(\cdot,X)\bullet dX^i\bullet dX^j \\ &= \partial_t\varphi(\cdot,X)dt + \sum_{i=1}^n \partial_i\varphi(\cdot,X)\bullet\Big[\,\mu_i(\cdot,X)dt + \sum_{j=1}^d \sigma_{ij}(\cdot,X)\bullet dB^j\Big] \\ &\qquad + \tfrac{1}{2}\sum_{i,j=1}^n \partial_i\partial_j\varphi(\cdot,X)\mathbf{a}_{ij}(\cdot,X)dt \\ &= (\partial_t+L_t)\varphi(\cdot,X)dt + \sum_{i=1}^n\sum_{j=1}^d \Big[\partial_i\varphi(\cdot,X)\sigma_{ij}(\cdot,X)\Big]\bullet dB^j.\end{aligned}$$

This proves the identity (12.1.2). As a stochastic integral, the right-hand side is an a.s. continuous local martingale. ∎

The connection between harmonic functions and martingales established in section 9.3 for the ordinary Laplacian now extends to the parabolic operator $\partial_u + L_u$. Frequently (at least in probabilistic texts), a function $\varphi \in \mathscr{C}^{1,2}(\mathbb{R}_+\times U)$ is called **space-time-harmonic** for this operator if

$$(\partial_t + L_t)\varphi = 0.$$

12.1.3 Corollary. If X is a solution of (12.1.1) and φ is a space-time-harmonic function for the corresponding operator then the process $(\varphi(t,X_t))_{t\geq 0}$ is a local martingale. If, moreover, the family $\{\varphi(S,X_S) : 0 \leq S \leq t, S \text{ stopping time}\}$ is uniformly integrable and $X_0 = x$ a.s. then

$$\varphi(0,x) = \mathbb{E}\big(\varphi(t,X_t)\big). \tag{12.1.3}$$

Proof. The first part is clear since the integral on the left-hand side in (12.1.2) vanishes. Under the additional assumption the process $(\varphi(s,X_s))_{s\geq 0}$ is a martingale up to time t by the criterion 4.2.3. Taking expectations thus proves the identity. ■

Remark. If the coefficients allow a solution of (12.1.1) for every initial condition then a converse of corollary 12.1.3 holds exactly as in theorem 9.3.2.

In remark 12.1.6 below we indicate sufficient criteria for the integrability conditions which ensure (12.1.3). First let us write down a similar statement for general test functions.

12.1.4 Theorem. Suppose that the semimartingale X solves the SDE

$$dX_t = \mu(t,X_t) + \sigma(t,X_t)\bullet dB_t$$

and that for some open subset U in $\mathbb{R}^n$

$$\mathbb{P}(X_t \in U) = 1 \text{ for every } t \geq 0.$$

Assume further that $\varphi \in \mathscr{C}^{1,2}(\mathbb{R}_+\times U)$ fulfills

$$\mathbb{E}\Big(\int_0^t\big[(\partial_i\varphi(s,X_s)\sigma_{ij}(s,X_s)\big]^2 ds\Big) < \infty \text{ for all } i, j = 1, \dots ,n \text{ and } t < \infty. \tag{12.1.4}$$

Then the process

$$X^\varphi = \varphi(\cdot,X) - \varphi(0,X_0) - \int_0^\cdot(\partial_s+L_s)\varphi(s,X_s)\, ds$$

is a martingale.

Proof. To apply the martingale criterion 5.6.3 we must have

$$\mathbb{E}(\langle X^\varphi\rangle_t) < \infty \text{ for all } t < \infty.$$

Since

$$\langle \int_0^{\cdot} \partial_i \varphi(\cdot,X)\sigma_{ij}(\cdot,X)dB^j \rangle_t = \int_0^t \big(\partial_i\varphi(s,X_s)\sigma_{ij}(s,X_s)\big)^2 ds.$$

and by the representation (12.1.2) this follows from the integrability condition.■

The following reformulation of the theorem sometimes is called **Dynkin's formula**.

12.1.5 Corollary. Under the hypothesis of the theorem,

$$\varphi(s,X_s) = \mathbb{E}\Big(\varphi(t,X_t) - \int_s^t (\partial_u + L_u)\varphi(u,X_u)du \;\Big|\; \boldsymbol{F}_s\Big)$$

whenever $s \le t$.

12.1.6 Remark. A sufficient condition for the integrability requirement (12.1.4) is the following one: The partial derivatives $\partial_i\varphi$ satisfy a linear growth condition

$$|\partial_i\varphi(s,x)| \le c(1 + |x|) \text{ on bounded t-intervals,}$$

the coefficients μ and σ are Lipschitz-continuous and finally, the initial condition X_0 has finite fourth moment. Then (12.1.4) is implied by theorem 11.2.3. In fact, the $\boldsymbol{M}_c$-part and the $\boldsymbol{A}_c$-part of the driving semimartingale of the SDE (12.1.1) are Brownian motion and dt, respectively, and therefore, the boundedness condition (11.2.6) for the corresponding quadratic (resp. total) variation on the stochastic interval $[S,T]$ amounts just to the condition that $T - S$ be bounded.

If the partial derivatives of φ are bounded then of course one needs only bounded second moments of X_0 for (12.1.4).

Let us gather the reward and show that solutions of the SDE (12.1.1) have the desired infinitesimal parameters μ and $\mathbf{a}$ in the sense of the previous section. The knack is Dynkin's formula which holds under the conditions discussed in the last remark.

12.1.7 Proposition. Suppose that the a.s. continuous semimartingale X takes a.s. values in an open subset U of $\mathbb{R}^n$ and fulfills the Dynkin formula for the function $\varphi \in \mathscr{C}^{1,2}(\mathbb{R}_+ \times U)$. Assume in addition that the infinitesimal parameters

μ and σ are jointly continuous in the variables t and x and that for some $\varepsilon > 0$ the family

$$\left((\partial_s + L_s)\,\varphi(s,X_s)\right)_{t\le s\le t+\varepsilon}$$

is uniformly integrable. Then

$$\lim_{h\to 0}\frac{1}{h}\,\mathbb{E}\left(\varphi(t+h,X_{t+h}) - \varphi(t,X_t)\,\middle|\,\mathcal{F}_t\right) = (\partial_t + L_t)\varphi(t,X_t)\ \text{a.s.}\,.$$

Proof. By continuity of φ, μ, σ and of the paths the map $t \longmapsto (\partial_t + L_t)\varphi(t,X_t)$ is continuous. The fundamental theorem of (classical) calculus yields

$$\lim_{h\to 0}\frac{1}{h}\int_t^{t+h}(\partial_s + L_s)\varphi(s,X_s)ds = (\partial_t + L_t)\varphi(t,X_t).$$

Moreover, the integral averages are uniformly integrable by our assumption. Thus the Dynkin formula implies

$$\begin{aligned}
&\lim_{h\to 0}\frac{1}{h}\mathbb{E}\left(\varphi(t+h,X_{t+h}) - \varphi(t,X_t)\,\middle|\,\mathcal{F}_t\right)\\
&= \lim_{h\to 0}\mathbb{E}\left(\frac{1}{h}\int_t^{t+h}(\partial_s + L_s)\varphi(s,X_s)ds\,\middle|\,\mathcal{F}_t\right)\\
&= \mathbb{E}((\partial_t + L_t)\varphi(t,X_t)\,|\,\mathcal{F}_t) = (\partial_t + L_t)\varphi(t,X_t).
\end{aligned}$$

This proves the proposition. ■

We discussed the question when a solution of the SDE (12.1.1) has the right infinitesimal parameters. We combine this with the considerations in section 11.3 to give sufficient conditions under which the SDE actually has a (unique) solution, when this solution enjoys the strong Markov property and hence is the desired diffusion process.

12.1.8 Theorem. Let μ and σ be jointly continuous and satisfy the Lipschitz conditions: for every $t^* < \infty$ there is a constant $c \in \mathbb{R}$ such that for all $t \in [0,t^*]$

$$\begin{aligned}
|\mu(t,x) - \mu(t,x')| &\le c\cdot|x - x'|\\
|\sigma(t,x) - \sigma(t,x')| &\le c\cdot|x - x'|.
\end{aligned}$$

Then for every initial condition X_0 the initial value problem associated to the SDE

$$dX_t = \mu(t,X_t)dt + \sigma(t,X_t)\bullet dB_t$$

has a unique solution X on $\mathbb{R}_+ \times \Omega$. This process is a diffusion if X_0 has finite second moments.

Proof. The SDE (12.1.1) is a special case of the SDE (11.3.1) where

$$f(t,x) = \left[\mu(t,x),\sigma(t,x)\right] \text{ and } Y = \binom{t}{B}.$$

The Lipschitz conditions on μ and σ imply the Lipschitz condition (11.3.2) on f. Therefore, the existence and uniqueness theorem 11.3.1 applies and theorem 11.3.3 yields the strong Markov property of the solutions.
In order to verify the infinitesimal condition (12.0.3) we check the assumptions of proposition 12.1.7. Choose φ from $\mathscr{C}_c^\infty(\mathbb{R}^n)$ and consider it as a function on $\mathbb{R}_+ \times \mathbb{R}^n$. Then the estimates in (12.1.4) are valid: since the partial derivatives of φ are bounded and X_0 has second moments the last part of remark 12.1.6 applies. Hence the Dynkin formula holds for φ. Similarly, the family $\left(L_s\varphi(X_s)\right)_{t\le s\le t+\varepsilon}$ is uniformly integrable. Thus proposition 12.1.7 can be applied to get (12.0.3). ■

Remark. Using essentially the same argument the condition (12.0.3) holds also if instead of the condition $X_0 \in \boldsymbol{L}^2$ the coefficients μ and σ are required to be uniformly bounded. This does not even need the moment estimates of section 11.2 which were used in remark 12.1.6.

12.2 On the Form of the Diffusion Operator

This section is not essential for the logical development of the remaining sections.

Let us have a closer look at the operators L_t. They are of the form

$$L_t\varphi(x) = \sum_{i=1}^{n} \mu_i(t,x) \frac{\partial\varphi}{\partial x_i}(x) + \tfrac{1}{2}\sum_{i,j=1}^{n} \mathbf{a}_{ij}(t,x) \frac{\partial\varphi^2}{\partial x_i\partial x_j}(x).$$

We answer two natural questions: (i) Can one read off the diffusion coefficient σ in the associated SDE

$$\text{(12.2.1)} \qquad dX_t = \mu(t,X_t)\,dt + \sigma(t,X_t)\bullet dB_t$$

from the matrix $\mathbf{a}$? (ii) Why is L_t elliptic?

Until now we have tacitly assumed that the SDE associated to the differential operators L_t is already at our disposal. More precisely, we have assumed that there is a "square-root" of the diffusion matrix $\mathbf{a}$, namely a $n\times d$-matrix σ with $\mathbf{a} = \sigma\sigma^{\top}$. In fact, such square-roots do always exist. The following lemma gives some informations about the question when the resulting SDE fulfills the hypothesis of the basic existence and uniqueness theorem.

12.2.1 Theorem. (a) For every positive definite matrix $\mathbf{a}$ there is a unique positive definite matrix σ such that $\sigma^2 = a$.
(b) For $c > 0$ let $\mathbf{S}_c$ be the set of all $\mathbf{a} \in \mathbf{S}_n^+$ such that $\Theta^{\top}\mathbf{a}\Theta \geq c|\Theta|^2$ for all $\Theta \in \mathbb{R}^n$. If $a, a' \in \mathbf{S}_c$ then the corresponding square roots σ and σ' satisfy

$$\|\sigma - \sigma'\| \leq \frac{1}{2\sqrt{c}} \|\mathbf{a} - \mathbf{a}'\| .$$

where $\|\cdot\|$ denotes any matrix norm which is invariant under orthogonal change of coordinates.

Remark. If the coefficient $\mathbf{a}$ in L_t satisfies for some $c > 0$ the inequality

$$c|\Theta|^2 \leq (\Theta^{\top}\mathbf{a}(t,x))\Theta \text{ for each n-vector } \Theta,$$

then L_t is said to be **uniformly elliptic**. The theorem implies that if L_t is uniformly elliptic and $\mu(t,\cdot)$ and $\mathbf{a}(t,\cdot)$ are Lipschitz continuous then there is an associated SDE with Lipschitz continuous μ and σ.

Proof. (a) Recall that there is an orthonormal base of eigenvectors of $\mathbf{a}$. Let $\lambda_1, \ldots, \lambda_r \geq 0$ denote the eigenvalues of $\mathbf{a}$. Set $\alpha_i = \sqrt{\lambda_i}$ and choose a polynomial P with $P(\lambda_i) = \alpha_i$. Then $\sigma = P(\mathbf{a})$ has the desired properties. As a polynomial in $\mathbf{a}$ the matrix σ is symmetric. If Θ is an eigenvector for λ_i then

$$\sigma\, \Theta = P(\mathbf{a})\, \Theta = P(\lambda_i)\Theta = \alpha_i\, \Theta,$$

hence $\sigma\sigma\,\Theta = \alpha_i^2\,\Theta = \mathbf{a}\,\Theta$ which proves that σ is a square-root of $\mathbf{a}$. Note that σ has eigenvalues α_i and that to each α_i belong as eigenvectors those of $\mathbf{a}$ belonging to λ_i. Hence σ has nonnegative eigenvalues and is nonnegative definite. To see that there is no other symmetric and nonnegative definite square-root, choose such a matrix σ' and let $\Phi_1, \ldots, \Phi_n$ be an orthonormal base of eigenvectors. If now γ is an eigenvalue corresponding to Φ_i then

$$\mathbf{a}\Phi_i = \sigma'\sigma'^{\top}\Phi_i = \gamma^2\Phi_i,$$

hence γ^2 is one of the eigenvalues λ_j of $\mathbf{a}$. We conclude that

$$\sigma\,\Phi_i = P(\mathbf{a})\Phi_i = P(\lambda_j)\Phi_i = \alpha_j\Phi_i = \gamma\Phi_i = \sigma'\Phi_i,$$

which proves $\sigma = \sigma'$.

(b) Connect the points $\mathbf{a}$ and $\mathbf{a}'$ by the straight line $\mathbf{a}(t) = \mathbf{a} + t(\mathbf{a}' - \mathbf{a})$, $0 \le t \le 1$. The set $\mathbf{S}_c$ is convex and therefore $\mathbf{a}(t) \in \mathbf{S}_c$ for all t. It is sufficient to show that the associated map $t \longmapsto \sigma(t)$ is differentiable and satisfies the estimate

$$\left\|\frac{d}{dt}\,\sigma(t)\right\| \le \frac{1}{2\sqrt{c}}\left\|\frac{d}{dt}\,\mathbf{a}(t)\right\| . \tag{12.2.2}$$

Fix t and let t' tend to t. Write σ for σ (t) and σ' for $\sigma(t')$. First we show $\sigma' \longrightarrow \sigma$.The set $\{\sigma(t) : 0 \le s \le 1\}$ is bounded and the map $\sigma \longmapsto \sigma^2$ is injective (by (a)) and continuous. Hence the inverse map is continuous and thus $t' \longrightarrow t$ implies $\mathbf{a}(t') \longrightarrow \mathbf{a}(t)$ and $\sigma' \longrightarrow \sigma$. Therefore

$$\|(\sigma' - \sigma)^2\| = o(\|\sigma' - \sigma\|). \tag{12.2.3}$$

Moreover

$$\sigma'^2 - \sigma^2 = (\sigma' - \sigma)\sigma + \sigma(\sigma' - \sigma) + (\sigma' - \sigma)^2 . \tag{12.2.4}$$

The matrix norm is invariant under orthogonal changes of coordinates. Thus for the proof of (12.2.2) we may assume that σ and $\mathbf{a} = \sigma^2$ are of diagonal form. Then (12.2.3) and (12.2.4) yield

$$(\sigma'^2 - \sigma^2)_{ij} = (\sigma' - \sigma)_{ij}(\sigma_{ij} + \sigma_{ij}) + o(\|\sigma' - \sigma\|)$$

for all i,j. This implies $\|\sigma' - \sigma\| = O(|t' - t|)$ and hence

$$(\sigma' - \sigma)_{ij} = \frac{(\sigma'^2 - \sigma^2)_{ij}}{\sigma_{ii} + \sigma_{jj}} + o(|t' - t|). \tag{12.2.5}$$

Since $\sigma_{ii} = \sqrt{\lambda_i}$ where λ_i is an eigenvalue of $\mathbf{a}(t)$we have $\sigma_{ii} + \sigma_{jj} > 2\sqrt{c}$. Thus (12.2.5) implies (12.2.2). This completes the proof of the theorem. ■

You may have wondered why we considered only differential operators of second order. Surprisingly, theorem 12.2.3 below states that L *has* to be of this form. The key observation is that an operator is elliptic if it fulfills the **weak maximum principle**, i.e. $L\varphi(x) \le 0$ if φ has a local maximum at x. Let us say that a function space ***D*** **contains locally a function** $\boldsymbol{\varphi}$ if for each argument x the function φ is equal to some element of ***D*** in a neighbourhood of x . The function φ itself is not necessarily contained in ***D***.

12.2.2 Proposition. Suppose that $L : \boldsymbol{D} \longrightarrow \mathscr{C}(\mathbb{R}^n)$ is a linear operator, where $\boldsymbol{D}$ is a sub*algebra* of $\mathscr{C}_c^2(\mathbb{R}^n)$ containing locally all functions $\varphi_{x_i}(y) = y_i - x_i$. If L satisfies the weak maximum principle then L is of the form

$$L\varphi(x) = \sum_{i=0}^{n} \mu_i(x) \frac{\partial \varphi}{\partial x_i}(x) + \tfrac{1}{2} \sum_{j,k=1}^{n} \mathbf{a}_{ij}(x) \frac{\partial^2 \varphi}{\partial x_i \partial x_j}(x)$$

for every $\varphi \in \boldsymbol{D}$ with mappings $\boldsymbol{\mu} : \mathbb{R}^n \longrightarrow \mathbb{R}^n$ and $\mathbf{a} : \mathbb{R}^n \longrightarrow \mathbf{S}_+^n$.

Proof. We claim that $L\varphi(x) = L\psi(x)$ as soon as the first and second partial derivatives of φ and ψ in x coincide. Then L is a linear combination of these derivatives.

Suppose that $\varphi, \psi \in \boldsymbol{D}$ with the same partial derivatives of first and second order. Choose $x \in \mathbb{R}^n$ and $\xi_x \in \boldsymbol{D}$ which is equal to $|\cdot - x|^2$ near x. Then each function

$$\eta_t(y) = t(\varphi(y) - \psi(y)) - \xi_x, \ t \in \mathbb{R},$$

has a local maximum in x. The weak maximum principle implies $L\eta_t(x) \le 0$ which is equivalent to

$$L\xi_x(x) \ge t\,(L\varphi(x) - L\psi(x)) \text{ for every } t \in \mathbb{R}.$$

We conclude that $L\varphi(x) = L\psi(x)$ as asserted. Therefore functions μ_i and $\mathbf{a}_{ij}$ on $\mathbb{R}^n$ exist such that L has the above representation.

To see that $\mathbf{a}(x) = (\mathbf{a}_{ij}(x)) \in \mathbf{S}_+^n$ choose $z_1, \dots, z_n \in \mathbb{R}$ and a function $\varphi_x \in \boldsymbol{D}$ such that $\varphi_x(y) = -\left[\sum_{i=1}^{n} z_i (y_i - x_i)\right]^2$ near x. Then φ_x has a local maximum in x and

$$\sum_{j,k=1}^{n} \mathbf{a}_{ij}(x) z_j z_k = - L\varphi_x(x) \ge 0.$$

Hence $\mathbf{a}$ is positive semidefinite which completes the proof. ■

We learned in the last section, that - besides the Markov property - the main probabilistic feature of a diffusion is that it satisfies Dynkin's formula. This fact already enforces L to be elliptic.

12.2.3 Theorem. Let $L: \boldsymbol{D} \longrightarrow \mathscr{C}(\mathbb{R}^n)$ be a linear operator where $\boldsymbol{D}$ is as in the proposition. Suppose that for each $x \in \mathbb{R}^n$ there is a process X with continuous paths such that $X_0 = x$ a.s. and the Dynkin formula 12.1.5 holds for every $\varphi \in \boldsymbol{D}$. Then L fulfills the weak maximum principle. In particular, L is an elliptic operator as stated above.

Proof. By way of contradiction suppose that $\varphi \in \boldsymbol{D}$ has a local maximum in x and $L\varphi(x) > 0$. Choose a strictly positive stopping time S such that $\varphi(X_s) \le \varphi(x)$ and $L\varphi(X_s) > 0$ for every $0 \le s < S$. Then the Dynkin formula yields

$$0 \ge \mathbb{E}\big(\varphi(X_{t\wedge S}) - \varphi(x)\big) = \mathbb{E}\Big(\int_0^{t\wedge S} L\varphi(X_s)\, ds\Big) > 0$$

which clearly is impossible. Hence L fulfills the weak maximum principle and application of the proposition completes the proof. ■

12.3 The Feynman-Kac Formula

In this section we give a very brief introduction to the Feynman-Kac formula. It relates the solutions of the equation

$$\frac{\partial}{\partial t}\psi = -L_t\psi + v\cdot\psi \tag{12.3.1}$$

to diffusion processes. Equation (12.3.1) is the generalization of the equation

$$\frac{\partial}{\partial t}\psi + L_t\psi = 0 \tag{12.3.2}$$

from the previous section: a multiplication operator is added to the operator L_t. Equation (12.3.1) is called Schrödinger equation. The original Schrödinger equation from quantum mechanics is obtained from (12.3.1) multiplying the left-hand side by $i\cdot\hbar$ and choosing $(\hbar/2m)\Delta$ for L_t:

$$i\hbar\frac{\partial}{\partial t}\psi = -\frac{\hbar^2}{2m}\Delta\psi + v\cdot\psi\,. \tag{12.3.3}$$

The complex solutions of (12.3.3) correctly describe the behaviour of an atomic particle in a potential field v. For us this physical interpretation is not important. The equation (12.3.1) is just another tool for the study of diffusion processes. However, today there is also much research on the connection between the probabilistic treatment of (12.3.1) and the quantum mechanical equation (12.3.3) (cf. e.g. Nagasawa (1989)).

Corollary 12.1.3 gave a probabilistic representation of the solutions of (12.3.2). The Feynman-Kac formula (12.3.5) does the same for the more general equation (12.3.1).

12.3.1 Theorem. Let G be an open subset of $\mathbb{R}^d$, let X be a solution of (12.1.1) and let T be a stopping time such that $X_{.\wedge T} \in G$ a.s.. Let $v : G \longrightarrow \mathbb{R}$ be a locally bounded Borel map and assume that $\psi \in \mathscr{C}^{1,2}(\mathbb{R}_+ \times G)$ is a solution of the equation (12.3.1). Then the process

$$(12.3.4) \qquad \left(e^{-\int_0^{t\wedge T} v(X_s)ds}\, \psi(t\wedge T, X_{t\wedge T})\right)_{t\geq 0}$$

is a local martingale. If, in particular, it satisfies integrability conditions such that it is a martingale on $[0,t]$ and $X_0 = x$ a.s. then

$$(12.3.5) \qquad \psi(0,x) = \mathbb{E}\left(e^{-\int_0^{t\wedge T} v(X_s)ds}\, \psi(t\wedge T, X_{t\wedge T})\right).$$

Proof. Since $(\frac{\partial}{\partial s} + L_s)\psi = v \cdot \psi$ we know from theorem 12.1.2 that

$$\psi(t, X_{t\wedge T}) - \int_0^{t\wedge T} v(X_s)\psi(s,X_s)\, ds$$

defines a local martingale. Writing $\Psi = (\psi(t\wedge T, X_{t\wedge T}))_{t\geq 0}$ and $V = (v(X_{t\wedge T}))_{t\geq 0}$ this means that

$$d\Psi - V\Psi ds \in d\boldsymbol{M}_c\,.$$

Since v is locally bounded on $\mathbb{R}^n$ and X has a.s. continuous paths the process V is a locally bounded process. Thus the process A given by $A_t = -\int_0^t V\, ds$ is in $\boldsymbol{A}_c$. Then $e^A \in \boldsymbol{A}_c$ as well and integration by parts yields

$$de^{A}\Psi = e^{A}\bullet d\Psi + \Psi\bullet de^{A} = e^{A}\bullet d\Psi + \Psi\, e^{A}\bullet dA = e^{A}\bullet\left(d\Psi - \Psi V ds\right).$$

Hence $e^A\Psi$ - i.e. the process in (12.3.4) - is in $\boldsymbol{M}_c$. If this local martingale is a martingale up to time t then (12.3.5) holds. This completes the proof. ∎

As you see this proof is quite short. However, if the function v takes negative values the verification of the the integrability condition is cumbersome. Therefore we restrict our attention to the simple case of nonnegative v. Then one has to check only the integrability of $\psi(t,X_t)$ and this can be done as in remark 12.1.6 using the moment estimates of section 11.2.

For $v \geq 0$ the exponential factor in (12.3.4) can be interpreted probabilistically as follows. The particles ω start at X_0 and move according to the dynamics prescribed by (12.1.1). But in contrast to the undisturbed case they suffer from the discounting influence of the potential v. They contribute to the expectation in (12.3.5) only with the weight $\exp\left(-\int_0^t v(X_s(\omega))\,ds\right)$. The following mechanism induces precisely this discounting factor: The particle ω is killed after an exponential waiting time whose intensity is spatially inhomogeneous and is given by the function v. In (12.3.5) the value $\psi(t,X_t(\omega))$ is counted only if the particle is still alive.

One straightforward application of the theorem is an alternative description of the Laplace transforms of various sojourn time distributions of the process X. In this case the function v is just a positive constant λ and therefore the integrability of the exponential factor causes no problem.

12.3.2 Corollary. Let G be an open subset of $\mathbb{R}^d$ and let X be a solution of (12.1.1) with the initial condition $X_0 = x$ for some $x \in G$. We assume that the coefficients μ and σ in the SDE and hence the operator $L = L_t$ are constant in time. Let S be the first exit time of X from G and let $\lambda > 0$. Suppose that φ_λ is a bounded solution of the equation

$$
(12.3.6) \qquad \begin{cases} L\varphi_\lambda - \lambda\varphi_\lambda = 0 & \text{on } G, \\ \varphi_\lambda = 1 & \text{on } \partial G. \end{cases}
$$

Then using the convention $e^{-\infty} = 0$ the following identity holds:

$$
\varphi_\lambda(x) = \mathbb{E}(e^{-\lambda S}).
$$

Proof. Let

$$
T_n = n \wedge \inf\left\{t \geq 0 : \operatorname{dist}(X_t, \mathbb{R}^d \backslash G) < \tfrac{1}{n}\right\}.
$$

Then $T_n \uparrow S$ a.s. by path continuity. The function $\psi(t,x) = \varphi_\lambda(x)$ solves (12.3.1) for $v(x) \equiv \lambda$. Therefore according to the theorem, the processes

$$
M_\lambda^{(n)} = \left(e^{-\lambda(t\wedge T_n)}\varphi_\lambda(X_{t\wedge T_n})\right)_{t\geq 0}
$$

are local martingales. Since $\lambda > 0$ and φ_λ is bounded on G they form a uniformly bounded sequence of martingales. Moreover

$$
\lim_{n\to\infty} (M_\lambda^{(n)})_{T_n} = \lim_{n\to\infty} e^{-\lambda T_n}\varphi_\lambda(X_{T_n}) = e^{-\lambda S} \text{ a.s..}
$$

In fact, on $\{S < \infty\}$ the sequence (X_{T_n}) converges a.s. to the point $X_S \in \partial G$ and hence $\varphi_\lambda(X_S) = 1$ a.s.; on $\{S = \infty\}$ the factor $e^{-\lambda T_n}$ enforces $(M_\lambda^{(n)})_{T_n} \longrightarrow 0$ since φ_λ is bounded. Hence by dominated convergence

$$\varphi_\lambda(x) = \lim_{n\to\infty} \mathbb{E}\left((M_\lambda^{(n)})_0\right) = \lim_{n\to\infty} \lim_{t\to\infty} \mathbb{E}\left((M_\lambda^{(n)})_{t\wedge T_n}\right)$$

$$= \lim_{n\to\infty} \mathbb{E}\left((M_\lambda^{(n)})_{T_n}\right) = \mathbb{E}\left(e^{-\lambda S}\right).$$

This completes the proof. ■

Example. (cf. Varadhan (1980), p.131) Choose $d = 1$ and $L\varphi = \frac{1}{2}\varphi'' + \mu\varphi'$, i.e. the process X is a Brownian motion with drift μ. Let G be the open interval $]-1,1[$. Then (12.3.6) becomes the second order differential equation

$$\tfrac{1}{2}\ddot{\varphi}_\lambda + \mu\dot{\varphi}_\lambda - \lambda\varphi_\lambda = 0$$

with the two additional conditions

$$\varphi_\lambda(1) = \varphi_\lambda(-1) = 1 .$$

The solution can be given explicitely as a trigonometric function. For $\mu \neq 0$ the expressions become somewhat heavy. In the case $\mu = 0$ one simply gets

$$\varphi_\lambda(x) = \frac{\cos(\sqrt{-2\lambda}\, x)}{\cos\sqrt{-2\lambda}} = \frac{e^{\sqrt{2\lambda}\,x} + e^{-\sqrt{2\lambda}\,x}}{e^{\sqrt{2\lambda}} + e^{-\sqrt{2\lambda}}} .$$

For $x = 0$ therefore the time S which a standard Brownian spends before first leaving $]-1,1[$ has the Laplace transform

$$\mathbb{E}^0(e^{-\lambda S}) = (\cos\sqrt{-2\lambda}\,)^{-1} = \frac{2}{e^{\sqrt{2\lambda}} + e^{-\sqrt{2\lambda}}} .$$

For a short discussion of the Feynman-Kac formula based on semigroup theory see Williams(1979), III.39. Many authors treat only the case of Brownian motion i.e. $\frac{1}{2}\Delta = L$. For a more recent paper on the integrability problems mentioned after the proof of theorem 12.3.1 see Pinsky (1986).

12.4 Weak Solutions and Uniqueness

Let us now go on from the development of diffusions in section 12.1. What is the state of affairs ? We fixed infinitesimal parameters $\boldsymbol{\mu}$ and $\mathbf{a}$. To them we associated a SDE of the diffusion type (12.1.1) which amounts to the choice of a Brownian motion. Theorem 12.1.8 then tells us that - provided smoothness of the coefficients - for given initial values and the given driving Brownian motion there is a unique solution of this SDE and that this solution is a diffusion. So uniqueness is connected to the particular choice of some Brownian motion and not only to the original objects μ and $\mathbf{a}$. But we cannot compare diffusions arising from different Brownian motions. *What* we may compare are the *distributions* of such diffusion processes. This leads to the concept of uniqueness in distribution or weak uniqueness. This concept is studied in some detail in this and in the last section. In generalization of section 12.1 the operators L_t may depend on the past.

We endow $\mathscr{C}(\mathbb{R}_+, \mathbb{R}^n)$ with the σ-field generated by the coordinate process Π given by

$$\Pi_t : \mathscr{C}(\mathbb{R}_+, \mathbb{R}^n) \longrightarrow \mathbb{R}^n, \ \omega \longmapsto \omega(t)$$

and with the natural filtration. The mappings

$$\mu : \mathbb{R}_+ \times \mathscr{C}(\mathbb{R}_+, \mathbb{R}^n) \longrightarrow \mathbb{R}^n,$$
$$\sigma : \mathbb{R}_+ \times \mathscr{C}(\mathbb{R}_+, \mathbb{R}^n) \longrightarrow \mathbb{R}^{n \times d}$$

are assumed to be product measurable and adapted. Since we work on the space of continuous paths *this is equivalent to assuming them to be either predictable or progressively measurable* (theorem 6.1.7). Further, we require the paths of μ and σ to be bounded on bounded time intervals. In section 12.1, we had the special case where $\mu(t,\Pi) = \mu(t,\Pi_t)$ and $\sigma(t,\Pi) = \sigma(t,\Pi_t)$ with measurable mappings

$$\mu : \mathbb{R}_+ \times \mathbb{R}^n \longrightarrow \mathbb{R}^n \text{ and } \sigma : \mathbb{R}_+ \times \mathbb{R}^n \longrightarrow \mathbb{R}^{n \times d}. \tag{12.4.1}$$

Note that $\mu(t,X)$ and $\sigma(t,X)$ are predictable if X is a continuous adapted process. The infinitesimal variance

$$\mathbf{a} = \sigma\sigma^{\top} : \mathbb{R}_+ \times \mathscr{C}(\mathbb{R}_+, \mathbb{R}^n) \longrightarrow \mathbf{S}^n_+$$

then is progressively measurable and locally bounded as well. The operators $L_t(\omega)$ are defined by

$$L_t(\omega)\varphi(x) = \sum_{i=1}^{n} \mu(t,\omega)\frac{\partial \varphi}{\partial x_i}(x) + \tfrac{1}{2} \sum_{i,j=1}^{n} \mathbf{a}_{ij}(t,\omega) \frac{\partial^2 \varphi}{\partial x_i \partial x_j}(x) \tag{12.4.2}$$

for $\varphi \in \mathscr{C}^2(\mathbb{R}^n)$.

We study weak and pathwise uniqueness of the associated SDE.

12.4.1 Definition. Write down formally the SDE

$$dX = \mu(\cdot,X)dt + \sigma(\cdot,X)\bullet dB. \tag{12.4.3}$$

This equation is said to have a **weak solution** X if there is an $(\boldsymbol{F}_t)$-Brownian motion B on some probability space and a process X on the same space fulfilling (12.4.3). **Weak uniqueness** or **uniqueness in distribution** or **law** holds if any two weak solutions with the same initial distribution (possibly associated to different Brownian motions) have the same distribution on $\mathscr{C}(\mathbb{R}_+,\mathbb{R}^n)$. The equation (12.4.3) is said to be **pathwise solvable** if it has a solution for *every* filtration $(\boldsymbol{F}_t)_{t\geq 0}$ and *every* $(\boldsymbol{F}_t)$-Brownian motion. **Pathwise uniqueness** holds if - given the initial distribution of X - it has only one solution for each $(\boldsymbol{F}_t)$-Brownian motion.

The hypotheses in the Picard-Lindelöf-Itô theorem 11.1.1 refer to the coefficients μ and $\mathbf{a}$ only. Hence it applies to all Brownian motions simultaneously or to neither of them and we always get (unique) pathwise solutions from there. In constructing weak solutions, one frequently starts with a candidate X for a solution and then tries to find a Brownian motion such that (12.4.3) holds. Typical is the following procedure which illustrates the Girsanov change of measure technique from section 10.2.

12.4.2 Example. (a) We construct a weak solution for the one-dimensional SDE

$$dX = \mu(t,X)\, dt + dB \tag{12.4.4}$$

by change of measure. On Wiener space $(\mathscr{C}(\mathbb{R}_+),\mathscr{B}(\mathscr{C}(\mathbb{R}_+)),\mathbb{W})$ with natural filtration $(\boldsymbol{F}_t)_{t\geq 0}$ the coordinate process Π is standard Brownian motion. Suppose that μ is in $\boldsymbol{B}$ and H = μ fulfills the assumptions of theorem 10.2.5. By that theorem

$$B_t^* = \Pi_t - \Pi_0 - \int_0^t \mu(s,\Pi)\, ds$$

defines a Brownian motion for some probability measure Q. Rewriting this identity in the differential form

$$d\Pi = \mu(t,\Pi)\, dt + dB^* \tag{12.4.5}$$

shows that Π is a weak solution of (12.4.4) on $(\mathscr{C}(\mathbb{R}_+), \mathscr{B}(\mathscr{C}(\mathbb{R}_+)),Q)$ with driving Brownian motion B^* (and initial condition $\Pi_0 = 0$ Q-a.s.).

(b) If μ is uniformly bounded then the solution of (12.4.4) is weakly unique. To see this, let us compute the law Q of a solution X on some probability space $(\tilde{\Omega},\tilde{\boldsymbol{F}},\tilde{\mathbb{P}})$ for a driving Brownian motion $\tilde{B}$. The transformulation

$$\left[\Pi - \Pi_0 - \int_0^{\cdot} \mu(s,\Pi)ds\right]\circ X = X - X_0 - \int_0^{\cdot} \mu(s,X)dx = \tilde{B}$$

shows that for the process B^* in the square-brackets the composition $B^*\circ X$ is a Brownian motion for $\tilde{\mathbb{P}}$ and hence B^* itself is a Brownian motion for $\mathbb{P}\circ X^{-1} = Q$ (and Π solves (12.4.4) for B^* and Q). Since μ is bounded the exponential process

$$Z_t = \exp\left(-\int_0^{\cdot} \mu(t,\Pi)dB^* - ½ \int_0^{\cdot} \mu^2(t,\Pi)dt\right).$$

is a martingale by proposition 9.2.4. Thus we may define consistent probability measures on the σ-fields $\boldsymbol{F}_t$ by $Z_t dQ$. Since we work on the Polish space $\mathscr{C}(\mathbb{R}_+)$ there is a probability measure $\mathbb{P}$ on $\mathscr{B}(\mathscr{C}(\mathbb{R}_+))$ such that $d\mathbb{P} = Z_t dQ$ on $\boldsymbol{F}_t$. Girsanov's theorem applied to the Q-Brownian motion B^* and $W_t = -\mu(t,\Pi)$ yields that the process

$$\Pi - \Pi_0 = B^* + \int_0^{\cdot} \mu(t,\Pi)dt$$

is a $\mathbb{P}$-Brownian motion. The law of Π_0 under $\mathbb{P}$ is the same as under Q since $Z_0 = 1$. Hence if $X_0 = 0$ in law then $\mathbb{P}$ is Wiener measure $\mathbb{W}$. Thus the law of Q of the solution X is $\frac{1}{Z_t} d\mathbb{W}$ in $\boldsymbol{F}_t$. Because Z depends only on μ all solutions of (12.4.4) have this law. If the initial condition is ν, then by the last formula $\Pi - \Pi_0$ is still a Brownian motion for $\mathbb{P}$. Hence $\mathbb{P} = \int \mathbb{W}^x d\nu(x)$. A modification of the last argument shows uniqueness in this more general situation.

(c) A typical application of the pathwise concept of solution is the following one. Suppose that (12.4.4) is pathwise uniquely solvable. We claim that then the assumption of proposition 10.2.7 holds, This result allows to compute explicitely the density of the law of B^* with respect to $\mathbb{W}$. Let us for simplicity assume that we have initial condition $X_0 = 0$ a.s. as in (a). Let $(\boldsymbol{F}_t)$ and $(\boldsymbol{G}_t)$ be the nullset augmentation of the natural filtrations of Π and B^*, respectively. Since B^* is $(\boldsymbol{F}_t)$-adapted we have $\boldsymbol{G}_t \subset \boldsymbol{F}_t$. For Q the process B^* is both a

$(\boldsymbol{F}_t)$-Brownian motion and a $(\boldsymbol{G}_t)$-Brownian motion. Pathwise existence gives a $(\boldsymbol{G}_t)$-adapted solution. Since $\boldsymbol{G}_t \subset \boldsymbol{F}_t$ pathwise uniqueness shows that Π is a.s. equal to this solution, i.e. Π is $(\boldsymbol{G}_t)$-adapted. This implies $(\boldsymbol{G}_t) = (\boldsymbol{F}_t)$ and shows that Π can be written in the form $\Pi_s = \psi_s(B^*)$ a.s. as required in proposition 10.2.7. An example of Tsirelson shows that even for uniformly bounded μ these two filtrations are different in general (cf. Rogers-Williams (1987), section V.18). This completes the discussion of the equation (12.4.4).

Before we prove that pathwise uniqueness is stronger than weak uniqueness we give an example showing that weak uniqueness does not imply strong uniqueness and weak existence does not imply strong existence either.

12.4.3 Example. (**Tanaka's SDE**) Consider the SDE

$$dX_t = \mathrm{sig}(X_t) \bullet dB_t \tag{12.4.6}$$

where $\mathrm{sig}(x) = 1$ for $x \geq 0$ and $\mathrm{sig}(x) = -1$ for $x < 0$. Then the following holds:
(i) there is a weak solution; for the natural filtration of the driving Brownian motion there is no pathwise solution;
(ii) all solutions are Brownian motions which implies weak uniqueness; for every solution X the process $-X$ is a solution too which implies that pathwise uniqueness fails.

Proof. (1) Each solution X is an a.s. continous local martingale whose quadratic variation according to corollary 7.4.3 has the form

$$\langle X\rangle_t = \int_0^t \big(\mathrm{sig}(X_s)\big)^2 \, d\langle B\rangle_s = \int_0^t 1 \, ds = t.$$

Thus it is a Brownian motion by Lévy's theorem. Hence irrespective of the particular choice of the driving Brownian motion B and of the filtration all solutions have the same law. This proves weak uniqueness. If in the SDE the function sig is replaced by the standard function sgn which vanishes at $x = 0$ then the zero process is also a solution (for the initial condition $x = 0$) and therefore weak uniqueness fails. This is the main reason for choosing sig instead of sgn.
(2) Suppose now that B is fixed and that X is a solution of (12.4.6). Because of (1), X is a Brownian motion and according to corollary 8.3.3 we have

(12.4.7) $$\lambda\otimes\mathbb{P}\big(X = 0\big) = 0.$$

Hence $\mathrm{sig}(-X) = -\,\mathrm{sig}(X)$ on a set of full $\lambda\otimes\mathbb{P}$-measure which yields

$$d(-X) = \mathrm{sig}(-X)\bullet dB,$$

i.e. $-X$ is also a solution. Thus the solution is not pathwise unique.

(3) Now we ask ourselves whether a given Brownian motion X can be represented as a solution of Tanaka's SDE . The answer is: Yes, indeed. We show that the driving Brownian motion B can be constructed (or reconstructed) from X via the formula

(12.4.8) $$B_t = \int_0^t \mathrm{sig}(X)\; dX.$$

In fact, assume that X is a Brownian motion and B is defined by (12.4.8). Since $\langle X\rangle_t = t$ we have

(12.4.9) $$\langle B\rangle_t = \int_0^t \mathrm{sig}(X_s)^2\, ds = t$$

and hence B is a Brownian motion by Lévy's theorem. Moreover, we have

$$\mathrm{sig}(X)\bullet dB = \mathrm{sig}(X)^2\bullet dX = dX\,,$$

i.e. X solves Tanaka's equation for this particular B. Conversely, assume (12.4.6) for any B. Then

(12.4.10) $$dB = \mathrm{sig}(X)^2\bullet dB = \mathrm{sig}(X)\bullet dX$$

which proves (12.4.8).

(4) Finally, we show that Tanaka's SDE is not pathwise solvable. More precisely, the process X is never adapted to the nullset augmentation $(\boldsymbol{F}_t)_{t\ge 0}$ of the natural filtration of the driving Brownian motion B. From (12.4.7) and corollary 8.3.4 we get applying (8.3.8) to $g(x) = \mathbf{1}_{\mathbb{R}\setminus\{0\}}(x)$,

$$B = \int_0^{\cdot} \mathrm{sig}(X)\; dX = \int_0^{\cdot} \mathrm{sgn}(X)\; dX = \int_0^{\cdot} \mathbf{1}_{\{|X|\neq 0\}}\; d|X|\,.$$

Hence B is adapted to the nullset augmentation of the natural filtration of $|X|$. But X itself is not adapted to this filtration (e.g. the sign of X_1 cannot be reconstructed from the path of $|X|$). This implies the assertion. Therefore X is not an adapted solution for the $(\boldsymbol{F}_t)$-Brownian motion B. ∎

Now we come to the main positive result: The theorem of Yamada-Watanabe (1971). It implies in particular that under the assumptions of theorem 12.1.8 weak uniqueness holds.

12.4.4 Theorem. For the SDE

$$dX = \mu(\cdot,X)dt + \sigma(\cdot,X) \bullet dB$$

pathwise uniqueness implies weak uniqueness.

In the proof of the theorem, we will use: whether a process X solves (12.4.3) for the Brownian motion B or not depends solely on the joint distribution of X and B.

12.4.5 Lemma. Let probability spaces $(\Omega, \boldsymbol{F}, \mathbb{P})$ and $(\Omega', \boldsymbol{F}', \mathbb{P}')$ with filtrations $(\boldsymbol{F}_t)_{t\geq 0}$ and $(\boldsymbol{F}'_t)_{t\geq 0}$ be given. Consider $(\boldsymbol{F}_t)$- and $(\boldsymbol{F}'_t)$-Brownian motions B and B' - each of dimension d. Assume further that X and X' are n-dimensional processes on Ω and Ω', respectively, with $\mathbb{P}\circ(B,X)^{-1} = \mathbb{P}'\circ(B',X')^{-1}$. Then either both X and X' solve the associated SDE or neither of them does.

Proof. We compare (B,X) and (B',X') with a fixed third process. Define a new probability space by

$$\Omega^* = \mathscr{C}(\mathbb{R}_+,\mathbb{R}^d)\times\mathscr{C}(\mathbb{R}_+,\mathbb{R}^n),$$
$$\boldsymbol{F}^* = \mathscr{B}(\mathscr{C}(\mathbb{R}_+,\mathbb{R}^d))\otimes\mathscr{B}(\mathscr{C}(\mathbb{R}_+,\mathbb{R}^n)),$$
$$\mathbb{P}^* = \mathbb{P}\circ(B,X)^{-1}.$$

Let Γ and Π be the projections onto the first and second component of Ω^*, respectively, and define a filtration by

$$\boldsymbol{F}^*_t = \sigma((\Gamma_s,\Pi_s) : s \leq t).$$

Then we have

$$\int_0^t \mu(s,X)\,ds = \int_0^t \mu(s,\Pi(B,X))\,ds = \left[\int_0^t \mu(s,\Pi)\,ds\right]\circ(B,X)\,. \tag{12.4.11}$$

We show that the analogous transformation formula

$$\int_0^t \sigma(s,X)\,dB = \left[\int_0^t \sigma(s,\Pi)d\Gamma_s\right]\circ(B,X) \tag{12.4.12}$$

holds for the stochastic integral. For the proof assume first that σ is an elementary process $\sigma = \sum H_i \mathbf{1}_{]t_i,t_{i+1}]}$ on $\mathscr{C}(\mathbb{R}_+, \mathbb{R}^d)$ where H_i is $\sigma(\Pi_t : t \leq t_i)$-measurable. Then

$$\sigma(s,X) = \sum H_i(X)\,\mathbf{1}_{]t_i,t_{i+1}]}(s)$$

and

$$\int_0^t \sigma(s,X)dB = \sum H_i(X)\ d(B_{t_{i+1}} - B_{t_i})$$
$$= \left[\sum H_i(\Pi)\left(\Gamma_{t_{i+1}} - \Gamma_{t_i}\right)\right] \circ (B,X)$$
$$= \left[\int_0^t \sigma(s,\Pi)d\Gamma\right] \circ (B,X)\ .$$

The general case of (12.4.12) follows by approximation of σ in the space $\boldsymbol{L}^{2,\mathrm{loc}}\left(\mathbb{R}_+ \times \mathscr{C}(\mathbb{R}_+, \mathbb{R}^d), \lambda \otimes \mathbb{W}^d\right)$ by elementary processes.

Because of (12.4.11) and (12.4.12) the following are equivalent:

(a) X solves the SDE w.r.t. B and $\mathbb{P}$,
(b) $X_t - X_0 = \int_0^t \mu(\cdot,X)\ ds + \int_0^t \sigma(\cdot,X)\ dB,$
(c) $\Pi_t(B,X) - \Pi_0(B,X) = \left[\int_0^t \mu(\cdot,\Pi)\ ds + \int_0^t \sigma(\cdot,\Pi)\ d\Gamma\right] \circ (B,X),$
(d) Π solves the SDE w.r.t. Γ and $\mathbb{P}^*$.

This proves the lemma since in (d) the pair (B,X) enters only via $\mathbb{P}^*$. ■

The idea for the proof of theorem 12.4.4 is to carry two weak solutions and their driving Brownian motions over to a common canonical setup - namely $\mathscr{C}(\mathbb{R}_+,\mathbb{R}^d) \times \mathscr{C}(\mathbb{R}_+,\mathbb{R}^n)$ and to use there pathwise uniqueness. As an important tool regular conditional distributions will be used which exist on the Polish spaces in question.

Proof of theorem 12.4.4. Assume that the SDE is pathwise uniquely solvable. Let $(\Omega',\boldsymbol{F}',\mathbb{P}')$ and $(\Omega'',\boldsymbol{F}'',\mathbb{P}'')$ be probability spaces with filtrations $(\boldsymbol{F}'_t)_{t\geq 0}$ and $(\boldsymbol{F}''_t)_{t\geq 0}$, adapted Brownian motions B' and B'' and solutions X' and X'' on Ω' and Ω'', respectively, such that the initial variables X'_0 and X''_0 have the same law ν. Let

$$Q : \left[\mathscr{C}(\mathbb{R}_+,\mathbb{R}^d) \times \mathbb{R}^n\right] \times \mathscr{B}(\mathscr{C}(\mathbb{R}_+,\mathbb{R}^n)) \longrightarrow [0,1]$$

be the regular conditional distribution of X'' given (B'',X''_0); it exists since all spaces in question are Polish (cf. Breiman (1968)). Define the new setting by

$$\Omega = \Omega' \times \mathscr{C}(\mathbb{R}_+,\mathbb{R}^n),\ \boldsymbol{F} = \boldsymbol{F}'\otimes\mathscr{B}(\mathscr{C}(\mathbb{R}_+,\mathbb{R}^n)),$$
$$\boldsymbol{F}_t = \boldsymbol{F}'_t \otimes \mathscr{B}_t(\mathscr{C}(\mathbb{R}_+,\mathbb{R}^n)),$$
$$\mathbb{P}(F' \times A) = \int_{F'} Q[B'(\omega'),X'_0(\omega');A]\ d\mathbb{P}'(\omega'),$$

where $\mathscr{B}_t(\ldots)$ denotes the natural filtration. A typical element of Ω will be written (ω',π). The process

$$B_t(\omega',\pi) = B'_t(\omega')$$

will serve as the common driving Brownian motion for two solutions Y' and Y" on Ω. To X' corresponds the process

$$Y'_t(\omega',\pi) = X'_t(\omega').$$

Plainly, the new pair (B,Y') has the same law as (B',X'). The counterpart of X" will be played by

$$Y''_t(\omega',\pi) = \pi_t.$$

Since B" is a $(\boldsymbol{F}''_t)$-Brownian motion and X''_0 is $\boldsymbol{F}''_0$-measurable the pair (B'',X''_0) has the law $\mathbb{W}\otimes\nu$. Similarly, the pair (B',X'_0) has the same law. Since for every set $C \in \mathscr{B}(\mathbb{R}^n)$

$$Q(\beta,x;\ \{\pi_0 \in C\}) = \mathbf{1}_C(x)$$

we have

(12.4.13) $$\mathbb{P}\Big(Y''_0(\omega',\pi) = X'_0(\omega')\Big) = 1$$

and hence also

$$\mathbb{P}\Big(B \in D, Y''_0 \in A\Big) = \mathbb{P}'\Big(B' \in D, X'_0 \in A\Big) = \mathbb{W}\otimes\nu(D \times A).$$

We conclude that (B'',X'') and (B,Y'') are equal in law since

(i) (B,Y''_0) and (B'',X''_0) both have $\mathbb{W}\otimes\nu$ as law,

(ii) the conditional distribution of Y" given (B,Y''_0) is the same as that of X" given (B'',X''_0).

Since (B',X') and (B,Y') on the one hand and (B'',X'') and (B,Y'') on the other hand are equal in law, we conclude from lemma 12.4.5 that both Y' and Y" solve (12.4.1) with common driving Brownian motion B. Since $Y'_0(\omega',\pi) = X'_0(\omega')$ the relation (12.4.13) shows $\mathbb{P}(Y'_0 = Y''_0) = 1$. By pathwise uniqueness $\mathbb{P}(Y' = Y'') = 1$ which implies that Y' and Y" have the same law. Hence X' and X" are also equal in law. This proves the theorem. ■

12.5 The Time - Homogeneous Case and Semigroups

Under weak uniqueness all solutions of the SDE (12.1.1) with the initial condition $X_0 = x$ a.s. have the same law - independent of the underlying probability space and the driving Brownian motion. This allows to speak of *the* law $\mathbb{P}^x$ and of *the* associated expectation operator $\mathbb{E}^x$ for functionals of the solution starting at x. The main purpose of this section is the reformulation of some of the previous results with the help of this family $(\mathbb{P}^x)_{x\in\mathbb{R}^n}$ of probability laws on $\mathscr{C}(\mathbb{R}_+,\mathbb{R}^n)$. We consider only the time homogeneous case which also can be described in terms of semigroups.

Now have another look at the end of section 2.1 where we explained why we did not assume the usual conditions on the filtrations !

In this section we always assume that

$$\mu : \mathbb{R}^n \longrightarrow \mathbb{R}^n, \ \sigma : \mathbb{R}^n \longrightarrow \mathbb{R}^{n\times d}$$

are coefficients such that the associated SDE

$$dX_t = \mu(X_t)\, dt + \sigma(X_t)\bullet dB_t \tag{12.5.1}$$

has a weakly unique solution on $[0,\infty[$ (for every initial condition $x \in \mathbb{R}^n$). Plainly, the associated operator L_t does no more depend on t and we denote it by L.

Using this notation we can formulate the following version of *Dynkin's formula* (12.1.5): For every function $\varphi \in \mathscr{C}^2(\mathbb{R}^n)$, every $x \in \mathbb{R}^n$ and every bounded stopping time T,

$$\mathbb{E}^x\Big(\varphi(X_T)\Big) - \varphi(x) = \mathbb{E}^x\Big(\int_0^T L\varphi(X_s)\, ds\Big). \tag{12.5.2}$$

Similarly, the *Feynman-Kac formula* gets for a $\mathscr{C}^2$-solution of $L\psi + v\psi = 0$ the form

$$\psi(x) = \mathbb{E}^x\Big(e^{-\int_0^T v(X_s)ds}\psi(X_T)\Big). \tag{12.5.3}$$

Now we give another approach to the (strong) Markov property which is based on weak uniqueness. A similar argument which uses the equivalent assumption of uniqueness of the associated martingale problem (cf. section 12.6) is given in Stroock-Varadhan(1979) or Rogers-Williams(1987), V.21.

12.5.1 Theorem. Let X be a continuous solution of the SDE (12.5.1) with respect to some $(\boldsymbol{F}_t)$- Brownian motion on some probability space $(\Omega, \boldsymbol{F}, \mathbb{P})$ and assume that Ω is a Polish space and $\boldsymbol{F} = \mathscr{B}(\Omega)$. Let S be a stopping time. Then for every $x \in \mathbb{R}^n$ and every Borel functional $f : \mathscr{C}(\mathbb{R}_+, \mathbb{R}^n) \longrightarrow \mathbb{R}$ one has

$$\mathbb{E}\left(f(X_{S+\cdot}) \;\middle|\; \boldsymbol{F}_S^+\right) = \mathbb{E}^{X_S}(f) . \tag{12.5.4}$$

Proof. Intuitively, the argument simply is this: The process $X_{S+\cdot}$ solves the SDE with initial condition X_S, whatever might have happened before time S. In order to give a precise meaning to the "whatever ... " we use the conditional distribution $(\mathbb{P}_\omega)_{\omega\in\Omega}$ on $\boldsymbol{F}$ given the σ-field $\boldsymbol{F}_S^+$. (We had to assume that $\boldsymbol{F}$ is the Borel-σ-field of a Polish space in order to make sure that this conditional distribution exists.) By definition of the conditional distribution, for every $\boldsymbol{F}$-measurable function g one has for $\mathbb{P}$-almost all ω that

$$\mathbb{E}(g \mid \boldsymbol{F}_S^+)(\omega) = \int g \, d\mathbb{P}_\omega. \tag{12.5.5}$$

In particular, a random variable (or a right-continuous process) which is independent of $\boldsymbol{F}_S^+$ has under $\mathbb{P}_\omega$ the same distribution as under $\mathbb{P}$ for almost all ω. Since X solves the SDE we have

$$X_{S+t} = X_S + \int_S^{S+t} \mu(X_s)\, ds + \int_S^{S+t} \sigma(X_s)\, dB_s .$$

Writing X^* for $X_{S+\cdot}$ and $B^* = B_{S+\cdot} - B_S$ this reads

$$X_t^* = X_S + \int_0^t \mu(X_s^*)\, ds + \int_0^t \sigma(X_s^*)\, dB_s^* . \tag{12.5.6}$$

The process B^* is a $(\boldsymbol{F}_{S+t}^+)_{t\ge 0}$-Brownian motion both for the original measure $\mathbb{P}$ and w.r.t. $\mathbb{P}_\omega$ for almost all ω by the strong Markov property of Brownian motion (corollary 9.1.11). Suppose now that the stochastic integral in (12.5.6) can be read also as one w.r.t. to the probability measure $\mathbb{P}_\omega$. Then (12.5.6) says that X^* solves the SDE also on the new probability space $(\Omega, \boldsymbol{F}, \mathbb{P}_\omega)$. For the measure $\mathbb{P}_\omega$ one has $X_S = X_S(\omega)$ a.s.. By weak uniqueness therefore, for such an ω, the law of X^* under $\mathbb{P}_\omega$ is the measure $\mathbb{P}^{X_S(\omega)}$. Hence using (12.5.6) we get $\mathbb{P}$-a.s.

$$\mathbb{E}\left(f(X_{S+\cdot}) \mid \boldsymbol{F}_S^+\right)(\omega) = \mathbb{E}\left(f(X^*) \mid \boldsymbol{F}_S^+\right)(\omega) = \int f(X^*)\, d\mathbb{P}_\omega = \mathbb{E}^{X_S(\omega)}(f) .$$

It remains to be shown that for a $(\boldsymbol{F}_{S+t}^+)_{t\ge 0}$-predictable integrand H a version of the stochastic integral process $\int_0^\cdot H\, dB^*$ w.r.t. $\mathbb{P}$ is also one for almost every $\mathbb{P}_\omega$. If H is an elementary process then the integral is an elementary integral and no measure is involved. Next we assume that $H \in \mathcal{L}^2(\lambda\otimes\mathbb{P})$. Note that $\mathbb{P} = \int \mathbb{P}_{\omega'}\, d\mathbb{P}$ according to (12.5.5). Thus there is a sequence $(H^n)_{n\geq 1}$ of elementary processes such that

$$\int\int\int |H - H^n|^2 d\lambda d\mathbb{P}_{\omega'} d\mathbb{P} = \int\int |H - H^n|^2 d\lambda d\mathbb{P} \longrightarrow 0.$$

Passing to a suitable subsequence, we may suppose that for $\mathbb{P}$-almost every ω' one has $|H - H^n| \longrightarrow 0$ in $\mathcal{L}^2(\lambda\otimes\mathbb{P}_{\omega'})$. Then for all $t \geq 0$

$$\sup_{0\leq s\leq t} \left| \int_0^s H\, dB^* - \int_0^s H^n\, dB^* \right| \longrightarrow 0$$

in $\mathbb{P}_{\omega'}$-probability. This implies the assertion for $H \in \mathcal{L}^2(\lambda\otimes\mathbb{P})$. The extension to $H \in \mathcal{L}^{2,loc}(\lambda\otimes\mathbb{P})$ is a straightforward localization argument. ■

The proof shows in particular that the right-hand side of (12.5.4) is a.s. equal to a $\mathcal{F}_S^+$-measurable function. This measurability statement can be improved: The measure $\mathbb{P}^x$ depends in a measurable way on x whenever weak uniqueness holds (cf. Stroock-Varadhan(1979); exercise 6.7.4). We show that under the more restrictive Lipschitz assumptions under which we proved pathwise (and thus weak) uniqueness this dependence is even continuous. More precisely, we show

12.5.2 Theorem. If in equation (12.5.1) the coefficients μ and σ are Lipschitz then the map

$$x \longmapsto \mathbb{E}^x(\varphi(X_t))$$

is continuous for every $\varphi \in \mathcal{C}_b(\mathbb{R}^n)$.

Proof. We use the continuous dependence on the initial conditions in theorem 11.2.3. Suppose that $x_k \longrightarrow x_\infty$ in $\mathbb{R}^n$. Let B be a Brownian motion on some probability space $(\Omega,\mathcal{F},\mathbb{P})$. For each $k \in \mathbb{N} \cup \{\infty\}$ let X^k be the solution of the SDE for the initial condition $X_0^k = x_k$. Then, plainly, $|X_0^k - X_0^\infty| \longrightarrow 0$ in $\mathcal{L}^2$ and hence by theorem 11.2.3 also $X_t^k \longrightarrow X_t^\infty$ in $\mathcal{L}^2$. For a bounded continuous function φ on $\mathbb{R}^n$ this implies $\varphi(X_t^k) \longrightarrow \varphi(X_t^\infty)$ in probability and even in $\mathcal{L}^1(\mathbb{P})$. The process X^k has the law $\mathbb{P}^{x_k}$ and hence

$$\mathbb{E}^{x_k}\big(\varphi(X_t)\big) = \mathbb{E}\big(\varphi(X_t^k)\big) \longrightarrow \mathbb{E}\big(\varphi(X_t^\infty)\big) = \mathbb{E}^{x_\infty}\big(\varphi(X_t)\big) .$$

This proves the lemma. ■

Combining this continuous dependence on the initial condition with the Markov property allows to establish a connection with semigroups on $\mathscr{C}_b(\mathbb{R}^n)$. Recall that $\mathscr{C}_b(\mathbb{R}^n)$ becomes a Banach space when endowed with the sup-norm. A **semigroup** on a Banach space **E** is a family of linear operators $P_t : \mathbf{E} \longrightarrow \mathbf{E}$, $t \geq 0$, such that P_0 is the identity and $P_s\, P_t = P_{s+t}$ for all s and t.

12.5.3 Proposition. Suppose that the coefficients μ and σ in the SDE (12.5.1) are Lipschitz continous. Then the relation

$$P_t\varphi(x) = \mathbb{E}^x(\varphi(X_t))$$

defines a semigroup of continuous linear operators on the Banach-space $\mathscr{C}_b(\mathbb{R}^n)$.

Proof. The semigroup property $P_{s+t} = P_s P_t$ follows from the Markov property: The latter implies that for every φ and all $s,t \geq 0$,

$$\mathbb{E}^x(\varphi(X_{s+t})|\, \boldsymbol{F}_s) = \mathbb{E}^{X_s}(\varphi(X_t)) .$$

Hence for the function $h(x) = P_t\varphi(x)\mathbb{E}^x(\varphi(X_t))$one has

$$P_{s+t}\varphi(x) = \mathbb{E}^x\Big(\varphi(X_{s+t})\Big) = \mathbb{E}^x\Big(\mathbb{E}^{X_s}(\varphi(X_t))\Big) = \mathbb{E}^x\Big(h(X_s)\Big) = P_s h(x) = P_s P_t \varphi(x)$$

completing the proof. ■

Rewriting the formula (12.0.3) which was verified in theorem 12.1.8 for t = 0 and the probability measure $\mathbb{P}^x$ we get for every φ in $\mathscr{C}_c^2(\mathbb{R}^n)$ as $h \longrightarrow 0$ that

$$\frac{1}{h}\Big(P_h\varphi - \varphi\Big)(x) = \frac{1}{h}\Big(\mathbb{E}^x(\varphi(X_h) - \varphi(x)\Big) \longrightarrow L\varphi(x).$$

Looking a little more carefully at the arguments in the proof of theorem 11.2.3 one can see that this convergence is even uniform in x. In terms of abstract semigroup theory this implies that these φ are in the domain of the generator of the semigroup and on the domain $\mathscr{C}_c^2(\mathbb{R}^n)$ the operator L and the generator of the semigroup coincide.

12.6 The Martingale Problem

Another approach to the description of diffusion processes was initiated (1969) and developed to a considerable extent (1979) by D.W. Stroock and S.R.S. Varadhan. It starts straight from the Dynkin formula or rather its preliminary form in theorem 12.1.2. This theorem implies that for a solution of a X associated to the differential operators L_t the processes

$$X^{\varphi} = \varphi(X) - \varphi(X_0) - \int_0^{\cdot} L_s\varphi(X_s)ds$$

are local martingales for all test functions φ. Stroock and Varadhan make - in their own words - this observation the cornerstone of their theory. They take $\Omega = \mathscr{C}(\mathbb{R}_+,\mathbb{R}^n)$ and the coordinate process Π given by

$$\Pi_t : \mathscr{C}(\mathbb{R}_+,\mathbb{R}^n) \longrightarrow \mathbb{R}^n, \ \omega \longmapsto \omega(t)$$

and look for a probability measure Q on Ω which makes each of the processes Π^{φ} a martingale. Such a measure Q is called a solution of the martingale problem associated to L. Under Q the process Π is a diffusion with infinitesimal parameters $\boldsymbol{\mu}$ and $\mathbf{a}$. Loosely spoken, the martingale problem is the probabilistic essence of diffusion theory. Its simple and elegant formulation has great appeal to probabilists.

Working on the Polish space of continuous paths has several technical advantages, also. So all the machinery from topological measure theory works, e.g. regular conditional probabilities exist and narrow convergence can be used in compactness arguments. The latter is essential in proving existence and uniqueness results in situations where the method based on SDEs does not apply. This part of the theory, however, is beyond the scope of the present text. What we can do is to accompany the reader a short piece of way and refer him then to the more advanced literature like Stroock and Varadhan (1979) and Stroock (1988). Rogers and Williams (1987), §§23ff also contains many helpful remarks.

To get a feeling that it might be reasonable to study the martingale problem we consider a simple example

Example. Let $L_t = \frac{1}{2}\frac{\partial^2}{\partial x^2}$ and consider the test functions $\varphi(x) = x$ and $\psi(x) = x^2$. Then for the coordinate process Π we have

$$\Pi^{\varphi}(t) = \Pi(t) \text{ and } \Pi^{\psi}(t) = \Pi(t)^2 - t.$$

Then Lévy's theorem tells us that Π is a Brownian motion if and only if Π^{φ} and Π^{ψ} are local martingales. Since Brownian motion has all moments they are even martingales. We conclude that - given the initial distribution of Π_0 - Wiener measure is the one and only solution of the martingale problem for this special L_t.

We work in the setting and continue with the notation from section 12.4. In the formal definiton of the martingale problem one has the choice of either allowing general test functions and to ask the associated processes X^{φ} only to be local martingales or to restrict the class of test functions and let the X^{φ} be martingales. We choose the latter alternative.

12.6.1 Definition. A probability measure Q on $\mathscr{C}(\mathbb{R}_+,\mathbb{R}^n)$ solves the **martingale problem** associated to **a** and μ if for every $\varphi \in \mathscr{C}_c^{\infty}(\mathbb{R}^n)$ the process

$$\Pi_t^{\varphi} = \varphi(\Pi_t) - \varphi(\Pi_0) - \int_0^t L_s\varphi(\Pi_s)ds$$

is a martingale.

Plainly, one could cancel the second term on the right-hand side.

The central result reveals a close connection between SDEs and the martingale problem (equivalence (a) $\Leftrightarrow$ (c) below). The equivalence (a) $\Leftrightarrow$ (b) yields a reformulation of the martingale problem in terms of space-time test functions.

12.6.2 Theorem. Let ν be a probability measure on $\mathbb{R}^n$ and Q a probability measure on $\mathscr{C}(\mathbb{R}_+,\mathbb{R}^n)$. Then for all μ and σ with $\mathbf{a} = \sigma\sigma^{\top}$ as in section 12.4 the following are equivalent:

(a) $Q\circ\Pi_0^{-1} = \nu$ and Q solves the martingale problem: the process Π^{φ} given by

$$\Pi_t^{\varphi} = \varphi(\Pi_t) - \varphi(\Pi_0) - \int_0^t L_s(\Pi)\varphi(\Pi_s)\, ds$$

is a martingale for every test function $\varphi \in \mathscr{C}_c^{\infty}(\mathbb{R}^n)$.

(b) $Q \circ \Pi_0^{-1} = \nu$ and Q solves a martingale problem for space-time test functions, i.e. the process given by

$$\psi(t,\Pi_t) - \psi(0,\Pi_0) - \int_0^t \Big[\frac{\partial}{\partial s} + L_s(\Pi)\Big]\psi(s,\Pi_s)ds$$

is a local martingale for every function $\psi \in \mathscr{C}^{1,2}(\mathbb{R}_+ \times \mathbb{R}^n)$.

(c) There is a probability space $(\Omega, \boldsymbol{F}, \mathbb{P})$ with a filtration $(\boldsymbol{F}_t)_{t \geq 0}$, a d-dimensional $(\boldsymbol{F}_t)$-Brownian motion B and a solution X of the SDE

$$dX_t = \mu(t,X)dt + \sigma(t,X) \bullet dB_t$$

with initial distribution $\mathbb{P} \circ X_0^{-1} = \nu$ and law $\mathbb{P} \circ X^{-1} = Q$.

12.6.3 Supplement. If $\mathbf{a}$ is never singular, i.e. if $\mathbf{a}(t,\Pi)$ is always positive definite and if $d = n$ we can take in (c)

$$\Omega = \mathscr{C}(\mathbb{R}_+, \mathbb{R}^n), \ \mathbb{P} = Q, \ X = \Pi$$

and

$$B_t = \int_0^t \sigma^\top \mathbf{a}^{-1}(s,\Pi)\, dY_s$$

where

$$Y_t = \Pi_t - \int_0^t \mu(s,\Pi)\, ds.$$

In particular, the theorem states that - given the initial distribution - the martingale problem has a solution if and only if the SDE has a weak solution.

For the proof, the results 12.6.4 and 12.6 .6 are useful. The first one is of independent interest. The martingale problem is stated there in terms of various classes of test functions.

12.6.4 Proposition. For a probability measure Q on $\mathscr{C}(\mathbb{R}_+, \mathbb{R}^n)$ the following are equivalent:

(a') Π^φ is a martingale for every $\varphi \in \mathscr{C}_c^\infty(\mathbb{R}^n)$.

(b') Π^φ is a martingale for every $\varphi \in \mathscr{C}_c^2(\mathbb{R}^n)$.

(c') Π^φ is a local martingale for every $\varphi \in \mathscr{C}^2(\mathbb{R}^n)$.

(d') Π^φ is a local martingale for all φ of the form $\varphi(x) = x_i$ or $\varphi(x) = x_i x_j$.

(e') $Y_t = \Pi_t - \Pi_0 - \int_0^t \mu(s,\Pi)ds$ is a local martingale with

$$dY(t) \bullet dY(t)^\top = \mathbf{a}(t,\Pi)dt.$$

We will not need the implication (e') → (a') and omit the proof; it is similar to the proof of (a) → (c) in the theorem. For the proof of (a') → (b') we use an analytical tool which should be known to most of the readers.

12.6 .5 Lemma. For every $\varphi \in \mathscr{C}_c^2(\mathbb{R}^n)$ there is a sequence $(\varphi_n)_{n\geq 1}$ in $\mathscr{C}_c^\infty(\mathbb{R}^n)$ converging uniformly to φ and such that the sequences of partial derivatives up to second order converge uniformly to the respective derivatives of φ as well.

A sketch of the proof is given as a series of exercises.

Proof of the lemma. If one believes that there are functions $\eta_\varepsilon \in \mathscr{C}_c^\infty(\mathbb{R}^n)$, $\varepsilon > 0$, vanishing outside an ε-ball $B(0,\varepsilon)$ around $0 \in \mathbb{R}^n$ and with $\int_{\mathbf{R}^n} \eta_\varepsilon(x)\,dx = 1$ then one is nearly done: Since φ is uniformly continuous, there is $\varepsilon_n > 0$ such that $|\varphi(x) - \varphi(y)| \leq \frac{1}{n}$ whenever $|x - y| \leq \varepsilon_n$. Then

$$\left|\int \eta_{\varepsilon_n}(x-y)(\varphi(y) - \varphi(x))\,dy\right| \leq \int_{B(x,\varepsilon_n)} \eta_{\varepsilon_n}(x-y)|\varphi(x) - \varphi(y)|\,dy \leq \frac{1}{n}.$$

Analoguous estimates hold for the partial derivatives of first and second order (use integration by parts). Hence $\varphi_n(x) = \int \eta_{\varepsilon_n}(x-y)\varphi(y)\,dy$ has the desired properties.

Those who do not believe have to work out some exercises:

(1) Show that the functions

$$\psi_n(t) = \begin{cases} t^{-n}e^{-1/t} & \text{for } t > 0 \\ 0 & \text{for } t \leq 0 \end{cases}, \quad n \geq 0,$$

are continuously differentiable with $\psi_n' = \psi_{n+2} - n\psi_{n+1}$ (prove first continuity at zero).

(2) Show by induction $\psi_0 \in \mathscr{C}^n(\mathbb{R})$ and $\psi_0^{(k)}(x) \longrightarrow 0$ as $|x| \longrightarrow \infty$ for all k.

(3) Show that the functions

$$\eta_\varepsilon(x) = \begin{cases} \dfrac{\psi_0(1 - \frac{|x|^2}{\varepsilon^2})}{\int \psi_0(1 - \frac{|y|^2}{\varepsilon^2})\,dy} & \text{if } |x| \leq \varepsilon \\ 0 & \text{if } |x| \leq \varepsilon \end{cases}$$

have the desired properties. ■

Proof of proposition 12.6.4. Suppose that (a') holds. Then (b') follows from the lemma.

Assume now (b') and choose $\varphi \in \mathscr{C}^2(\mathbb{R}^n)$. The exit times

$$T_m = \inf\{t \geq 0 : |\Pi_t| > m\}$$

from closed balls with radius m are stopping times. Choose $\varphi_m \in \mathscr{C}_c^\infty(\mathbb{R}^n)$ which is equal to 1 on such a ball, vanishes outside a m+1-ball and has uniformly bounded partial derivatives of first and second order. Then $\varphi_m \cdot \varphi$ is uniformly bounded and

$$\Pi^\varphi_{\cdot\wedge T_m} = \Pi^{\varphi_m \cdot \varphi}_{\cdot\wedge T_m}$$

is a uniformly integrable martingale by (b'). Since $T_m \uparrow \infty$ and Π^φ starts at zero the latter is a local martingale by definiton which proves (c').

(d') is a special case of (c').

Suppose now that (d') holds. Taking $\varphi(x) = x_i$ in (d') shows that each component of the process Y from (e') is a local martingale. Proving the identity in (e') is a rather cumbersome job (if we knew already that Π solves the SDE then the identity could easily be deduced from lemma 12.1.1; but we do not.). We assume for simplicity of notation that Π starts at zero. We must show

$$\langle Y^i, Y^j \rangle_t = \int_0^t \mathbf{a}_{ij}(s,\Pi)\, ds .$$

By the characterization 7.3.15 of mutual variation this amounts to check that the a.s. continuous processes

$$Z^{ij} = Y^i Y^j - \int_0^\cdot \mathbf{a}_{ij}(s,\Pi)\, ds$$

$$= \Pi^i \Pi^j - \Pi^i \int_0^\cdot \mu_j(s,\Pi) ds - \Pi^j \int_0^\cdot \mu_i(s,\Pi)\, ds$$

$$+ \int_0^\cdot \mu_i(s,\Pi)\, ds \cdot \int_0^\cdot \mu_j(s,\Pi)\, ds - \int_0^\cdot \mathbf{a}_{ij}(s,\Pi)\, ds$$

are local martingales. By assumption, Π^φ is a local martingale for $\varphi(x) = x_i x_j$. We are going to verify that the difference $\Pi^\varphi - Z^{ij}$ is a local martingale and thus prove the desired result. Write Π^φ down explicitely:

$$\Pi^\varphi = \Pi^i \Pi^j - \int_0^\cdot \Pi^i(s)\mu_j(s,\Pi)\, ds - \int_0^\cdot \Pi^j(s)\mu_i(s,\Pi)\, ds - \int_0^\cdot \mathbf{a}_{ij}(s,\Pi)\, ds .$$

Differential calculus gives for the difference in question (recall $d\boldsymbol{A}_c \bullet d\boldsymbol{S} = \{0\}$):

$$(\Pi^{\varphi} - Z^{ij})(t)$$

$$= \Pi^i \int_0^t \mu_j(s,\Pi)\, ds - \int_0^t \Pi^i \mu_j(s,\Pi)\, ds + \Pi^j \int_0^t \mu_i(s,\Pi)\, ds - \int_0^t \Pi^j \mu_i(s,\Pi)\, ds$$

$$- \int_0^t \mu_i(s,\Pi) ds \cdot \int_0^t \mu_j(s,\Pi)\, ds$$

$$= \int_0^t \Big[\int_0^s \mu_j(u,\Pi) du\Big] d\Pi^i(s) + \int_0^t \Big[\int_0^s \mu_i(u,\Pi) du\Big] d\Pi^j$$

$$- \int_0^t \Big[\int_0^s \mu_j(u,\Pi) du\Big] \mu_i(s,\Pi) ds - \int_0^t \Big[\int_0^s \mu_i(u,\Pi) du\Big] \mu_j(s,\Pi) ds$$

$$= \int_0^t \Big[\int_0^s \mu_j(u,\Pi) du\Big]\Big[d\Pi_j(s) - \mu_j(s,\Pi)\, ds\Big]$$

$$+ \int_0^t \Big[\int_0^s \mu_i(u,\Pi) du\Big]\Big[d\Pi_i(s) - \mu_i(s,\Pi) ds\Big].$$

The integrators $\Pi^j - \int_0^{\cdot} \mu_j(s,\Pi)\, ds$ are local martingales and consequently so are the integrals themselves. Hence $\Pi^{\varphi} - Z^{ij}$ is a local martingale which completes the proof of (e'). ■

The next lemma is merely technical and adresses a measurability question concerning the coefficients. It is borrowed from Stroock (1987). In a first reading you might take it on trust.

12.6.6 Lemma. Let $\mathbf{a} \in \mathbf{S}^n_+$ and denote by ρ the orthogonal projection of $\mathbb{R}^n$ onto Range($\mathbf{a}$). Then

$$\rho = \lim_{\varepsilon \downarrow 0} (\mathbf{a} + \varepsilon E_n)^{-1} \mathbf{a}.$$

Let further

$$\mathbf{a}' = \lim_{\varepsilon \downarrow 0} (\mathbf{a} + \varepsilon E_n)^{-1} \rho.$$

Then $\mathbf{a}' \in \mathbf{S}^n_+$ and

$$\mathbf{a}'\mathbf{a} = \mathbf{a}\mathbf{a}' = \rho.$$

Next suppose that σ is a $n \times d$-matrix with $\mathbf{a} = \sigma\sigma^{\top}$. Then

$$\text{Range}(\mathbf{a}) = \text{Range}(\sigma) \text{ and } \sigma^{\top}\mathbf{a}'\sigma = \rho_{\sigma}$$

where ρ_{σ} denotes the orthogonal projection of $\mathbb{R}^d$ onto Range($\sigma^{\top}$). In particular, the mappings

$$\mathbf{a} \longmapsto \rho,\ \mathbf{a} \longmapsto \mathbf{a}',\ \sigma \longmapsto \rho_{\sigma}$$

are measurable.

Proof. Set

$$\mathbf{a}_\varepsilon = (\mathbf{a} + \varepsilon E_n)^{-1}$$

which exists and is positive definite. If $\Theta \in \text{Range}(\mathbf{a})^\perp$ (where the $\text{Range}(\mathbf{a})^\perp$ is the orthogonal complement of Range ($\mathbf{a}$)) then $\mathbf{a}\Theta = 0$ and $\mathbf{a}_\varepsilon\mathbf{a}\Theta = 0$. Each $\Theta \in \text{Range}(\mathbf{a})$ has the form $\Theta = \mathbf{a}\Phi$ and we conclude from the obvious identity $\mathbf{a}_\varepsilon\, \mathbf{a} = E_n - \varepsilon\, \mathbf{a}_\varepsilon$ that

$$\mathbf{a}_\varepsilon\mathbf{a}\Theta = \Theta - \varepsilon\mathbf{a}_\varepsilon\mathbf{a}\Phi \longrightarrow \Theta \text{ as } \varepsilon \downarrow 0.$$

This shows that $\lim_{\varepsilon\downarrow 0} (\mathbf{a} + \varepsilon E_n)^{-1}\, \mathbf{a}$ exists and is equal to the projection ρ onto Range ($\mathbf{a}$). Again, if $\Theta \in \text{Range}(\mathbf{a})^\perp$ then $\mathbf{a}_\varepsilon\rho\Theta = 0$ and if $\Theta \in \text{Range}(\mathbf{a})$ then there is $\Phi \in \text{Range}(\mathbf{a})$ such that $\Theta = \mathbf{a}\Phi$ and so

$$\mathbf{a}_\varepsilon\rho\Theta = \mathbf{a}_\varepsilon\Theta = \mathbf{a}_\varepsilon\mathbf{a}\Phi \longrightarrow \Phi \text{ as } \varepsilon \downarrow 0.$$

This proves $\mathbf{aa'} = \rho$. Plainly, $\mathbf{a'} \in \mathbf{S}_+$.

From the smoothness of $\mathbf{a} \longmapsto \mathbf{a}_\varepsilon$ follows the measurability of the maps

$$\mathbf{a} \longmapsto \rho \text{ and } \mathbf{a} \longmapsto \mathbf{a'}.$$

Suppose $\mathbf{a} = \sigma\, \sigma^\top$. Plainly, Range ($\mathbf{a}$) $\subset$ Range(σ). Conversely, if $\Theta \in \text{Range}(\sigma)$ then $\Theta = \sigma\, \Phi$ with $\Phi \in \text{Null}(\sigma)^\perp = \text{Range}(\sigma^\top)$. We find Φ' with $\Phi = \sigma^\top\Phi'$ and have $\Theta = \sigma\sigma^\top\Phi'$, i.e. $\Theta \in \text{Range}(\mathbf{a})$. In summary, $\text{Range}(\mathbf{a}) = \text{Range}(\sigma)$.

We show now $\sigma^\top\mathbf{a'}\sigma = \rho_\sigma$. If $\Theta \in \text{Range}(\sigma^\top)^\perp$ then $\Theta \in \text{Null}(\sigma)$ and so $\sigma^\top\mathbf{a'}\sigma\, \Theta = 0$. On the other hand, if $\Theta \in \text{Range}(\sigma^\top)$ then $\Theta = \sigma^\top\Phi$ for some

$$\Phi \in \text{Null}(\sigma^\top)^\perp = \text{Range}(\sigma) = \text{Range}(\mathbf{a})$$

and therefore

$$\sigma^\top\mathbf{a'}\sigma\, \Theta = \sigma^\top\mathbf{a'}\mathbf{a}\Phi = \sigma^\top\rho\, \Phi = \sigma^\top\Phi = \Theta.$$

This proves the last identity and also the measurability of the map $\sigma \longmapsto \rho_\sigma$. The proof is now complete. ■

Proof of theorem 12.6.2. The implication (c)→(b) follows for the special case $\mu(t,X(\omega)) = \mu(t,X(t,\omega))$ and $\sigma(t,X(\omega)) = \sigma(t,X(t,\omega))$ from theorem 12.1.2. The general case can be treated similarly. If (b) holds then all processes in (a) are local martingales; since each of them is uniformly bounded on bounded time intervals they are even martingales by criterion 4.2.3 which proves (a). So the proof reduces to the implication (a)→(c). We follow the lines in Stroock (1987). We continue with the notation introduced in the lemma.

If $\mathbf{a}$ is never singular and n = d (as is the case when we take square roots according to lemma 12.2.1) then $\boldsymbol{\rho} = E_n$ and $\boldsymbol{\rho}_\sigma^\perp \equiv 0$ and we could construct a Brownian motion from the coordinate process on $\Omega = \mathscr{C}(\mathbb{R}_+,\mathbb{R}^n)$. Suppose on the other hand that $\mathbf{a}$ - and hence also $\boldsymbol{\sigma}$ - vanishes and say the initial measure ν is a point measure ε_y . Then the process X we are looking for must for each ω satisfy the ordinary differential equation

$$dX_t(\omega) = \boldsymbol{\mu}(t,X(\omega))\, dt,\ X_0(\omega) \equiv 0,$$

and Q is the degenerate measure concentrated on the solution (which is the same for all ω). Hence $(\mathscr{C}(\mathbb{R}_+,\mathbb{R}^n),Q)$ is not rich enough to carry a Brownian motion. We must compensate this deficiency by adding an independent Brownian motion.

We start our construction with the probability space $(\Omega,\boldsymbol{F},\mathbb{P})$:

$$\begin{aligned}
\Omega &= \mathscr{C}(\mathbb{R}_+,\mathbb{R}^n\times\mathbb{R}^d) = \mathscr{C}(\mathbb{R}_+,\mathbb{R}^n) \times \mathscr{C}(\mathbb{R}_+\times\mathbb{R}^d),\\
\boldsymbol{F} &= \mathscr{B}(\mathscr{C}(\mathbb{R}_+,\mathbb{R}^n))\otimes\mathscr{B}(\mathscr{C}(\mathbb{R}_+,\mathbb{R}^d)) = \mathscr{B}(\mathscr{C}(\mathbb{R}_+,\mathbb{R}^n\times\mathbb{R}^d)),\\
\mathbb{P} &= Q\otimes\mathbb{W}^d,
\end{aligned}$$

where $\mathbb{W}^d$ is d-dimensional Wiener-measure. Write each $\omega \in \Omega$ in the canonical form

$$\omega = \begin{bmatrix} X(\cdot,\omega) \\ W(\cdot,\omega) \end{bmatrix}$$

with a d-dimensional Brownian motion independent of the n-dimensional process X. As the filtration we take the natural filtration of the process $\begin{bmatrix} X \\ W \end{bmatrix}$.
We conclude from lemma 12.6.4 that

$$Y_t = X_t - X_0 - \int_0^t \boldsymbol{\mu}(s,X)\, ds$$

is a local martingale with

$$dY(t)\bullet dY^\top(t) = \mathbf{a}(t,X)dt. \tag{12.6.1}$$

From these ingredients we construct the Brownian motion B. Let now $\boldsymbol{\rho}(t,X)$ and $\boldsymbol{\rho}_\sigma(t,X)$ be the orthogonal projections of $\mathbb{R}^n$ and $\mathbb{R}^d$ onto Range($\mathbf{a}(t,X)$) and Range($\boldsymbol{\sigma}(t,X)$), respectively. Set

$$\boldsymbol{\rho}_\sigma^\perp = E_d - \boldsymbol{\rho}_\sigma$$

and let $\mathbf{a}'$ be the element of $\mathbf{S}_+^n$ with

$$\mathbf{a}'\mathbf{a} = \mathbf{a}\mathbf{a}' = \boldsymbol{\rho}.$$

Define the n-dimensional process

$$B(t) = \int_0^t \left[\sigma^{\top}\mathbf{a}',\rho_\sigma^{\perp}\right](\cdot,X)\, d\begin{bmatrix} Y(s) \\ W(s)\end{bmatrix}.$$

By the lemma, this integral is well-defined. To make sure that we really have found a Brownian motion we compute the mutual variation:

$$dB\bullet dB^{\top} = \left[\left[\sigma^{\top}\mathbf{a}',\ \rho_\sigma^{\perp}\right](\cdot,X)\bullet d\begin{bmatrix} Y \\ W\end{bmatrix}\right]\bullet\left[d\left[Y^{\top},W^{\top}\right]\bullet\begin{bmatrix}\mathbf{a}' \\ (\rho_\sigma^{\perp})^{\top}\end{bmatrix}(\cdot,X)\right]$$

$$= \left[\left[\sigma^{\top}\mathbf{a}',\rho_\sigma^{\perp}\right](\cdot,X)\bullet\begin{bmatrix}\mathbf{a}(\cdot,X) & 0 \\ 0 & E_d\end{bmatrix}dt\ \bullet\ \begin{bmatrix}\mathbf{a}'\sigma \\ \rho_\sigma^{\perp}\end{bmatrix}\right](\cdot,X)$$

In the diagonal we have substituted (12.6.1) and mutual variation of Brownian motion. Since Y and W are independent local martingales the mutual variations $\langle Y^i,W^j\rangle$ vanish and thus the zeroes enter the matrix. By lemma 12.6.6 and in particular, because of Range($\mathbf{a}$) = Range(σ) we have

$$\sigma^{\top}\mathbf{a}'\mathbf{a}\mathbf{a}'\sigma = \sigma^{\top}\rho(\mathbf{a}'\sigma) = \sigma^{\top}\mathbf{a}'\sigma = \rho_\sigma.$$

Because of this identity we may proceed with our calculation and write:

$$dB\bullet dB^{\top} = [\sigma^{\top}\mathbf{a}'\mathbf{a}\mathbf{a}'\sigma + \rho_\sigma^{\perp}\rho_\sigma^{\perp}](\cdot,X)\ dt = [\rho_\sigma + \rho_\sigma^{\perp}](\cdot,X)\ dt = E_d\ dt.$$

Hence B is a Brownian motion on the probability space $(\Omega,\boldsymbol{F},\mathbb{P})$. To complete the argument we must verify that X solves the differential equation associated with B. Since

$$\sigma\sigma^{\top}\mathbf{a}' = \mathbf{a}\mathbf{a}' = \rho$$

we see that

$$\sigma(\cdot,X)\bullet dB = \sigma\left[\sigma^{\top}\mathbf{a}',\ \rho_\sigma^{\perp}\right](\cdot,X)\bullet d\begin{bmatrix} Y \\ W\end{bmatrix} = \sigma\sigma^{\top}\mathbf{a}'(\cdot,\ X)\bullet dY + \sigma\rho_\sigma^{\perp}\bullet dW$$

$$= \rho(\cdot,X)\bullet dY = dY - \rho^{\perp}(\cdot,X)\bullet dY$$

where $\rho^{\perp} = E_n - \rho$. By the lemma, the second term vanishes since

$$\rho^{\perp}(\cdot,X)\bullet dY\bullet\left[\rho^{\perp}(\cdot,X)\bullet dY\right]^{\top} = \rho^{\perp}(\cdot,X)\bullet dY\bullet dY^{\top}\bullet\rho^{\perp}(\cdot,X)$$

$$= \rho^{\perp}(t\ ,X)\mathbf{a}(t,X)\ \rho^{\perp}(\cdot,X)\ dt = 0,$$

and we may rewrite the identity as $dX(t) = \mu(t,X)\ dt + \sigma(t,X)\bullet dB(t)$. That X has initial distribution ν is clear. This proves the implication (a)→(c) and therefore completes the proof of the theorem. ■

Proof of the supplement 12.6.3. If **a** is never singular and $n = d$ then $\rho_\sigma^\perp \equiv 0$ and we could have carried out the whole procedure on $\mathscr{C}(\mathbb{R}_+,\mathbb{R}^n)$. ■

The equivalence of weak uniqueness and uniqueness in the martingale problem is a straightforward consequence of the main theorem.

12.6.7 Theorem. For the SDE

$$dX = \mu(\cdot,X)\,dt + \sigma(\cdot,X)\bullet dB \tag{12.6.2}$$

weak uniqueness holds if and only if the associated martingale problem has a unique solution with prescribed initial condition.

Let us combine theorems 12.4.4 and 12.6.7. In the following scheme (SDE) stands for the SDE (12.6.2) and (MP) for the associated martingale problem.

(SDE) is pathwise uniquely solvable

$\Downarrow$

(SDE) is uniquely solvable in the weak sense

$\Updownarrow$

(MP) is uniquely solvable

and by theorem 12.6.2

(SDE) has a weak solution

$\Updownarrow$

(MP) has a solution.

In particular, the law of a solution X for (SDE) does not depend on the choice of the square root σ of **a**, as soon as we have weak uniqueness. In fact, different square roots σ lead to the same martingale problem, hence the law of X - being the solution of (MP) - only depends on **a**.

References

Arnold L. (1974): Stochastic differential equations. Theory and applications. John Wiley & Sons

Asmussen S. (1987): Applied probability and queues. John Wiley & Sons: Chichester · New York · Brisbane · Toronto · Singapore

Bachelier L. (1900): Théorie de la spéculation. Thesis. Ann. Ecole Norm. Sup. 17, 21-88

Billingsley P. (1968): Convergence of probability measures. John Wiley & Sons, Inc., New York · London · Sydney · Tokyo

Billingsley P. (1979): Probability and measure. John Wiley & Sons: Chichester · New York · Brisbane · Toronto · Singapore

Blankenship G.L. and **Li C.W.** (1986): Almost sure stability of linear stochastic systems with Poisson process coefficients. Siam J. Appl. Math., 46, No.5, 875-911

Breiman L. (1968): Probability. Addison-Wesley Company: Reading, Menlo Park, London, Ontario

Cameron R.H. and **Martin W.T.** (1944): Transformation of Wiener integrals under translations. Ann. Math. 45, 386-396

Chung K.L. and **Williams R.J.** (1983): Introduction to stochastic integration. Progress in Probability and Statistics Vol. 4, Birkhäuser: Boston, Basel, Stuttgart

Cocozza C. and **Yor M.** (1980): Démonstration d'un théorème de F. Knight à l'aide de martingales exponentielles. Sém. Prob. XIV. Lecture Notes in Mathematics 784, 496-499. Springer, Berlin

Dambis K. (1965): On the decomposition of continuous submartingales. Teor Verojatnost., Primenen. 10, 438-448

Davis B. (1975): Brownian motion and analytic functions. Ann. Prob. 7, 913-932

Dawson D.A. and **Gärtner J.** (1989): Large deviations, free energy functional and quasi-potential for a mean field model of interacting diffusions. Mem. Am. Math. Soc. 398

Dellacherie C. and **Meyer P.A.** (1978): Probabilities and potential. North Holland Publ.: Amsterdam-New York-Oxford

Dellacherie C. and **Meyer P.-A.** (1980): Probabilities and potential (B). Theory of Martingales. North-Holland Publ.: Amsterdam - New York - Oxford

Doob J.L. (1953): Stochastic processes. John Wiley & Sons, Inc.: New York; Chapman & Hall, Limited: London

Dubins L.E. and **Gilat D.** (1978): On the distribution of maxima of martingales. Proc. Am. Math. Soc. 68, 337-338

Dubins L.E. and **Schwarz G.** (1965): On continuous martingales. Proc. Nat. Acad. Sci. USA 53, 913-916

Durrett R. (1984): Brownian motion and martingales in analysis. The Wadsworth Mathematics Series, Wadsworth Advanced Books & Software: Belmont

Einstein A. (1905): Über die von der molekularkinetischen Theorie der Wärme geforderte Bewegung von in ruhenden Flüssigkeiten suspendierten Teilchen. Ann. Physik 17, 549-560

Elliott R.J. (1982): Stochastic calculus and applications. Applications of Mathematics 18, Springer-Verlag: New York, Heidelberg, Berlin

Engel D. (1982): The multiple stochastic integral. Mem. Am. Math. Soc. 265

Feller W. (1966): An introduction to probability theory and its applications. vol II. John Wiley & Sons, Inc.: New York

Freedman D. (1971): Brownian motion and diffusion. Holden-Day: San Francisco, Cambridge, London, Amsterdam

Gard C.Th. (1988): Introduction to stochastic differential equations. Marcel Dekker, Inc.: New York and Basel

Girsanov I.V. (1960): On transforming a certain class of stochastic processes by absolutely continuous change of measure. Teor. Verojatnost. i Primenen. 7, 285-301

Ikeda N. and **Watanabe S.** (1981): Stochastic differential equations and diffusion processes. North-Holland Publishing Company: Amsterdam, Oxford, New York

Itô K. (1944): Stochastic integral. Proc. Imp. Acad. Tokyo 20, 519-524

Itô K. (1951): On stochastic differential equations. Mem. Am. Math. Soc. 4

Itô K. (1975): Stochastic differentials. Appl. Math. Optim. 1, 374-381

Itô K. and **McKean Jr. H.P.** (1974): Diffusion processes and their sample paths. Second Printing, Die Grundlehren der mathematischen Wissenschaften in Einzeldarstellungen Band 125, Springer Verlag: Berlin, Heidelberg, New York

Kakutani S. (1961): Spectral analysis of stationary Gaussian processes. Proceedings of the Fourth Berkely Symposion. University of California: Berkely

Karatzas I. and **Shreve S.E.** (1988): Brownian motion and stochastic calculus. Springer-Verlag: New York, Berlin, Heidelberg, London, Paris, Tokyo

Karlin S. and **Taylor H.M.** (1981): A second course in stochastic processes. Academic Press: New York, London, Toronto, Sidney, San Francisco

Knight F.B. (1970): A reduction of continuous square integrable martingales to Brownian motion. in: Dinges H. (ed). Lecture Notes in Mathematics 190, 19-31. Springer, Berlin

Kolmogorov A.N. (1931): Über die analytischen Methoden in der Wahrscheinlichkeitsrechnung. Math. Ann. 104, 415-458

Kolmogorov A.N. (1933): Zur Theorie der stetigen zufälligen Prozesse. Math. Ann. 108, 149-160

Lenglart E. (1983): Sémi-martingales et intégrales stochastiques en temps continue. Rev. CETHEDEC 75 (1983) 91-160

Letta G. (1984): Martingales et intégration stochastique. Scuola Normale Superiore, Pisa

Lévy P. (1965): Processus stochastique et mouvement Brownien. Deuxième édition revue et augmentée, Gauthier-Villars & Cie. Paris

Liptser R.S. and **Shiryayev A.N.** (1977): Statistics of random processes I, General theory. Springer Berlin

Mandelbrot B. (1983): The fractalgeometry of nature. W. H. Freeman and Company, New York

Maruyama G. (1954): On the transition probability functions of the Markov processes. Natur. Sci. Rep. Ochanomizu Univ. 5,10-20

McKean H.P. (1969): Stochastic integrals. Academic Press: New York, London

Meyer P.-A. (1963): Decomposition of supermartingales; the uniqueness theorem. Illinois J. Math. 7, 1-17

Mohammed S.E.A. (1984): Stochastic functional differential equations. Pitman: Boston, London, Melbourne

Mohammed S.E.A. (1987): Non linear flows for linear stochastic delay equations. Stochastics, 17. 207-213

Nagasawa M. (1989): Transformations of diffusion and Schrödinger processes. Prob. Th. Rel. Fields 82, 109-136

Nelson E. (1967): Dynamical theories of Brownian motion. Mathematical Notes (3), Princeton University Press

Ornstein L.S. and **Uhlenbeck G.E.** (1930): On the theory of Brownian motion. Phys. Rev. 36, 823-841

Pinsky R. (1986): A spectral criterion for the finiteness or infiniteness of stopped Feynman-Kac functionals of diffusion processes. Ann. Prob. 14, 1180-1187

Pitman J. and **Yor M.** (1986): Asymptotic laws of planar Brownian motion. The Ann. Prob. Vol. 11, 733-779

Protter, Ph. (1986) Stochastic integration without tears. Stochastics Vol. 16, 295-325

Riesz F. (1914): Les opérations fonctionnelles linéaires. Ann. Sci. Ecole Norm. Sup., sér. 3, 28, 9-214

Rogers L.C.G. and **Williams D.** (1987): Diffusions, Markov processes and martingales, Volume 2: Itô-calculus. John Wiley & Sons. Chichester · New York · Brisbane · Toronto · Singapore

Skorokhod A.V. (1989): Asymptotic methods of the theory of stochastic differential equations. Published by the Am. Math. Soc.

Stratonovich R.L. (1964): A new form of representation of stochastic integrals and equations. Vestnik Moskow Univ., Ser. I, 1, 3-12. Repr. in SIAM J. Control 4, 362-371

Stroock D.W. (1987): Lectures on stochastic analysis: diffusion theory. London Mathematical Society, Student Texts 6

Stroock D.W. and **Varadhan S.R.S.** (1979): Multidimensional diffusion processes. Springer Verlag: New York

Sussmann H.J. (1978): On the gap between deterministic and stochastic ordinary differential equations. Ann Prob. 6,19-41

Turelli M. (1977): Random environments and stochastic calculus. Theoret. Population Biol. 12, 140-178

Varadhan S.R.S. (1980): Lectures on diffusion problems and partial differential equations. Tata Institute. Springer: Berlin, Heidelberg, New York

Wentzel A.D. (1980): A course in the theory of stochastic processes. Mc Graw Hill: New York, London

Wiener N. (1923): Differential space. Journal of Math. and Physics 2, 131-174

Wiener N. (1930): The homogeneous chaos. Amer. J. Math. 80, 897-936

Williams D. (1979): Diffusions, Markov processes, and martingales. John Wiley and Sons: Chichester, New York, Brisbane, Toronto

Wong E. and **Zakai M.** (1969): Riemann-Stieltjes approximations of stochastic integrals. Z. Wahrsch. verw. Gebiete. 12,87-97

Yamada T. and **Watanabe S.** (1971): On the uniqueness of solutions of stochastic differential equations. J. Math. Kyoto Univ. 11, 155-167

Index of Common Notation

Standard Notation

$\mathbb{N} = \{1, 2, \ldots\}$	
$\mathbb{N}_0 = \{0, 1, 2, \ldots\}$	
$\mathbb{Z}$	set of integers
$\mathbb{Q}$	set of rational numbers
$\mathbb{R}$	set of real numbers
$\mathbb{Q}_+, \mathbb{R}_+$	resp. set of nonnegative elements
$\mathbb{R}^n$	Euclidean n-space
$\mathbb{C}$	set of complex numbers
$i, \ldots, q$	nonnegative integers
$a, \ldots, h; r, \ldots, z$	real numbers
$\overline{A}$	closure of set A
A^c	complement of set A
$B \setminus A$	rel. complement
$A \triangle B$	symmetric difference
δ_{ij}	equals 1 if i = j and else 0
f'	derivative of a real function in one variable
∂_i	i-th partial derivative
$\frac{\partial}{\partial x_i}$	same
$\mathbf{1}_A$	characteristic function equals 1 on A and 0 outside A
$S \wedge T$	pointwise minimum
$S \vee T$	pointwise maximum of functions
$f\vert A$	restriction of function f to set A
E_d	d-dim unit matrix
$\sigma(\boldsymbol{A})$	σ-field generated by $\boldsymbol{A}$
$\boldsymbol{F} \otimes \boldsymbol{G}$	product of σ-fields
$\mathscr{B}(\boldsymbol{E})$	Borel-σ-field of a metric space
$\mathscr{C}(\boldsymbol{E}), \mathscr{C}_b(\boldsymbol{E}), \mathscr{C}^2(\boldsymbol{E})$	spaces of continuous, bounded continuous, twice cont. diff. functions
$\mathbb{E}$	expectation
$\boldsymbol{L}^p$	space of p-integrable functions
λ	Lebesgue measure
$\lvert\cdot\rvert$	Euclidean norm
$\lVert\cdot\rVert_p$	$\boldsymbol{L}^p$-norm
$\lVert\cdot\rVert_\infty$	sup-norm
$i = \sqrt{-1}$	
$\triangle$	Laplacian
∇	gradient

Specific Notation

Subject Index

123